EXTENDED ATMOSPHERES AND CIRCUMSTELLAR MATTER IN SPECTROSCOPIC BINARY SYSTEMS

PROF. OTTO STRUVE

(1897–1963)

INTERNATIONAL ASTRONOMICAL UNION
UNION ASTRONOMIQUE INTERNATIONALE

SYMPOSIUM No. 51

(STRUVE MEMORIAL SYMPOSIUM)

HELD AT PARKSVILLE, B.C., CANADA, 6–12 SEPTEMBER, 1972

EXTENDED ATMOSPHERES AND CIRCUMSTELLAR MATTER IN SPECTROSCOPIC BINARY SYSTEMS

EDITED BY

A. H. BATTEN

Dominion Astrophysical Observatory, Victoria, B.C.

D. REIDEL PUBLISHING COMPANY

DORDRECHT-HOLLAND / BOSTON-U.S.A.

1973

Published on behalf of
the International Astronomical Union
by
D. Reidel Publishing Company, P.O. Box 17, Dordrecht, Holland

Sold and distributed in the U.S.A., Canada, and Mexico
by D. Reidel Publishing Company, Inc.
306 Dartmouth Street, Boston,
Mass. 02116, U.S.A.

Library of Congress Catalog Card Number 72–97942

ISBN-13:978-90-277-0362-0   e-ISBN-13:978-94-010-2614-7
DOI: 10.1007/978-94-010-2614-7

This volume and the symposium of

which it is a record are dedicated to the memory of

OTTO STRUVE (1897–1963)

a former president of the International Astronomical Union

(1952–1955)

whose pioneer research created the field of study

to which the symposium was devoted

INTRODUCTION

The proposal to organize a Symposium on circumstellar matter and extended atmospheres in binary systems was first made by the Dominion Astrophysical Observatory to the Executive Committee of the International Astronomical Union in the summer of 1969. It received the support of the presidents of Commissions 29 (Stellar Spectra), 30 (Radial Velocities), 36 (Stellar Atmospheres), and 42 (Photometric Double Stars). Approval in principle was given by the Executive Committee almost immediately, and the Committee further suggested that the Symposium be officially designated the Struve Memorial Symposium. Final approval was given at the time of the 1970 General Assembly of the Union, when the dates of the Symposium were set for August or September, 1972. The Organizing Committee set up consisted of K.O. Wright (Chairman), A. H. Batten, K.-H. Böhm, A. A. Boyarchuk, G. Larsson-Leander, and M. Plavec. In addition, J. Sahade and F. B. Wood acted as advisory members. Local organization was entrusted to a committee consisting of A. H. Batten, E. K. Lee, and C. D. Scarfe. The final dates selected were September 6–12, 1972, and the Symposium was held at the Island Hall Hotel, Parksville, B.C., on Vancouver Island some 90 miles from Victoria.

The Organizing Committee attempted to arrange a Symposium of the type in which no contributed papers would be presented and discussion would range as widely as possible over the field covered by the six invited review papers. Inevitably some of the longer contributions to discussion bear a strong resemblance to papers, but they have all been presented in this volume as a record of continuous discussion. Because different review papers provoked differing amounts of discussion, it has been impossible to maintain a one-to-one correspondence between sessions devoted to review papers and sessions devoted to discussion. In general, the discussion session immediately following a review was concerned with that review, but both the Saturday sessions, and the Tuesday morning session, tended to be free discussion ranging over the whole field of the Symposium. Parts of other sessions were similarly free and a connection between the discussion and the review paper is not always apparent.

All discussion was recorded on tape, and all speakers had a chance to correct virtually all the transcripts. Further editing of the corrected transcripts was frequently necessary. Written summaries were provided for many of the longer contributions. These were frequently shortened and edited also. Some speakers wished to have their contributions bodily moved from the record of one session to that of another. This was rarely possible, but in the editing of the volume, a few cross references have been introduced. Within sessions, some departures have been made from the original order of presentation, and some stretches of dialogue have been reduced or eliminated by making a speaker's first statement more explicit.

Participants at the Symposium were welcomed by Dr. J. L. Locke, on behalf of the National Research Council of Canada, the Mayor of Parksville, Mr. R. G. Young, and Dr. J. Sahade, as Vice-President of the International Astronomical Union. Greetings were also received from the Canadian Astronomical Society. Dr. Sahade also spoke of his "Remembrances of Otto Struve" at the closing banquet of the Symposium on Monday, September 11. He said:

The kind of meeting that we are having here at Parksville is the kind of meeting that Professor Otto Struve would have enjoyed the most and therefore it is fitting to devote it to his memory.

When I think of the new knowledge we have acquired in the field of close binaries in the past few years through the developments of high resolution photometry, radio interferometry and space technology and through a more thorough analysis of objects like the β Canis Majoris stars, for instance, the figure of Otto Struve always comes to my mind. And I recall then his untiring enthusiasm and strong passion for Astronomy his wide field of interest, his open mind, his sense of humour.

He liked to talk about Astronomy practically all the time, he liked to talk with his colleagues and friends about their work and to hear comments on his ideas and thoughts. Struve was always an open-minded person ready to offer an explanation for an observational fact or to accept an alternative suggestion or to change his mind without much hesitation if necessary.

Let me mention as an example that he had worked on β Lyrae for many years and devoted much of his thoughts to find an interpretation for the system that would account for all the observational facts. He was somehow linked with the conclusions that considered that the mass of the primary component was larger than that of the secondary and had been accepted for about fifteen years. In spite of this Struve was receptive, without any conservative feelings, even with joy, to the possibility that the situation in regard to the masses was just the reverse. For him it was always more important to find a more adequate model than to stick to long-accepted ideas.

Struve was working on SX Cassiopeiae at the time I came to Yerkes to work with him. I had just finished my graduate work at La Plata and had no experience in close binary problems. One cloudy evening at the McDonald Observatory he described to me the observational facts and asked me for an idea that would explain them, an idea that, of course, I could not offer. This illustrates the way Struve always acted, he was ready to talk even to a student without feeling any sense of superiority – trying to spread his knowledge around, trying to arouse other people's interest in his own line of interest and trying to find solutions to the problems in a sort of a team work for which Struve was no doubt a leader. While in Berkeley he used to have afternoon tea with Su-Shu Huang and me and during tea we would talk about our research and he would also relay to us the queries he had in mind. One of the problems that concerned Struve then was gravity and how to plan observations that would disclose its nature.

Actually Struve was interested in a wide variety of problems and this is shown by his very extensive work and also by his 155 articles in *Sky and Telescope* where he practically always made a suggestion or submitted an idea.

I still remember Struve's excitement when shown the radial velocity results of AU Monocerotis, which confirmed his interpretation of the discrepancy between the photometric and the spectrographic results of U Cephei, and when shown the radial velocity curve of XZ Sagittarii, which implied a large mass ratio between the two components at variance with the then accepted idea that in close binaries the mass ratio must be always close to unity. I still remember Struve's excitement when shown the emission lines found in Algol or the largely violet-displaced lines of He I $\lambda\lambda$ 3888 and 5876 in HD 47 129. And I could imagine how excited Struve would feel now when we are able to find such far-reaching evidence of interaction between gaseous streams and circumstellar envelopes in close binaries.

Struve was a very hard worker who worked practically without rest and in this context we can say that his life was one of constant and indefatigable endeavour. But in his task as an organizer, an administrator and a teacher, we can distinguish three distinct epochs. The first one covers his years at the Yerkes Observatory when his energies were devoted to the erection of the McDonald Observatory and to the aim of having of a brilliant and homogeneous staff of research workers. Struve gave dimension to the place, thus creating at Yerkes an atmosphere that was felt as soon as one entered the building.

The second epoch refers to Struve's years in Berkeley. He had then no telescope of his own at his disposal and had to do his observing at Mount Wilson. His efforts were devoted then to the students and to the goal of making the School of Astronomy at Berkeley the best in the United States.

Struve's third epoch as an organizer covers the last four years of his life. He had ambitious plans and was very optimistic about the role that the presence of a stellar spectroscopist could play in a radioastronomical observatory. Let me mention that among his plans he thought of extending the range of observations of β Lyrae to the radio region, something that was possible to accomplish successfully only recently. However, the administrative duties at Green Bank took most of his energies and what he could actually do was something different from what he had envisaged. Part of his time was going to be devoted to the writing of books; unfortunately there was no time for Struve to write except his *Astronomy in the 20th Century*.

Struve was characterized by his sense of humour. He would introduce all speakers at colloquia and the like by telling the audience any funny stories he would know about the speakers. And he would change an announced topic of a lecture of his on the β Canis Majoris stars to a paper on β Lyrae, stating that that was all right because the two subjects had something in common: the letter β.

Everybody respected Struve because he was a very kind, unselfish and modest person. He would never give orders, he would always suggest or, at most, strongly suggest. In his talks and reviews he would naturally mention his results, but without mentioning his name among those who contributed to the subject. He was always trying to help and had a deep feeling of loyalty to his friends and colleagues.

The contributions that Struve made in the field of close binaries stand today as the foundation of everything that was done afterwards. I am sure that without his work we would not be dealing here in Parksville with a subject in such a flourishing and exciting state as we are finding it today.

The admiration of all of us for Struve and his work is at the same time a stimulus, a challenge and a responsibility. Our Symposium shows that it is so.

In addition, informal reminiscences were shared with participants by D. M. Popper, A. D. Thackeray, O. C. Wilson, D. B. Wood, F. B. Wood and K. O. Wright. Greetings from the banquet were sent to Professor P. Swings, a long-time friend and colleague of Struve's who was unfortunately unable to be present; Dr. A. H. Joy, and Dr. L. H. Aller.

Other social events included excursions to Long Beach, Vancouver Island, to Little Qualicum Falls, and (for the ladies) to the Fish Hatchery on Qualicum River. A member of the staff of the Hatchery also showed participants a movie of his own making about the Pacific Salmon. There was a swimming race between photometrists and spectroscopists which was clearly won by the astronomers! The good offices of the Parksville and District Chamber of Commerce in helping to organize some of these activities are gratefully acknowledged.

Financial support for the symposium in the form of travel grants came from the International Astronomical Union and the National Research Council of Canada. The latter body also bore all the running expenses of the Symposium, and much of the social expenditure. We record with gratitude a generous donation from the Nicolaus Copernicus Observance Committee of Manitoba Inc. which helped to ensure that the exceedingly active group of astronomers in Warsaw was represented at the Symposium. Thanks are due to the chairmen of the various sessions, whose names appear at the heads of the respective records of discussion. Miss E. M. Edmond (Mrs. Cole) and Miss H. D. Mann undertook the taxing labour of transcription of the recorded discussion. Without their help, the production of this volume would have been impossible. Thanks are also due to Mrs. I. McColl, who did most of the final typing. Mr. E. K. Lee, Mr. B. W. Baldwin, and Mr. W. A. Fisher rendered invaluable service in the practical organization of the sessions. The courtesy and help of the management and staff of the Island Hall Hotel are also gratefully acknowledged. Finally the Editor wishes to thank all participants who have co-operated so well in correcting transcripts and supplying summaries.

The photograph of Otto Struve used as the frontispiece was taken by Mr. S. H. Draper when Struve visited the Dominion Astrophysical Observatory in 1955. The original print was autographed by Dr. Struve, and has been copied by Mr. Draper for this volume.

A. H. BATTEN

October, 1972

TABLE OF CONTENTS

LIST OF PARTICIPANTS

J. Andersen, Copenhagen University Obervatory, Denmark.

B. W. Baldwin, University of Victoria, Canada.

G. T. Bath, University of Oxford, England.

A. H. Batten, Dominion Astrophysical Observatory, Victoria, Canada.

P. Biermann, Universitäts Sternwarte, Göttingen, F.R.G.

K.-H. Böhm, University of Washington, Seattle, U.S.A.

C. T. Bolton, David Dunlap Observatory, Richmond Hill, Canada.

B. W. Bopp, University of Texas, Austin, U.S.A.

S. Catalano, Osservatorio Astrofisico, Catania, Italy.

K.-Y. Chen, University of Florida, Gainesville, U.S.A.

M. L. Cooper, U.S. Naval School Monterey, U.S.A

A. Cowley, University of Michigan, Ann Arbor, U.S.A.

E. J. Devinney, University of S. Florida, Tampa, U.S.A.

T. H. H. Lloyd Evans, Royal Observatory, Edinburgh, Scotland.

R. Faraggiana, Osservatorio Astronomico, Trieste, Italy.

M. G. Fracastoro, Osservatorio Astronomico, Pino Torinese, Italy.

B. A. Goldberg, University of British Columbia, Vancouver, Canada.

M. de Groot, European Southern Observatory, Santiago, Chile

D. S. Hall, Dyer Observatory, Nashville, U.S.A.

H. K. Hansen, Brigham Young University, Provo, U.S.A.

T. Herczeg, University of Oklahoma, Norman, U.S.A.

R. W. Hilditch, Dominion Astrophysical Observatory, Victoria, Canada.

G. Hill, Dominion Astrophysical Observatory, Victoria, Canada.

S.-S. Huang, Dearborn Observatory, Evanston, U.S.A.

D. G. Hummer, Joint Institute for Laboratory Astrophysics, Boulder, U.S.A.

J. B. Hutchings, Dominion Astrophysical Observatory, Victoria, Canada.

M. Kitamura, Tokyo Astronomical Observatory, Japan.

G. Larsson-Leander, Lund Observatory, Sweden.

E. K. Lee, Dominion Astrophysical Observatory, Victoria, Canada.

K.-C. Leung, University of Nebraska, Lincoln, U.S.A.

A. P. Linnell, Michigan State University, East Lansing, U.S.A.

C. Magnan, Institut d'Astrophysique, Paris, France.

J. M. Marlborough, University of Western Ontario London,, Canada.

D. H. McNamara, Brigham Young University, Provo, U.S.A.

E. F. Milone, University of Calgary, Canada.

T. J. Moffett, University of Texas, Austin, U.S.A.

B. Nordström, Stockholm Observatory, Sweden

J. P. Oliver, University of Florida, Gainesville, U.S.A.

G. J. Peters, University of California, Los Angeles, U.S.A.

M. Plavec, University of California, Los Angeles, U.S.A.

R. S. Polidan, University of California, Los Angeles, U.S.A.

D. M. Popper, University of California, Los Angeles, U.S.A.

E. L. Robinson, University of Texas, Austin, U.S.A.

J. Sahade, Instituto de Astronomia y Fisica del Espacio, Buenos Aires, Argentina.

C. D. Scarfe, University of Victoria, Canada.

J. Smak, Institute of Astronomy, Warsaw, Poland.

A. D. Thackeray, Radcliffe Observatory, Pretoria, S. Africa.

A. B. Underhill, Goddard Space Flight Center, Greenbelt, U.S.A.

F. Van 't Veer, Institut d'Astrophysique, Paris, France.

K. Walter, University of Tübingen, F.R.G.

J. A. J. Whelan, Joint Institute of Laboratory Astrophysics, Boulder, U.S.A.

O. C. Wilson, Hale Observatories, Pasadena, U.S.A.

R. E. Wilson, University of S. Florida, Tampa, U.S.A.*

D. B. Wood, Goddard Space Flight Center, Greenbelt, U.S.A.

F. B. Wood, University of Florida, Gainesville, U.S.A.

K. O. Wright, Dominion Astrophysical Observatory, Victoria, Canada.

Secretariat:

Miss E. M. Edmond, Dominion Astrophysical Observatory, Victoria, Canada.

Miss H. D. Mann, Dominion Astrophysical Observatory, Victoria, Canada.

* Now at NASA Institute of Space Studies, New York, U.S.A.

DISCUSSION OF OBSERVATIONS OF THE FLOW OF
MATTER WITHIN BINARY SYSTEMS

ALAN H BATTEN

Dominion Astrophysical Observatory, Victoria B. C.

Abstract. Observational determinations of density, dimensions, temperatures, and velocities of circumstellar features are surveyed and discussed, with a view to establishing limiting values that could be useful in any theoretical treatment of circumstellar structure. Densities of the order of 10^{13} particles cm^{-3} are found for streams and disks, although there is evidence for regions of much lower density in many systems, and some systems may have much denser circumstellar matter.

Dimensions of disks seem to be remarkably constant (fractional radius ~ 0.3) in a wide variety of systems. The total mass of circumstellar matter is always a small fraction of the mass of the system. Temperatures are usually similar to the temperatures of the stars in the system, although hot spots certainly exist in some systems, and 'flare' activity is evidence of localized very high temperatures. On the other hand, some systems may contain solid circumstellar matter. Observed velocities are usually several hundred km s^{-1}. The empirical relation between the velocity of rotation of disks and the orbital period is described. Finally, the stability of these features is briefly discussed.

1. Introduction

The observational evidence for the existence of circumstellar matter in close binary systems was reviewed recently (Batten, 1970) and this review has been updated, amended, and extended in a book to be published shortly (Batten, 1973). Despite great activity in the field within the past eighteen months, which has made this Symposium seem even more timely than it appeared to be when it was first proposed, a second updating now of my 1970 review would contain more repetition than could be reasonably justified. Therefore, in this review, I shall try to examine both new and old papers in order to find out what is known about the physical properties of circumstellar material in binary systems and to determine what constraints the presently available observations place on any theory of the structure and motions of this circumstellar matter.

In the 1970 paper, I introduced the concept of three elements in the circumstellar matter: disk, streams, and cloud (see Figure 1 of that paper). This concept still seems to me to be useful, and I shall use the same terminology in this review. There are not of course, sharp density boundaries between the different elements, as is inevitably suggested by the diagram: the real distinctions between elements are probably kinematic, and even these cannot be sharp. Nevertheless, densities within a region several stellar radii above the surface of one component of a binary system are likely to be very much less than those within a stream flowing between the components. Even within a stream, there may be great variations in density, depending on the velocity of the material within the stream (Korsch and Walter, 1969). The distinction between these three elements is certainly important in discussions of circumstellar density, however. Unfortunately, not every investigator has made it clear to which part of a binary system his estimate of circumstellar density applies. One recent review included erroneous conclusions about the capacity of the observed streams for mass transfer

Batten (ed.), Extended Atmospheres and Circumstellar Matter in Spectroscopic Binary Systems, 1–21.

partly because a 'mean stream density' was derived from figures referring to all three elements. From the title of this paper, which is based on that laid down by the Organizing Committee, you may expect me to concentrate on a discussion of the streams, but it seems to me to be necessary to consider all three elements of the circumstellar matter.

New observational tools are now at our disposal that will undoubtedly prove powerful aids in the study of circumstellar matter, and greatly contribute to our understanding of it. The first observations of a close binary system (β Lyrae) made from above the Earth's atmosphere have been published (Kondo, 1971; Kondo *et al.*, 1972) and many more have been obtained by the OAO. Observations of the far ultraviolet region of the spectrum have proved of great help in understanding the loss of mass from stars of early spectral type (Morton 1967a, b; Morton *et al.*, 1968; Hutchings, 1970) and they may also be expected to contribute decisively to our knowledge of mass exchange within binary systems. Unfortunately, many of the most interesting systems are rather faint for the instruments presently above the atmosphere of the Earth. There is also increasing evidence for the identification of several galactic X-ray sources with close binary systems (Burbidge *et al.*, 1967; Kristian *et al.*, 1967; Kraft and Demoulin, 1967; Murdin and Webster, 1971, 1972; Bolton, 1972a; and – less certainly – Braes and Miley, 1972; van den Bergh, 1972). Prendergast and Burbidge (1968) have shown that X-rays could be generated by streaming circumstellar matter – a hypothesis that has been further elaborated by Kraft (1972). Finally perhaps the greatest innovation has been the detection of radio waves from four close binary systems: Antares B, β Lyrae, β Persei, and HD 226 686 (probably Cyg X-1). The clearest observational picture presently available is that for β Persei (Wade and Hjellming, 1972; Hughes and Woodsworth, 1972). This indicates that a kind of flaring activity takes place at times within the system, and this may be paralleled by activity in the optical spectrum (Bolton, 1972b). These are exciting developments indeed, but it is still difficult to integrate them into our already existing corpus of knowledge. There is no doubt that these new results will figure prominently in our discussions, but I shall base my review primarily on the results of ground-based, conventional observations which have so far contributed the greatest part of our knowledge of circumstellar matter in binary systems. The last word, however, at least for some systems, may come from these new observations in spectral regions very different from those employed in conventional spectroscopy and photometry, or from observations in the normal spectral regions made with very much higher time resolution than is now possible.

2. Location and Dimensions of the Elements of Circumstellar Matter

Disks are usually detected by their emission spectra which, in many systems, are best seen at primary eclipse (when the star within the disk is hidden). The behavior of these emission lines during eclipse clearly reveals that they arise in some structure that surrounds the eclipsed (usually primary, or more massive) star. The classic example is RW Tauri, but disks are known in systems of widely different sorts, from short-period systems containing high-temperature stars, through the Algol-type systems (including RW Tauri) to systems like SX Cassiopeiae and VV Cephei that contain giant or supergiant stars. For the Algol-type systems, crude estimates of the radii of the disks can be made from the duration of the visibility of the emission lines during eclipse. Many of these have been quoted elsewhere (Batten, 1970). Plavec (1968) developed a more refined method for determining the size of the disk from the width of the line profile. This method is based on the assumption that individual particles within the ring rotate in Keplerian orbits. He found that the disk in RW Tauri must have a radius of about 2.6 times the radius of the star that it surrounds. Baldwin (1972) applied the same analysis to observations of U Cephei, and found that the disk had a radius of 2.8 stellar radii (or about 8.4 solar radii).

In the much larger system of SX Cassiopeiae, Günther (1959) has estimated that the disk has a radius of about 4.6 times that of the primary star. He did this by comparing the durations of primary and secondary eclipse, assuming that the disk is opaque enough to eclipse the secondary star, but that its own eclipse (in the light of the continuous spectrum) is not perceptible. Another long-period system, AR Pavonis, has been discussed by Thackeray (1959). From the variation of the emission-line profile during eclipse, he finds that the primary star must be surrounded by an extended disk, about twice the radius of the *secondary* star, that is both expanding and rotating.*

The spectrum of VV Cephei shows very strong emission lines of hydrogen which are again interpreted as arising in a structure surrounding the (invisible) hotter component, a star probably of late O or B spectral type. Model computations by Hutchings and Wright (1971) suggest that this the denser part of the structure has a radius of about 50 times the radius of the hot star, which they believe to have a radius of $13 R_\odot$, but the whole structure extends to about 150 times the radius of the hot star (the separation of the two stars appears to be about $5400 R_\odot$, and the radius of the M-type star to be about $1800 R_\odot$). This structure perhaps should not be called a disk, since they estimate that it fills about 85 per cent of the spherical region of radius $1950 R_\odot$ that surrounds the primary star.

Disks are also observed in the eruptive binaries where they may account for a significant fraction of the total light. These systems, for the most part, have very short periods and very small orbits. Disks in these systems seem to have radii approximately equal to the Sun's radius (Smak, 1971a; Kraft, 1959). The central stars in the disks are probably white dwarfs, at least in the systems DQ Herculis and U Geminorum. The separation of the two stars in the latter system, although very uncertain, is estimated to be about $3 R_\odot$ (Smak, 1971a).

If the radius of the disk is denoted by ϱ, and the major semi-axis by a, then we can use the quotient ϱ/a as a measure of the size of the disk relative to that of the system. If U Geminorum is typical of eruptive binaries, then ϱ/a is approximately 0.3 in these systems. In the Algol-type systems it appears to be somewhat larger, namely about 0.6 – again if U Cephei and RW Tauri can be considered typical, but the method of

---

* See, however, his own comments on p. 98.

determination used for these two systems will necessarily give a larger result than would be obtained by other methods. In VV Cephei, ϱ/a is approximately 0.35. Brown *et al.* (1970) have measured, with their stellar interferometer, the angular diameter of the emitting region that surrounds the Wolf-Rayet component of γ^2 Velorum. They find $\varrho/a = 0.24$. Thus ϱ/a seems to be nearly constant at about 0.3 in many different kinds of systems. Only in the Algol-type systems does the disk seem to be larger relative to the size of the system. Even in these systems, the disks are larger only by a factor of two that may well be accounted for by observational uncertainty.

Even less is known about the extent of these disks above or below the orbital plane. The convention, at least in the Algol-type systems, has been to assume that the circumstellar matter is highly concentrated toward the orbital plane. The famous representation of the 'ring' in RW Tauri (Skilling and Richardson, 1947) shows a very thin ring indeed – but this was only ever intended as a rough qualitative representation. Unfortunately, it has become fixed in many people's minds as a real indication of the nature of these rings or disks. Batten (unpublished) has found evidence for the existence of fairly substantial disk, in U Cephei, which scatters an appreciable amount of the light of the primary star and must therefore cover a considerable portion of its surface. These disks, therefore, must have a thickness (perpendicular to the orbital plane) that is at least comparable to the diameter of the central star. On the other hand, since the disks are rotating rapidly (except in the long-period systems, see Section 6) they are likely to be very appreciably flattened, and therefore to extend much less above and below the orbital plane than they do within it. Figure 1, although still schematic, probably gives a better idea of the sort

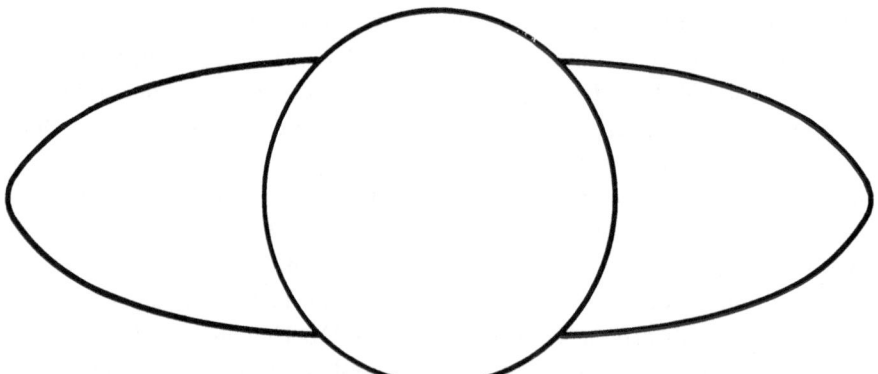

Fig. 1.   Schematic representation of a disk in a typical Algol system.

of structure that seems to exist in many Algol-type systems. Disks of similar size, although possibly of somewhat different character (Section 5), have been proposed for a number of systems by Hall and his colleagues (e.g. Hall, 1969; Hall and Taylor, 1971). In the eruptive binaries, according to Smak (1971a) the disks are 'quite extended in the z-coordinate', but little is known. In VV Cephei, as we have already seen, the model by Hutchings and Wright indicates that the 'disk' nearly fills a large spherical volume, and therefore must extend quite far out of the orbital plane. It would seem most natural to assume that the structure is an oblate spheroid of revolution. Its rotation is much less than that of disks in short-period systems, and therefore the structure is much less flattened. Coyne (1970) has found from the variable polarization of light in β Lyrae that the disk in that system must be appreciably flattened. None of these estimates is directly based on physical measurements and the extension of circumstellar disks outside the orbital plane is but poorly known. It is, indeed, difficult to determine this observationally, and the uncertainty in this quantity must reflect on any attempt to determine the masses of circumstellar disks.

Even fewer determinations have been made of the *precise* course and dimensions of streams. Many schematic diagrams have been published for particular systems, but they are nearly all qualitative, even including those that resemble, in appearance, the precision and detail of an anatomical drawing! Our ideas about the courses of streams are based very largely on the results of particle-dynamics calculations in which pressure effects within the streams, collisions between individual ions, and any possible magnetic forces are ignored. The most comprehensive theoretical calculations of trajectories yet made are based on these assumptions of particle dynamics (Plavec *et al.*, 1964; Plavec and Kříž, 1965). Prendergast (1960), Sobouti (1970), and Biermann (1971) have studied the hydrodynamical behaviour of circumstellar gases in binaries, but they have computed velocity fields rather than trajectories. The observer who wishes to compare his results with theory can make the comparison only with particle-dynamics trajectories. Surprisingly the comparison is not bad, although Kopal (1971a) argues that particle dynamics is insufficient even for a useful first approximation. The matter can only be settled by determining the entire trajectories (course and velocity) of streams in several systems. This is very difficult and perhaps impossible to do, but until it has been done arguments about whether particle dynamics is useful at all will remain academic. No doubt both hydrodynamic and magnetic effects are present in many systems (Walter, 1971a, 1972) and may dominate in some, but in most systems there seems to be little or no conflict between the observations of streams – as far as they have been interpreted – and the predictions of particle dynamics. Observations of disks are in a different category. Some disks have been observed which ought not to exist if the assumptions of particle dynamics were rigorously valid. The point of this argument is not to try to defend a theory that *must* be wrong in its predictions of details, but to urge that it is a useful first approximation.

Circumstellar matter is to be found not only around one of the components of a binary system, but also between the two components. This is attested by observations of emission at Hα, in the spectrum of Algol (Struve and Sahade, 1957) and possibly also in the spectrum of U Coronae Borealis (Struve *et al.*, 1957). This is often attributed to a concentration of matter near the Lagrangian point, but the emission could equally well arise from the whole length of the stream between the two components. In the case of Algol, however, variations in the intensity of the emission do seem to indicate an accumulation of matter near the Lagrangian point (Andrews, 1967).

Other observations of the stream in Algol have been made by Fletcher (1964) who has detected its effect on the He I line at $\lambda 4471$ in the spectrum of the system. The observations show an increase in the amount of absorption in this line about seven hours before the middle of primary eclipse (or about 2 h before the eclipse begins). The absorption reaches a maximum about three hours before mid-eclipse and thereafter declines steeply to a sharp minimum at mid-eclipse, rising again to a secondary maximum about four hours later. The sharp decrease, of course, is caused by the eclipse of the brighter star by the fainter. The other variations were explained by Fletcher as the result of the eclipse of a gaseous stream, also by the fainter star. The stream absorbs extra light in the wavelength $\lambda 4471$, and therefore modifies the limb-darkening coefficient in that wavelength so that it is different both from that in the neighbouring continuum and from the value it would have in a pure stellar line. The observations can be represented by a tongue of matter contiguous with the secondary component. The extent of the stream above and below the orbital plane is shown by Fletcher's analysis to be comparable to its width within the orbital plane. Here is an attempt to map the stream in Algol at least semi-quantitatively, but the results do raise some questions. First, we expect on theoretical grounds that the stream does not originate from a whole hemisphere of the secondary component, but from a small area around the Lagrangian point of the, presumably, distorted secondary star. Second, we may question whether circumstellar matter so close to the G-type component of the system would be hot enough to absorb the radiation of a helium line in a detectable quantity – the observation of additional absorption at $\lambda 4471$ suggests that the bulk of the matter producing the absorption is in fact fairly close to the B-type

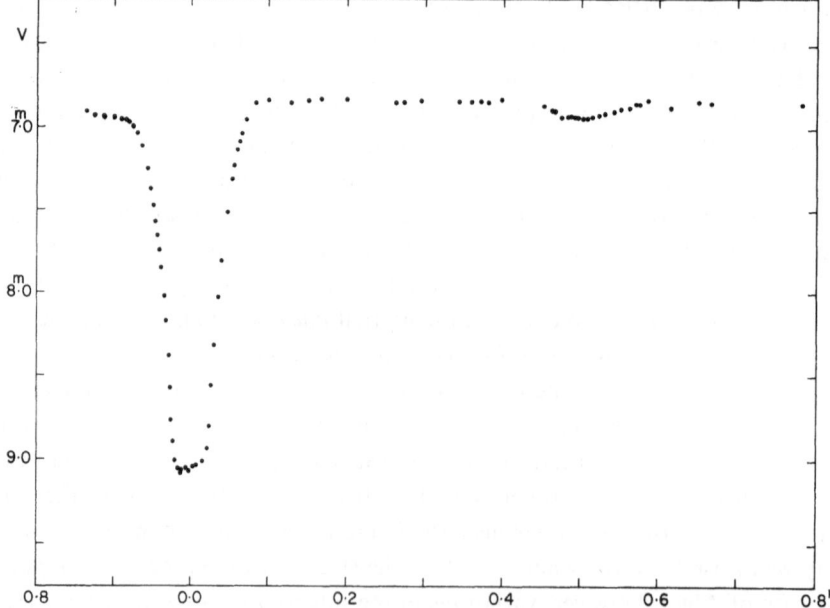

Fig. 2. The light curve of U Cep in V light, as determined by Broglia (unpublished).

component. Third, it is exceedingly implausible that a tongue of matter emerging from the G-type star should just stop. The matter must either fall into the B-type star, or go around it, and in either case it should still be detectable. These last two arguments suggest that the matter producing the extra absorption may be in a tongue extending from the B-type star, rather than from the G-type star, and it might be worthwhile to reconsider the same observations on the basis of a somewhat different model.

The light curve of U Cephei provides a similar clue to the course of the stream in this system (Figure 2). Between phase $0^{P}.8$ and $0^{P}.9$ the total light of the system decreases by about $0^{m}.1$, and the shoulder preceding eclipse is lower than that following it by this amount (eclipse begins at $0^{P}.92$). The disturbance of the light curve is the same in all colours, and can be traced right into eclipse, until phase $0^{P}.95$. The similarity of the disturbance in all colours suggests that it is caused by electron scattering in a stream that wholly or partially obscures the primary component at these phases, until it is itself eclipsed by the secondary star. The phases at which the disturbance appears and disappears define boundaries in the orbital plane between which the obscuring matter must lie (Figure 3). The extent of the stream perpendicular to the orbital plane cannot be derived from this information, but a plausible assumption is that the stream is roughly conical, the axis of the cone being aligned somewhere within the triangle of Figure 3. The observed diminution of $0^{m}.1$ in the light of the system would require

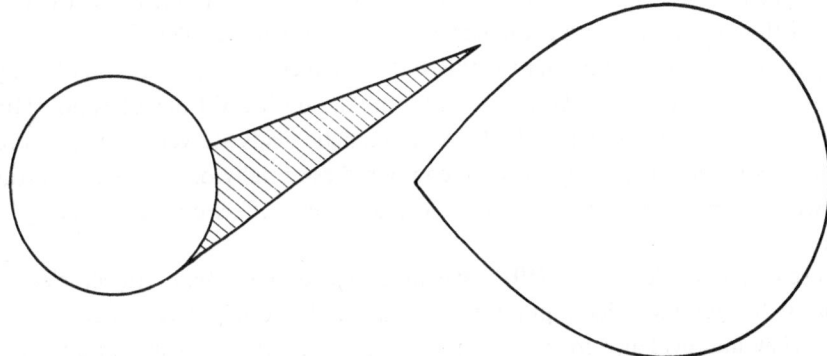

Fig. 3.   Location of the stream in U Cep, as indicated by the light curve.

1.4×10^{23} scattering electrons above each square centimetre of the star's surface, if the absorption is taking place uniformly over the entire stellar disk. More probably only a portion of the stellar disk is covered. If we knew the length of the stream and its diameter, we could compute its density. The length cannot be much more than the distance between the surfaces of the two stars, say about twice the radius of the primary star, and a reasonable guess at the area of the stream where it meets the surface of the star is that it is approximately equal to the area of the projected disk of the Sun. The primary star has a radius of about three solar radii, so that these assumptions lead to a density for the stream of about 3×10^{12} electrons cm^{-3} – somewhat lower than I have previously estimated (Batten, 1970, 1973) but still within the range found for stream densities. There is evidence that some of this stream goes right round the B-type star, and returns to the G-type star.

If the course deduced for the stream in U Cephei is correct, and if the matter does originate from the Lagrangian point of the secondary component, then somewhere it must make a sharp turn. Such a turn is compatible with particle-dynamics trajectories (Plavec *et al.*, 1964) for particles that have been ejected tangentially to the surface of the secondary star, at fairly high velocities. It is not necessary to propose magnetic fields, although, of course, they may be present. Direct evidence of magnetic fields would be hard to obtain, because the spectrum of U Cephei and those of most Algol-type systems, have such broad lines that Zeeman splitting would be almost impossible to detect. Indirect evidence might come from the careful determination of trajectories. Walter (1971b) has considered the trajectory of the stream in another Algol-type system, SW Cygni, and more recently (1972) in the system TW Andromedae. He finds that the light curve of this system cannot be interpreted without the assumption of additional light, and he supposes this light to arise in bright spots on the surface of the primary star, where the stream meets the surface. If this assumption is granted, then he finds that the bright spots are in quite high latitudes of the primary star (the equator being supposed to coincide with the orbital plane) and he suggests that the stream has been bent out of the orbital plane by magnetic fields. This hypothesis might not be necessary, however, if more were known about the density distribution within the stream, perpendicular to the orbital plane. Hall and Garrison (1972) have proposed a different model for this system. Other indirect evidence of the presence of magnetic fields may be provided by the radio observations of Algol (Hughes and Woodsworth, 1972). Appreciable local magnetic fields are very probably present within the streams, since they must be composed essentially of plasma. It is still not clear whether these local fields would have a large-scale effect on the trajectory of the stream.

Hansen and McNamara (1959) have investigated the stream trajectories in the system of RZ Scuti and have published a detailed, but qualitative, scheme of trajectories. They find evidence for a stream circulating the whole system, and for a large counter-rotating eddy in the neighbourhood of the following hemisphere of the primary star. This eddy is proposed to explain changes in the asymmetry of the helium-line profiles observed at phases 9 and 13 days (the orbital period is just over 15 days). These changes are in turn held responsible for step-like distortions of the velocity curve. Hansen and McNamara were the first to point out this step-like feature, although there are also hints of it in the velocity curve of U Cephei (Struve, 1960) and similar features are to be found in the velocity curve of VV Cephei (Hutchings and Wright, 1971). There is insufficient room in the system of U Cephei for a counter eddy of the size proposed by Hansen and McNamara for RZ Scuti, and the velocity curve of VV Cephei would require several eddies for its explanation. These considerations throw some doubt on the explanation offered by Hansen and McNamara, and yet the steps do seem to be associated in some way with the presence of gas streams. These step-like features may be a useful clue to the study of the nature and structure of the streams.

Most of the little quantitative work that has been done on gas streams has been concerned with the Algol-type systems, but a model devised for UX Ursae Majoris – a system that has much in common with the eruptive binaries – by Walker and Herbig (1954) should be mentioned. They found evidence for a stream completely detached from the hotter component, approximately half circling it (visible from phases $0\overset{P}{.}4$ to $0\overset{P}{.}8$) from which matter is falling on to the hot star. Its width is comparable to the radius of that star, as is also the separation of the inner boundary of the stream from the surface of the star. The portion of the stream projected on the stellar surface between phases $0\overset{P}{.}7$ and $0\overset{P}{.}8$ is believed to be hotter than the rest.

The dimensions of clouds can often be roughly estimated from the dilution effects in their spectra. Clouds are the elements in circumstellar matter that display shell spectra. Although they are often drawn as detached shells, they are probably more nearly tenuous clouds that surround the system, or perhaps spirals of matter escaping the system, as Kuiper (1941) envisaged for β Lyrae, or Herbig, Preston, Smak, and Paczyński (1965) have suggested for V Sagittae. They have traced out a spiral jet to some ten times the separation of the component stars, and Figure 15 of their paper indicates both the location and the width (in the orbital plane) that they have deduced for the jet from their measurements of the velocities and intensities of absorption reversals in the hydrogen emission lines of this complex spectrum. As the authors themselves point out, the reliability of the radial scale of the model depends strongly on the accuracy with the radial velocity can be determined from the absorption reversal. The maximum width of the jet within the orbital plane appears to be somewhat greater than the separation of the two stars.

Thackeray (1971) has studied the emission lines in the spectrum of HD 72 754 – a system that seems to resemble β Lyrae. The hydrogen and helium emission lines are cut into by absorption, and Thackeray finds that the combined profiles can be explained if they arise from a narrow, detached ring rotating around the whole system, and having a radius of four to five times the separation between the two stars. These features in the spectrum resemble the spectrum of W Crucis. Thackeray explicitly states that the ring is not expanding as the similar features in β Lyrae and W Crucis seem to be.

In some systems that show shell spectra, the shell component seems not to arise from a shell, in any ordinary sense of that word, but from portions of streams that, because of the size of the system, are far enough from either of the stellar components to display a shell-type spectrum. This seems to be the case, for instance, in AX Monocerotis (orbital period $232\overset{d}{.}5$ – see Plavec and Harmanec, 1972). Shell features appear in the spectrum only at certain orbital phases, when, presumably ,the relevant part of the stream is exposed to view. A similar situation is encountered in the system of V 367 Cygni (Heiser, 1961) in which the shell component of the spectrum also seems to arise in a stream. The pattern of streaming in this system, however, is quite different from anything encountered in the Algol-type systems. The streams are centred on the primary component (of spectral type A2), apparently falling in on one side, and expanding outwards on the other.

3. Density and Mass of Circumstellar Features

The available evidence on the density of disks, streams, and clouds has already been summarized (Batten, 1970, 1973). Only a few new estimates have been published since the latter compilation. Walter (1971a) estimates an electron density of 10^{12} cm^{-3} for the stream in SW Cygni, and this fits well with my earlier conclusion that electron densities in the streams are of the order of 10^{13} cm^{-3}, with an uncertainty of at least one in the exponent (which may be due either to mere uncertainty, or possibly to a real variation from system to system). Disks probably have similar but slightly lower densities, and clouds seem usually to have electron densities of 10^{11} cm^{-3}, or less.

Warner and Nather (1971), however, propose a much higher density for the streams in U Geminorum – namely, 1.5×10^{16} electrons cm^{-3}. This estimate is based on the observed period change, which is in turn interpreted as being caused by mass transfer between the components. They also estimate the density of the disk as 6×10^{17} electrons cm^{-3}. This is derived by assuming its mean lifetime, in the absence of any supply of matter, to be of the order of 10^5 s, as computed theoretically by Gorbatskij (1969) and by balancing this against the amount of matter supplied to the disk by the stream to obtain the mass of the disk. If a value is then assumed for the volume of the stream (they assume 10^{30} cm^3) then the density can easily be derived. These densities derived by Warner and Nather are considerably higher than any other empirical estimates. While U Geminorum is undoubtedly an unusual system, these high densities have not yet been found in any other eruptive binary. Such estimates as have been made are more in line with those made for 'ordinary' binaries, e.g. DQ Herculis 3×10^{13} electrons cm^{-3} (Kraft, 1959), AE Aquarii 6×10^{10} particles cm^{-3} (Crawford and Kraft, 1956), WZ Sagittae 4.5×10^{12} electrons cm^{-3} (Krzeminski and Kraft, 1964). The estimate of stream density depends entirely on the period change being correctly identified as due to mass transfer. The rate of change is derived from observations extending over five years, or 10000 cycles. It is not clear whether the period may not be steadily increasing, or whether abrupt changes have occurred. Warner and Nather interpret the period change as a steady increase corresponding to a rate of mass transfer of 3×10^{-7} $M_\odot$ y^{-1} or approximately 2×10^{19} gm s^{-1} although even a steady change of period could be partly due to mass loss. Now Kraft (1972) in his discussion of X-ray radiation from close binaries suggests that X-rays are not observed from U Geminorum stars because the rate of mass transfer is too low, and the rate he quoted is about 10^{16} gm s^{-1}. Warner and Nather may have overestimated the mass transfer rate, and hence the density of the stream because the period change is still poorly determined (and perhaps misinterpreted). The high density of the disk depends on the density determined for the stream, and is therefore probably similarly overestimated.

Smak (1971a, b) has suggested that the observed period changes in eruptive binaries are not caused by mass transfer, but by interaction between the rotating disk and the star that it surrounds. Some eruptive binaries show irregular period changes, in either direction, of approximately $(d \ln P/dt) = 10^{-10}$ or 10^{-8} d^{-1}. To produce changes of this magnitude in the way he has proposed, Smak finds that the disks must have at least 10^{-6}, and possibly 10^{-4}, of the mass of their central stars (generally assumed, in the eruptive binaries, to be white dwarfs of approximately solar mass). Even with the exceptionally high densities derived by Warner and Nather, and with the volume they assumed, the mass of the disk in U Geminorum comes to less than 10^{-9} $M_\odot$. It therefore seems that the mechanism suggested by Smak cannot apply to eruptive binary systems.

The volumes of the disks in Algol-type systems are naturally larger – by a factor of the order of 10^5. The estimated mean density of these disks, however, is of the order of 10^{13} electrons cm^{-3}, and if ionization within the disk is complete, or nearly so, the mass of the disk again works out to be of the order of 10^{-9} or 10^{-8} $M_\odot$. The mass of the disk in SX Cassiopeiae has recently been estimated at 10^{-9} to 10^{-6} $M_\odot$ (Koch, 1972).

In the larger, long-period systems the disks are larger still. In VV Cephei the 'disk' is very large, but the density within it falls off very steeply (Hutchings and Wright, 1971). Its total volume is of the order of 10^{43} cm^3. If it had a uniform density of 10^{11} particles cm^{-3}, and if these particles were all hydrogen atoms, the total mass of the disk would be about 10^{30} gm or 10^{-3} $M_\odot$ (of the order of 10^{-4} of the mass of its central star). Because the density is increasing steeply inwards and is probably much higher in the central regions, the total mass of the disk must be greater than this. In this system, therefore, the disk is massive enough to interact gravitationally with the components of the binary. Its effect on period changes would be hard to detect, however, since the period is 20 y.

A very massive disk indeed was proposed for β Lyrae (Woolf, 1965) comparable in mass to the central star itself. It is very doubtful whether so massive a disk would be stable. (In one sense no disks are stable: Gorbatskij's (1969) computations indicate disks would speedily collapse if no mass were supplied to them. A very massive disk, however, presents special problems of short-term stability.) Shulov (1967) attempted to derive the density and mass of the disk in β Lyrae from his polarization measurements and found a total mass for the disk of 10^{-7} $M_\odot$. In view of the strength of the emission lines in the spectrum of β Lyrae, which are usually attributed to the disk, this may seem a rather low mass, but even if the disk is a scaled down version of that found in VV Cephei, it must be several orders of magnitude less massive than was suggested by Woolf.

Hall (1971) has proposed that the secondary component of BM Orionis is a disk of total mass one to three solar masses. There are difficulties in reconciling the light curve with the fact that only a B-type spectrum is visible if the system is supposed to consist of two normal spherical stars. Hall deduces a mass for the disk of from one to three solar masses, but it is not clear whether the disk is homogeneous or contains a normal central star surrounded by a not very massive, but optically thick disk.

As was apparent in Section 2, the dimensions of streams are very much less well-known than those of disks, and the masses are correspondingly uncertain. In most systems, however, the volume of the stream seems to be much less than that of the disk, and since the densities are similar, the masses of the streams must be small compared with those of the disks. Little, if anything is known about the masses of clouds Although they probably contain relatively little mass at any one time, because of their low densities, they may be the means by which some of the mass of a binary system is completely lost into space. The available data indicate that the amount of mass between the components of a binary system is nearly always a very small fraction of the total mass of the system.

4. Ionization and Temperature of Circumstellar Matter

Circumstellar matter must be very far from local thermodymanic equilibrium, and therefore its temperature, in any given system, cannot be uniquely defined. In any system in which the matter shows an identifiable spectrum, however, some estimate of the degree of ionization, and hence of the ionization temperature, can be made, although this will necessarily be different from the temperature defined by the mean energy of individual atoms. For example, in β Lyrae, different portions of the stream appear to have different temperatures. Thus the expanding shell (cloud) is given a spectral type of B2 or B5 compared with the central star's type of B8. The red-displaced satellite lines are described as resembling an A2 spectrum (Sahade et $al.$, 1959), or more precisely, mainly on the basis of the ratio of the intensities of the lines of Fe I to those of Fe II, intermediate in type between α Cygni (A2) and α Persei (F5) (Struve and Zebergs, 1959), while the violet-displaced satellite lines correspond to a distinctly higher stage of ionization. Thus different ionization temperatures can be assigned to the various parts of the circumstellar matter in β Lyrae, although the different sets of lines do show anomalous intensities. For example, the red-satellite lines include lines of helium that should scarcely be visible at all in an A2 spectrum, and the early type of the shell lines no doubt represents an effect of the reduced electron pressure, rather than of higher temperature. Thus, the example of β Lyrae both illustrates what can be done in this way, and shows the difficulty of assigning a meaningful temperature to circumstellar matter.

Heiser (1961) has tried to estimate the excitation temperature of the stream in V 367 Cygni at different phases. At phases $0^p.18$ and $0^p.46$ he finds an excitation temperature of 6700 K, while at phase $0^p.78$ it is only 5400 K. It is quite conceivable that the stream should appear to have different temperatures at different phases, for one may be looking at different parts of the stream which may be at different distances from the exciting star (probably about A2 in V 367 Cygni), but it is not clear whether this reported difference is significant. In U Cephei, also, there is some evidence that a part of the stream has a 'temperature' similar to that of the G-type component, while most of it seems to display the same ionization as does the B-type component.

Not much work has been done on the temperatures of streams in other Algol-type systems, although, as already pointed out in Section 2, the presence of a helium line in the spectrum of the stream in Algol itself suggests an ionization in the stream similar to that in the atmosphere of the primary component.

Hutchings and Wright (1971) derive values for the excitation temperature at different levels in the envelope surrounding the secondary star of VV Cephei. They believe this star to be of late O or early B spectral type. They find an excitation temperature that varies from 16000 K at a distance from the secondary star equal to 0.01 times the separation of the two stars, to 7000 K at a distance of 0.12 times the separation. These figures are not obtained directly from observation, but by fitting a model to the emission-line profiles.

Kraft (1959) attempted to determine temperatures for the assumed nova component of DQ Herculis and the surrounding disk, from their combined colour. He found that the best fit was obtained if he assumed the nova component (which contributed 25% of the visual light) has a temperature of 80000 K and an electron temperature of 40000 K for the disk. Other arguments (from the ionization equilibrium of helium and hydrogen) indicate even higher temperatures. Similar high temperatures were found for circumstellar matter in T Coronae Borealis (Kraft, 1958).

Warner and Nather (1971) have also estimated a temperature of 50000 K for the circumstellar matter in U Geminorum, but this is confined to the hot spot that they believe is formed by the collision of the stream and the disk, and which they estimate to be confined to an area of about 3×10^9 cm diameter. Their estimate of temperature, however, depends partly upon the very high densities they have assumed for the stream and disk, although not sensitively so since it is the ratio of the two densities that is important. Nevertheless, Smak (1971a) has attempted to estimate the temperatures of hot spots in several eruptive binaries from their colours (which can be determined from the colour change during eclipse if the eclipse is essentially of the hot spot) and finds the surprisingly low range 7000 K to 18000 K. From this result, he deduces equivalent black-body temperatures for the disks of 12000 K or less. Thus there is much uncertainty in the estimated temperatures of both hot spots and disks in eruptive binaries.

Another attempt to estimate the ionization temperature of a disk in a long-period system has been made by Günther (1959) for SX Cassiopeiae. He has attempted to derive an ionization temperature by means of Saha's law, and finds 11500 K although the effective temperature of the central star is only 8400 K. As he estimated a fairly high density for the disk (about 10^{12} electrons cm^{-3}) this temperature seems to be anomalously high. This, however, probably again indicates the difficulty of defining the temperature of circumstellar matter in a meaningful way.

Related to questions of temperature and ionization is the occurrence of flaring in close binary systems. 'Flares' have been observed in W Ursae Majoris systems (Huruhata, 1952; Eggen, 1948) and in YY Geminorum (Bopp and Moffet, 1971) who now relate them to circumstellar matter in this latter system (private communication). Changes in the equivalent width of Hα in the spectrum of U Cephei also suggest some sort of flare activity in that system. The most important recent evidence, however, is undoubtedly the radio observations of Algol, cited in Section 1, that strongly suggest a flare-like process is taking place in that system.

Hjellming (1972) has made some estimates of the densities and temperatures needed to produce the observed radio fluxes. He finds that the emitting region must have a diameter in the range 0.1 AU to 13 AU, a temperature between 2.5×10^5 K and 2×10^9 K and electron densities between 6×10^7 cm^{-3} and 9×10^{10} cm^{-3}. These figures suggest that the 'flares' do not occur in the stream or disk, but in the cloud.

There is a need to be careful about terminology here. The word 'flare' immediately suggests a solar flare, and there is a temptation to link the transfer of matter within these systems to solar-type activity on the mass-losing component. Judged as a means of mass transfer, however, a normal solar flare is an inefficient mechanism. It accelerates a very small amount of matter to a much higher velocity than is needed to effect mass transfer, and gives it a large amount of radiant energy as well. An ideal mass-transfer mechanism would expend the same amount of energy on very much more matter. The activity on a stellar surface that leads to mass transfer is probably different in kind from solar activity, but it may well take place in surges that give rise to some of the phenomena associated with solar flares. It may be these phenomena that have been observed. The situation is complicated by the use of the same term to describe flare stars which may also be quite different phenomenon. It might be well to use the term 'flare' circumspectly, and perhaps in our discussions we can think of a better term.

5. Solid Circumstellar Matter

Some binary systems may contain solid circumstellar matter. Clouds of solid particles have been proposed to explain light variation of the type found in R Coronae Borealis, and there seems no fundamental reason why they should not exist in binary systems, especially in those containing only cool stars. Kopal (1954, 1971b) has suggested that a ring of solid particles surrounds the invisible secondary component of ε Aurigae (although a *disk* of such particles was suggested even earlier – Lüdendorff, 1924; Schoenberg and Jung, 1938). Wilson (1971) developed this idea further, particularly with reference to Cameron's (1971) proposal that the system contains a black hole. This latter proposal is not necessary for an adequate explanation of ε Aurigae (Demarque and Morris, 1971), and the scientific attitude toward it would seem to be to maintain a cautious economy of hypotheses. Demarque and Morris do not even favour the hypothesis of solid particles, preferring instead a hot gaseous ring. If the secondary component really is a hot star, as they propose, then solid particles within the system would be unlikely although it has recently been suggested that they formed around Nova Serpentis 1970 (Kleinmann *et al.*, private communication to J. B. Hutchings). The matter cannot yet be regarded as resolved. Evans (1971) has suggested that there are clouds of solid particles in the system of CC Eridani (and in other similar dwarf, late-type spectroscopic binaries) although an alternative explanation in terms of star spots has been proposed (Krzeminski, 1969). A disk of solid particles has also been proposed for another system containing late-type dwarfs, namely RT Lacertae (Hall and Taylor, 1971). Further investigation of many of these systems seems desirable.

6. Velocities Within Circumstellar Matter

Velocities are the only clue we have to the dynamics of circumstellar matter. Since they must be determined spectroscopically, we can know only the line-of-sight component of the velocity at any given orbital phase of a binary system. As the orbital phase changes, the observed change in radial velocity is produced not only by the changing aspect of the system, but also, often, by a change in the portion of the stream exposed to view.

The velocities of rotation of the disks can be completely observed. Struve (1950) frequently drew attention to the relation

$$V^3 \propto 1/P,$$

between the rotational velocity of a disk, V, and the orbital period, P. Huang and Struve (1956) showed that a relation of the form

$$V^3 = (1/P) f(m_1, m_2, a, \varrho)$$

was to be expected from particle dynamics, where m_1 and m_2 are the masses of the two stars in the system, and a and ϱ have already been defined in Section 2. The function f is given by

$$f(m_1, m_2, a, \varrho) = 2\pi G m_1 \left(\frac{m_1}{m_1 + m_2} \right)^{1/2} (a/\varrho)^3$$

which is not expected to vary widely from system to system (i.e. m_1, a/ϱ, and $\{m_1/(m_1 + m_2)\}^{1/2}$ are supposed to be similar in all systems) so the observed relation appeared to have some theoretical basis. Most of the observational data available to Struve concerned Algol-type systems, in which, as far as we know, these three quantities can be assumed to be similar. We now have data for eruptive binaries, and very large systems like ε Aurigae and VV Cephei, and, surprisingly, they all fit a similar, but different, relation, namely

$$V^{4.6} \propto 1/P$$

(see Figure 4). The rings in eruptive binaries rotate more slowly than Struve's formula predicted. In Section 2, it was shown that ϱ/a is fairly constant from system to system, over this wide range of systems that has now been observed, but of course there is a wide variation of m_1. It is surprising that the results from such diverse systems should fit *any* relation, and this suggests that the relation actually found has some significance. Smak's (1969) result that the observed rotational velocities of circumstellar disks are underestimates of the true velocities will not affect the form of the relation between V and P. This observed relation may be a definite indication of the failure of particle dynamics within disks, since it cannot be derived under the assumptions of particle

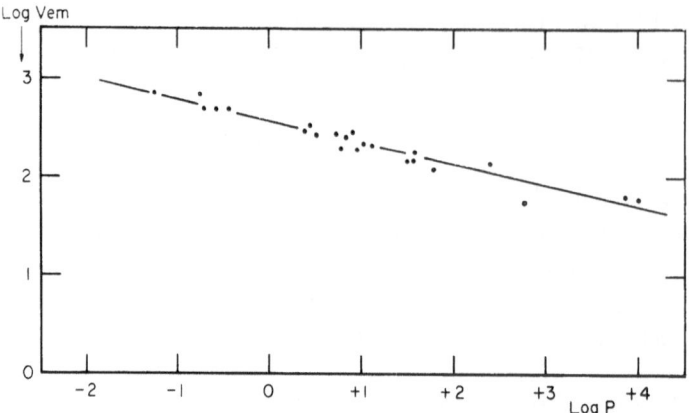

Fig. 4.   Empirical relation between the velocity of rotation of the disk (V_{em})
and the orbital period

dynamics. The failure of these assumptions within disks is hardly surprising, especially since some disks are observed in systems (U Cephei, U Sagittae, RW Tauri) in
which, according to those assumptions, disks should not form at all (Kruzewski, 1967).

The existence of this apparent relation between V and P in systems of remarkably
different characteristics is one aspect of a surprising uniformity within these systems.
Circumstellar matter is found in systems of vastly different scale, from the eruptive
binaries with orbital periods of a few hours, to supergiant systems like VV Cephei
and ε Aurigae with their periods of 20 y and 27 y respectively. The linear dimensions
of the systems vary correspondingly. The short-period systems contain white dwarfs,
or very hot subdwarfs, the Algol-type systems contain main-sequence B-type stars,
while other systems contain enormously large, late-type supergiants. There seems to
be no *a priori* reason to suppose that circumstellar matter will behave in the same way
in all these diverse systems, and yet the models proposed to explain the phenomena
have many points of similarity, of which the relation between V and P is only the most
striking. A possible exception to this rule is the systems containing very massive,
early-type stars, in which the motion of circumstellar matter may be dominated by
mass loss from one or both components (Sahade, 1962).

The velocities of streams are harder to deduce, but at some phases (particularly just
before primary eclipse, at least in the Algol-type systems) the stream can be observed
in projection on the primary star, with its velocity virtually entirely in the radial
direction. Typical values for the velocity then seem to be several hundred km s^{-1}
(about 450 km s^{-1} in U Cephei). These values coincide with the expected terminal
velocity of matter falling on to the primary star. This terminal velocity is fairly in-
sensitive to the assumed velocity of ejection from the secondary star, and we cannot
deduce very much about the ejection velocity, unless the whole trajectory can be
determined. It is for this reason that the question of the value of particle-dynamics
trajectories is hard to settle observationally. There is no evidence from the streams
(as distinct from the disks) to show that such trajectories are inadequate.

Most of the material on velocities of clouds has been summarized in the earlier sections. Some clouds seem to be expanding, and some do not. Others seem to be rotating. Further observations are desirable to try to establish the rate of mass loss from systems containing clouds.

7. Stability of Circumstellar Features

An almost untouched question is that of the stability of streams, disks, and clouds. If few systems have been examined in sufficient detail for a proper determination of the circumstellar properties at one epoch, none at all, as far as I know, has been observed in necessary detail at more than one epoch.

Three kinds of changes may be expected in circumstellar features:

  (i) apparent changes with changing aspect of the system,

  (ii) short-term changes (probably fluctuations about a mean) within a few orbital periods, or less,

  (iii) long-term changes within the system, possibly of evolutionary interest.

Changes of type (i) are only apparent, and should be periodic. In practice, however, they may be difficult to separate from some changes of type (ii). Changes of type (iii) would be difficult to observe, requiring observations from at least *three* epochs. Nevertheless, Sahade has suggested that changes in the streams of Plaskett's star may be of this type. Larsson-Leander (1969) has reported that β Lyrae appeared to become slightly fainter and slightly redder over an interval of ten years – a change that would fit in with our evolutionary ideas, if β Lyrae is indeed in the rapid phase of mass exchange. Guinan (private communication) believes changes have occurred in the emission in the spectrum of β Lyrae even within the past year. Walter (1970) has reported real changes in the light curve of RV Ophiuchi, over an interval of 50 yrs. These could be of type (iii), but Walter himself seems to prefer to regard them as changes of type (ii). Small changes in the size and location of a stream could, he believes, have a considerable effect on the light curve.

This section, however, is primarily concerned with changes of type (ii), so far as they can be distinguished from those of type (i). The reality of changes of type (iii) must for a long time remain speculative. The emission observed in RW Tauri at primary minimum is variable in intensity, and sometimes absent (Joy, 1942, 1947). Similar variations are found in U Cephei and U Sagittae, in each of which the emission has been looked for several times, and found only once. This is so even though observations have been carefully timed to achieve the maximum probability of detection of any ring. An obvious explanation is that the disk producing the emission is sometimes absent. This is backed up by theoretical investigations that make the formation of a disk seem rather unlikely in these particular systems (Kruzewski, 1967) and also suggest that disks will speedily collapse if the streams that feed them are cut off (Gorbatskij, 1969). Gorbatskij's calculations referred to U Geminorum systems, and whether they can also be applied to Algol-type systems I do not know. They do, however, point to a problem in the interpretation of U Cephei, and that is the contrast between the apparent permanence of photometric effects, and the transience of spectroscopic effects, of circumstellar matter. If the disturbances of light and velocity curves are correctly interpretated as the result of circumstellar matter, we have here a paradox since the light curve shows that the stream has been fundamentally constant for at least fifty years, while the velocity curve shows variations from epoch to epoch and a disk has been seen only once in the system. Spectroscopic observations, of course, are more senstive to changes in the stream than broad-band photometric observations are, but the complete disappearance of the disk is hard to explain if the stream has indeed been virtually constant. One possible explanation is that changes occur in the ionization of the stream and disk that would greatly affect the observable emission, but hardly effect the efficiency of the circumstellar matter in electron scattering – which is what determines its effect on the light curve. The observed emission lines could be produced by about one atom in every 10^4. If the normal state of the matter were complete ionization, the disk would be invisible in emission. If occasionally a relatively few ions recombined with electrons, then emission lines would become briefly visible. A difficulty is to imagine a mechanism that would achieve this. Flaring activity would be expected to operate the other way.

Other systems in which the disk is permanently present, usually of long period, (e.g. SX Cassiopeiae) may well exhibit changes of extent and structure of the disks, but insufficient observations have yet been made to establish any. One of my present aims is to build up a bank of observations of such systems that may be useful in the future elucidation of this problem.

Some possible changes in eruptive binaries have been mentioned by Kraft (1958, 1959) but the general situation remains that not enough observations have yet been made. Heiser (1961) also discusses changes in V 367 Cygni, that may be real changes of type (ii).

8. Conclusions

The aim of this review has been to investigate what is already known about the physical properties of circumstellar matter in order to find out what constraints are placed on any theory by the observational results at present aviable. This has been shown to be a difficult task, because quantitative observational estimates are few, and not always accordant. I have deliberately avoided any detailed discussion of the dynamics of circumstellar matter, believing this to be properly a part of the area that Dr. Huang is to review. The following conclusions summarize the result of our inquiry.

(a) The density of circumstellar matter is probably about the best determined quantity at present. Several arguments lead to densities for the streams of about 10^{13} particles cm^{-3}. There is no doubt a real variation between systems, as well as observational uncertainty in this figure. Estimates as high as 10^{16} or 10^{17} particles cm^{-3} have been suggested for U Geminorum. While it cannot be ruled out that the circumstellar matter in this system is exceptional, these figures are so much higher than the average, even in similar systems, that their derivation should be scrutinized very carefully before they are accepted. It seems likely that their derivation is based on a misinterpretation of the observed period change. Disks around one component of the binary probably have densities similar to those of the streams in the same systems. Clouds are less dense, being more remote from the stellar components in the binary system. Their densities appear to be 10^{11} particles cm^{-3}, or less.

(b) Velocities of the matter in streams can usually be observed only in the neighbourhood of the primary star, where they appear to be several hundred km s^{-1}. Such values are consistent with what would be expected if mass is being transferred from one star to the other, largely or solely under the control of gravitation (that is, hydrodynamic and magnetic effects appear not to be detectable above the limits set by observational uncertainty). The very existence of disks, however, requires a modification of particle dynamics, at least in some systems. The observed velocities within the disks (also several hundred km s^{-1} – except in the very long-period systems) may not be explicable simply in terms of particles describing orbits around the primary star.

(c) Dimensions of streams are very hard to estimate. The disks appear to have radii of several times the radius of their central star, or (in Algol-type systems) about half the separation between the stars. The masses of disks are very small, about $10^{-9} M_\odot$. Sometimes they may be larger, but seldom if ever massive enough for period changes to be explicable as an interaction between rotating disk and central star, as proposed by Smak (1971b). This kind of interaction is ruled out by the observations, even if the high circumstellar densities proposed by Warner and Nather (1971) for U Geminorum are accepted.

(d) Temperatures of circumstellar features are hard either to define or determine. Some estimate can often be made of the approximate degree of ionization. These estimates show that streams have temperatures less than or equal to that of the hotter star in the system. Although this conclusion is to be expected, it is perhaps of some value to show that it has observational support. Where two streams, or a stream and a disk collide, a hot spot may be formed. Some estimates of the temperatures of these spots make them very hot – about 50000° K – but there is considerable uncertainty just how hot they are. There is both optical and radio evidence for flare-like activity in systems containing circumstellar matter. It is not yet clear whether the seat of the activity is the matter itself, or one of the component stars. In either case, it could be associated with temporary changes in the degree of ionization of the circumstellar matter. Finally, under this head, the possibility (and significance) of solid circumstellar matter perhaps deserves more theoretical attention.

(e) The stability of many circumstellar features is unknown. The available observational evidence is inadequate and confusing.

This is the best that the observers have done so far. Many of the observations required are difficult to make and present a real challenge. The interesting binary systems are very faint, at least at the phases that are of most importance for the study of circumstellar matter. Good high-dispersion spectrograms of many of these systems have been almost impossible to obtain, until recently, and even now the radio observations of binary systems are being made at the limit of the detecting apparatus. Broad-band photometry, while useful, is not sensitive to the effects of circumstellar material. New observational techniques providing high time-resolution, and observation from above the Earth's atmosphere will certainly improve our knowledge, as the first X-ray and far UV observations have already shown. They may even revolutionize our concepts. Nevertheless the conventional techniques have not yet been utilized to their full, and are capable of giving us much more information about many systems. This field of circumstellar matter is challenging to observers of all kinds, and there is much work to be done.

9. Acknowledgments

I am grateful to my colleagues Drs. R. W. Hiltditch, J. B. Hutchings, and G. J. Odgers for critical comments on an earlier draft of this paper. Figures 2 and 4 are reproduced by kind permission of Pergamon Press.

References

Andrews, P. J.: 1967, *Astrophys. J.* **147**, 1183.
Baldwin, B. W.: 1972, M.Sc. Thesis, University of Victoria (unpublished).
Batten, A. H.: 1970, *Publ. Astron. Soc. Pacific* **82**, 574.
Batten, A. H.: 1973, *Binary and Multiple Systems of Stars*, Pergamon Press, Oxford, pp. 166–221.
Bierman, P.: 1971, *Astron. Astrophys.* **10**, 205.
Bolton, C. T.: 1972a, *Nature* **235**, 271.
Bolton, C. T.: 1972b, *J. Roy. Astron. Soc. Can.* **66**, 219.
Bopp, B. W. and Moffet, T. J.: 1971, *Astrophys. J. Letters* **168**, L117.
Braes, L. L. E. and Miley, G. K.: 1972, *Nature* **235**, 273.
Brown, R. H., Davis, J., and Herbison-Evans, D.: 1970, *Monthly. Notices Roy. Astron. Soc.* **148**, 103.
Burbidge, E. M., Lynds, C. R., and Stockton, A. N.: 1967, *Astrophys. J. Letters* **150**, L95.
Cameron, A. G. W.: 1971, *Nature* **229**, 178.
Coyne, G. V.: 1970, *Ric. Astron. Specola Astron. Vatic.* **8**, 85.
Crawford, J. A. and Kraft, R. P.: 1956, *Astrophys. J.* **123**, 44.
Demarque, P. and Morris, S. C.: 1971, *Nature* **230**, 516.
Eggen, O. J.: 1948, *Astrophys. J.* **108**, 15.
Evans, D. S.: 1971, *Monthly Notices Roy. Astron. Soc.* **154**, 329.
Fletcher, E. S.: 1964, *Astron. J.* **69**, 357.
Gorbatskij, V. G.: 1969, *Astrophys. Space Sci.* **3**, 179.
Günther, O.: 1959, *Astron. Nachr.* **285**, 97 and 105.
Hall, D. S.: 1969, *Bull. Am. Astron. Soc.* **1**, 345.
Hall, D. S.: 1971, *Veröffentl. Remeis-Sternw. Bamberg* **9**, 217.
Hall, D. S. and Garrison, L. M.: 1972, *Publ. Astron. Soc. Pacific* **84**, 552.
Hall, D. S. and Taylor, M. C.: 1971, *Bull. Am. Astron. Soc.* **3**, 12.
Hansen, H. K. and McNamara, D. H.: 1959, *Astrophys. J.* **130**, 791.
Heiser, A. M.: 1961, *Astrophys. J.* **134**, 568.
Herbig, G. H., Preston, G. W., Smak, J., and Paczynski, B.: 1965, *Astrophys. J.* **139**, 306.
Hjellming, C.: 1972, *Nature Phys. Sci.* **238**, 52.
Hughes, V. A. and Woodsworth, A.: 1972, *Nature Phys. Sci.* **236**, 42.
Huang, S.-S. and Struve, O.: 1956, *Astron. J.* **61**, 300.
Huruhata, M.: 1952, *Publ. Astron. Soc. Pacific* **64**, 200.
Hutchings, J. B.: 1970, *Monthly Notices Roy. Astron. Soc.* **147**, 367.
Hutchings, J. B. and Wright, K. O.: 1971, *Monthly Notices Roy. Astron. Soc.* **155**, 203.
Joy, A. H.: 1942, *Publ. Astron. Soc. Pacific* **54**, 35.
Joy, A. H.: 1947, *Publ. Astron. Soc. Pacific* **59**, 171.
Koch, R. H.: 1972, *Astron. J.* **77**, 500.

Kondo, Y.: 1971, *Veröffentl. Remeis-Sternw. Bamberg* **9**, 298.
Kondo, Y., Giuli, R. T., Modisette, J. L., and Rydgren, A. E.: 1972, *Astrophys. J.* **176**, 153.
Kopal, Z.: 1954, *Observatory* **74**, 14.
Kopal, Z.: 1971a, *Publ. Astron. Soc. Pacific* **83**, 521.
Kopal, Z.: 1971b, *Astrophys. Space Sci.* **10**, 332.
Korsch, D. and Walter, K.: 1969, *Astron. Nachr.* **291**, 231.
Kraft, R. P.: 1958, *Astrophys. J.* **127**, 625.
Kraft, R. P.: 1959, *Astrophys. J.* **130**, 110.
Kraft, R. P.: 1972, *Bull. Am. Astron. Soc.* **4**, 219.
Kraft, R. P. and Demoulin, M.-H.: 1967, *Astrophys. J. Letters* **150**, L183.
Kristian, J., Sandage, A. R., and Westphal, J. A.: 1967, *Astrophys. J. Letters* **150**, L99.
Kruszewski, A.: 1967, *Acta Astron.* **17**, 297.
Krzeminski, W.: 1969, in S. S. Kumar (ed.), *Low Luminosity Stars*, Gordon and Breach, New York, p. 57.
Krzeminski, W. and Kraft, R. P.: 1964, *Astrophys. J.* **140**, 921.
Kuiper, G. P.: 1941, *Astrophys. J.* **93**, 133.
Larsson-Leander, G.: 1969, *Arkiv Astron.* **5**, 253.
Lüdendorff, H.: 1924, *Sitzber. Deut. Akad. Wiss. Berlin* **49**.
Morton, D. C.: 1967a, *Astrophys. J.* **147**, 1017.
Morton, D. C.: 1967b, *Astrophys. J.* **150**, 535.
Morton, D. C., Jenkins, E. B., and Bohlin, R. C.: 1968, *Astrophys. J.* **154**, 661.
Murdin, P. and Webster, B. L.: 1971, *Nature* **233**, 110.
Murdin, P. and Webster, B. L.: 1972, *Nature* **235**, 37.
Plavec, M.: 1968, *Bull. Astron. Inst. Czech.* **19**, 11.
Plavec, M., Sehnal, L., and Mikuláš, J.: 1964, *Bull. Astron. Inst. Czech.* **15**, 171.
Plavec, M. and Kříž, S.: 1965, *Bull. Astron. Inst. Czech.* **16**, 297.
Plavec, M. and Harmanec, P.: 1972, *Comm. 27 IAU, Inf. Bull. Var. Stars*, No. 163.
Prendergast, K. H.: 1960, *Astrophys. J.* **132**, 162.
Prendergast, K. H. and Burbidge, G. R.: 1968, *Astrophys. J. Letters* **151**, L83.
Sahade, J.: 1962, in J. Sahade (ed.), *Symposium on Stellar Evolution*, La Plata, p. 185.
Sahade, J., Huang, S.-S., Struve, O., and Zebergs, V.: 1959, *Trans. Am. Phil. Soc. New Series* **49**, Part I.
Schoenberg, E. and Jung, B.: 1938, *Astron. Nachr.* **265**, 221.
Shulov, O. S.: 1967, *Astrofizika* **3**, 233.
Skilling, W. T. and Richardson, R. S.: 1947, *Astronomy*, Henry Holt and Co., New York.
Smak, J.: 1969, *Acta Astron.* **19**, 155.
Smak, J.: 1971a, *Veröffentl. Remeis-Sternw. Bamberg* **9**, 248.
Smak, J.: 1971b, *Acta Astron.* **22**, 1.
Sobouti, Y.: 1970, *Astron. Astrophys.* **5**, 149.
Struve, O.: 1950, *Stellar Evolution*, Princeton Univ. Press, Princeton.
Struve, O.: 1960, *Sky Telesc.* **19**, 276.
Struve, O. and Sahade, J.: 1957, *Publ. Astron. Soc. Pacific* **69**, 41.
Struve, O., Sahade, J., and Huang, S.-S.: 1957, *Publ. Astron. Soc. Pacific* **69**, 342.
Struve, O. and Zebergs, V.: 1959, *Astrophys. J.* **130**, 817.
Thackeray, A. D.: 1959, *Monthly Notices Roy. Astron. Soc.* **119**, 629.
Thackeray, A. D.: 1971, *Monthly Notices Roy. Astron. Soc.* **154**, 103.
Van den Bergh, S.: 1972, *Nature* **235**, 271.
Wade, C. M. and Hjellming, C.: 1972, *Nature* **235**, 270.
Walker, M. F. and Herbig, G. H.: 1954, *Astrophys. J.* **120**, 278.
Walter, K.: 1970, *Astron. Astrophys.* **5**, 140.
Walter, K.: 1971a, in *Analytical Procedures for Eclipsing Binary Light Curves*, IAU Colloquium No. 16 (Philadelphia); in press.
Walter, K.: 1971b, *Astron. Astrophys.* **13**, 249.
Walter, K.: 1972, in press.
Warner, B. and Nather, R. E.: 1971, *Monthly Notices Roy. Astron. Soc.* **152**, 219.
Wilson, R. E.: 1971, *Astrophys. J.* **170**, 529.
Woolf, N. J.: 1965, *Astrophys. J.* **141**, 155.

PROBLEMS OF GASEOUS MOTION AROUND STARS

SU-SHU HUANG

Dept. of Astronomy, Northwestern University

Abstract. A distinction is drawn between radial and tangential modes of ejection from stars, and the possible flow patterns are described. They are: expanding streams, falling streams, jet streams, circulatory streams, and gaseous envelopes. Motion around Be stars is discussed at some length, as a preliminary to studying more complicated flow in binary systems. The rotational velocity of the Be star is insufficient to form the ring. It appears likely that radial instability is temperature sensitive. Rings and disks in binary systems are discussed from the point of view of periodic orbits for particles within the gravitational field of such a system. The formation of these rings is discussed. The expected relation between rotational velocity of the ring and the orbital period is discussed. The relation of circumstellar streams to period changes is mentioned. Finally, the influence of magnetic fields on the circumstellar material and the system is discussed.

1. Introduction

Gaseous motion around stars is a very broad problem that can be studied from several points of view. In the past decade numerous papers have been presented in journals and at conferences that have been devoted solely to this subject. It is not my purpose to give a general review of the entire field. Indeed so many papers have appeared that a review by a single person cannot do justice to them all. Rather I will limit myself only to those topics within the general scope of gaseous motion around stars that I myself have been interested in since the days of my close association with the late Dr. Struve who had greatly influenced me in putting the emphases of theoretical studies on the observational relevance, a practice I have followed in this presentation.

In what follows we will first classify in Section 2 both ejection modes and flow patterns according to a kinematic consideration because what can be directly observed is nothing more than kinematics of gaseous motions. Three problems will then be discussed: (1) the emission problem of Be stars, (2) rings and disks in binary systems, and (3) magnetic braking and its consequences respectively in Sections 3, 4, and 5.

2. Kinematic Modes of Ejection and Flow Patterns

2.1. EJECTION MODES

One of the important problems associated with mass loss from stars is the ejection mechanisms. We shall divide them into three kinematic modes: (1) the pure mode of radial ejection, (2) the pure mode of tangential ejection and (3) the mixed mode.

2.1.1. *Radial Mode of Ejection*

The radial mode of ejection extends over a wide range from the solar wind to nova and supernova outbursts. It may be caused by several factors, like (1) a high temperature and pressure layer such as the solar corona in the case of a solar wind (Parker, 1963) or for that matter the stellar wind, (2) a hot and expanding photosphere as in the case of the

nova explosion (McLaughlin, 1943, 1950; Pottasch, 1959; *et seq.*), (3) radiative instability as has been suggested by Lucy and Solomon (1970), (4) magnetic and other kinds
of instability that may occur in the outer layers of the star, (5) overshots of current
elements in the convective zone in stellar atmospheres, or (6) the simple process of
evaporation from the surface.

We have considered the radial ejection only in a statistical sense. For example take
the simple case of evaporation, which has been invoked by Jeans (1925; also Spitzer,
1952; Öpik, 1963) in explaining the escape of mass from planetary atmospheres. In
the very top layers of the atmosphere where collisions are infrequent, particles with
velocities greater than the escape velocitiy will get away from the atmosphere for good.
Since the thermal velocities of particles are randomly oriented in space, few trajectories of the escaped particles actually follow the radial path. However, when all escaped
particles are considered together, the motion is statistically radial if the planet neither
rotates nor possesses a magnetic field.

Of the six cases of the radial mode of ejection all except the last three cases can be
treated by Parker's theory of the solar wind or its variations.

2.1.2. *Tangential Mode of Ejection*

The pure mode of mass ejection through the tangential motion can be envisaged in
connection with the gravitational contraction of a rapidly rotating body. It was pointed
out by Laplace in his theory of formation of the solar system that a nebula with a net
angular momentum will flatten itself first in a lenticular form in the process of gravitational contraction. Eventually the outer ring of this lenticular body will detach itself
from the central part because of the latter's further contraction. In this way the matter
left the central body without the aid of radial motion. Hereafter we shall call it the
Laplace mode of mass ejection. Obviously the Laplace mode of detachment is related
only to the gravitational contraction of a rapidly rotating body. It does not involve
any change of kinematic state of the ejected mass at the equator. Hence no ejection
theory is required.

Another kind of the pure tangential mode of ejection is through the action of viscosity which transports the angular momentum from the inner layer of the atmosphere
to the outer region, thereby creating rotational instability at the equator. The viscosity
may be molecular, photon, turbulent or magnetic. A general review of the effectiveness of these different causes of viscosity has been given by Limber and Marlborough
(1968).

2.1.3. *Mixed Mode of Ejection*

The two pure modes described previously represent only an idealization. Actually most
cases of ejection have both radial and tangential velocity-components at the time of
ejection. In the first place ejection of mass by stars in the binary system follows necessarily the mixed mode simply because of lack of spherical symmetry. Also, in the
presence of rotation and/or magnetic field, the solar and stellar winds are no longer
purely radial. They represent interesting cases of mass ejection in astrophysics (e.g.

Limber, 1967). We shall make no attempt to present a general survey of this mixed mode
of ejection but will select a few relatively simple cases in this category for discussion
here, because it is only through a complete understanding of simple cases that the
more complicated cases can be meaningfully tackled.

2.2. FLOW PATTERNS

Gaseous motion in an extended medium can take place in an infinite variety of ways.
Look at the wind on the surface of the Earth. At any time there are global wind systems, local wind systems and what may be called microwind systems that you find in
the neighborhood of tall buildings. It is hopeless to examine them in all their details.
This has made meteorology one of the most frustrating branches of science and the
weather forecasting one of the unreliable scientific predictions. But the meteorologist
who can set up stations at different locations for direct measurements is lucky when
compared with the astrophysicist who can only observe the entire system of gaseous
motion around the star from a single remote station many parsecs away. That we have
learned something about geaseous motions around the star must be attributed to the
tenacity of astrophysicists, like Dr. Otto Struve, who has persistently searched for the
mystery of the stellar world in spite of this handicap.

I cite this meteorological example in order to emphasize the simple truth that our
knowledge of the gaseous motion around the star cannot be complete. We will be
lucky if we can discern some general patterns of motion. With this point of view in
mind we can discuss three basic modes of motion of gases after they have been ejected from the star. They are:

(1) General expanding streams characterized by the P Cygni profile, WR emissions,
and the solar wind.

(2) General falling streams to the star from which the matter originally came, or to
the companion in the case of a binary system such as that suggested by Huang (1963a)
for the observed emission in β Lyrae.

(3) Jet streams that are confined to a limited region of space and have perhaps much
higher densities than the general medium, such as those suggested by Struve (1941)
for interpreting the satellite lines observed in β Lyrae and by Batten (1969, 1970;
also Batten and Plavec, 1971) for interpreting the observed emission in U Cephei.

(4) Circulatory streams characterized by the gaseous ring in Be stars – a problem
that will be extensively discussed later in this paper.

(5) Envelopes characterized by random motions without systematic motions. Because there is no containing wall in channeling the motion of ejected matter, the latter will
inevitably diffuse into the entire vicinity around the star, creating what we may call an envelope. In addition to diffusion the collisions of streams also help the creation of the envelope. This situation may be seen by considering a room full of smoking people, with each
puff producing a little stream. Soon the entire room is filled with smoke through diffusion of particles as well as mixing of different streams. However one can still discern the
little streams of smoke coming out of each smoker's mouth even when the room has
already been filled with smoke, provided that the density of smoke particles in the

general background is less than that in the stream. In such a case we must distinguish two substrata – those in the general background and those in the streams when we study the motion of smoke particles in the room. Similarly when we discuss the gaseous motion around the star, we must also consider in most cases these two substrata: the envelope characterized by the random motion and the streams characterized by the systematic motion.

Under this consideration the evelope may be identified with what Batten (1970) calls 'clouds'. However if we imagine in our example of smoking that the density of smoke particles streaming out from the smoker's mouth is much less than that already present in the room, the stream would hardly be distinguishable. While such an imagined case does not actually happen in the smoke-filled room because someone will open the window long before this state of affairs would occur, it is quite possible that the density of the stellar envelope becomes so high that the streams can no longer be traceable. They mix with the envelope immediately after being ejected from the star. This situation can happen when the star is massive and its strong gravity holds the matter there. In this way the star develops a dense envelope. Perhaps the extended atmospheres such as in Zeta Aurigae stars are envelopes in this sense. The solar chromosphere and the inner part of the solar corona may also belong to this classification, because some kind of mass ejection, in addition to the well-known mechanism of simple energy transfer, must be taking place below the chromosphere in order to replenish the mass that is being continuously lost from the outer corona in the form of the solar wind.

In summary the motion of ejected matter in the stellar neighborhood can be basically divided into two substrata (1) one with completely random motion and (2) one with somewhat systematic motion or motions. The great variety of cases of gaseous motion that will be actually observed in stars arises from (1) densities and flow velocities in the streams with respect to the density and random velocities in the envelope and (2) the different nature of streams. Also these two substrata – occupying the same region in the stellar neighborhood are continuously interacting with each other. Such an involved state is further complicated by the fact that the ejection of mass by the star may not be a continuous and steady process. All these factors contribute to the time and spatial variations of observed motions.

Around single stars the envelope may have an axial symmetry even perhaps only in a statistical sense. But in the binary system the envelope must be quite complicated. Since space around a binary system can be naturally divided into three regions by the innermost contact surface (Kuiper, 1941) – one around each of the two component stars and one around the entire system, three envelopes can in principle exist with each being confined to one region of space. However that does not mean that the envelopes are separated by empty space. Rather they denote only regions of relatively high densities separated by regions of relatively low densities. Because of the geometrical and physical nature of the stars some binaries may have less than three such envelopes or none at all. In any case the non-spherical distribution of matter in the envelopes in the binary system may produce phenomena that change with phase, although such changes could be caused by many other factors, such as gravity, reflection, etc. recently considered by Buerger (1969) that are not related to the envelopes at all.

In what follows we often discuss the problems of the gaseous stream for the sake of simplicity without considering the presence of the envelope although the motion is actually observed in the general background of the envelopes as has been already emphasized. We shall consider mainly some problems connected with the circulatory mode of motion – topics that are closely related to the transport of angular momentum from the star to the gaseous stream. In addition we will also discuss the reversed problem of the effect of angular-momentum transport in the ejection process, on the star itself especially in the presence of stellar magnetic field.

3. Gaseous Motion Around Be Stars

It is our point of view that in order to understand the complicated cases of the gaseous motion in the binary system, it is helpful to learn first the relatively simple case of the gaseous motion around the single star. For this reason we shall consider here one of the simplest cases of the gaseous motion, namely the circulatory motion around single Be stars.

3.1. MODE OF EJECTION

While it appears to be trivial to divide the ejection modes into the three kinds mentioned previously, it nevertheless gives us a new insight to some interesting problems. Struve (1931; also Crampin and Hoyle, 1960) has proposed that the gaseous ring is associated with many a Be star and is the source of its emission lines and further assumed that the ring was formed according to the Laplace mode of mass ejection. Actually it is easy to see that this cannot be so. Let us first consider the time scale. The gravitational contraction of a star is a slow process. The time scale is measured in terms of thousands of years. The behavior of the emission ring is not consistent with such a slow process. It appears and disappears in terms of decades and its intensity varies within hours or even minutes. Such rapid changes suggest that the ring cannot be formed by the slow and gentle detachment of matter according to the pure Laplace mode.

Secondly, Slettebak (1966) has found that the largest observed rotational velocities for stars of spectral types from O9.5 to F0 of luminosity classes IV–V are always below the computed equatorial breakup velocites. How, then, could the gaseous ring be formed purely by the Laplace mode of ejection. Slettebak mentioned radiation pressure and macroturbulence as the possible cause for this discrepancy. Of course, the presence of a stellar magnetic field can also change the picture (Hazlehurst, 1967; Limber and Marlborough, 1968). However, because of the high efficiency of angular momentum transport through magnetic fields, it is likely to help dissipate the gaseous ring instead of maintaining it. Another way of looking at the discrepancy has been advanced by Hardrop and Strittmatter (1968) who gave reasons to suggest that the observed rotational velocities might have been underestimated. According to them the most rapidly rotating stars are indeed at the point of rotational breakup at the equator. However, this explanation cannot quantitatively remove the discrepancy. Therefore we are inclined to agree with Limber (1970) that a star need not be rotationally unstable at its photosphere in order to show the Be phenomenon. In other words, the ring is not formed by the pure tangential mode.

Finally we may point out that many single A-type stars rotate as rapidly as B-type stars, but few, if any, show the emission feature that can be attributed to the gaseous ring. Of course one may argue that this could be due to a difference in excitation by B- and A-type stars. However, such as argument meets a serious difficulty in the fact that the emission rings have been found around A-type, and occasionally even F-type stars in binary systems. Among 15 such systems collected by Sahade (1960), twelve stars around which the gaseous ring revolves are of spectral type A and one of spectral type F. Only two are B stars. This shows not only that A stars but also F stars can easily excite the emitting atoms in the gaseous ring. It follows that the deficiency of gaseous rings around the single and rapidly rotating A stars is intrinsic and deserves a close examination (Huang, 1972).

From what has been said, it is clear that formation of the gaseous ring around Be stars is not due purely to rapid rotation of the star. There must be involved some radial instability in addition to rapid rotation. Whatever is the cause of the instability, it creates a radial component in the ejection velocity that facilitates the formation of rings.

From the observed fact that few rings have been found around single A-stars in rapid rotation, and may around single B stars, we may suggest that the radial instability is temperature sensitive. If we measure such an instability, whatever is its nature, by assigning a radial component for the ejection velocity, it may be said that this component increases with the temperature of the star. Hence in the A-stars this radial component must be very small or non-existent so that it playes little part in influencing the rotating matter at the stellar equator. Hence the ring is not formed. But in the B stars the velocity component contrï:buted by radial instability must be large enough in lifting up the matter in the equator to become a detached ring that produces emission lines. As the temperature within the range of B-type stars increases, the frequency of occurrence of emission lines increases correspondingly. While only a fraction of one percent of B9 and early A-type stars show emission lines, this figure increases to 1% for B8 stars, 5% for B5 stars and 10–15% for early B-type stars (Curtiss, 1926; Merrill and Burwell, 1933). Although these data are more than 30 yrs old, and a new and refined statistical study of the percentage of Be stars in different sub-divisions is highly recommended, we believe that this general trend will not be changed by any new statistical study. Thus such a general trend confirms the temperature-sensitiveness of the process of ring formation.

When the temperature of the star further increases, the radial component of the ejection velocity will become even greater. Eventually this would make the resultant velocity so high that the ejected matter can escape from the gravitational attraction of the star. If we assume that the star rotates so rapidly that the matter at its equator is on the verge of rotational breakup, the condition of matter escaping into infinity will

be met when the radial component of ejection velocity is equal in magnitude to the rotational velocity at the equator. For stars which rotate with less rapidity, the radial component will be somewhat higher than this value before the condition of escape is
reached. Thus the range of the radial component of ejection that may lead to the formation of gaseous rings is not very wide. This may explain why the gaseous rings around
single stars exist dominantly around those of spectral type B and the frequency of
occurrence of rings declines when the star becomes hotter. That does not mean, of
course, that emission will disappear in very hot stars. On the contrary a higher percentage of O stars shows emission than that of B stars. Only the nature of the emission
is changed. For O stars emission will dominantly come from the amorphous gas that
moves away from the star instead of arising from the rotating ring. Indeed S Pup,
an O5f star, shows the absorption line at λ 1175 of C III shifted to shorter wavelengths
by about 5 Å with an emission edge in the normal position (Carruthers, 1968).

In summary the formation of gaseous rings around Be stars results from two events
that simultaneously go on in the star, rapid rotation and the radial instability. In A
stars either radial instability is weak or non-existent, one condition for ring formation
is missing. Therefore few gaseous rings have been observed around single A stars.
In O stars the radial instability is so strong that the resultant velocity of ejection
becomes greater than the escape velocity. Therefore we see the decline of gaseous
rings but at the same time a rise of the frequency of emission in hotter and hotter stars.
From this view the emission resulting from the ring and that resulting from the expanding gas form a group differentiated only by the relative importance of the tangential
component versus the radial component of the ejection velocity. The temperature
dependence of the radial instability as inferred from this discussion seems to agree with
what has been suggested by Lucy and Solomon (1970) as due to the radiation pressure.

Lucy and Solomon's theory of mass flow is based on a consideration of radiation
pressure. The absorption in the ultra-violet resonance lines of ions such as Si IV,
C IV, N V, and S VI that are located in the intensity peak of stellar continuous radiation creates negative effective gravities in the outer parts of the reversing layers of hot
and luminous stars. As a result a static reversing layer is untenable and a radial flow
results. Their theory applies to Of and early supergiant stars and explains satisfactorily the P Cygni profiles observed in the far-ultraviolet spectra of these stars. According to their calculation B stars close to the main-sequence are located just near the
edge of the instability region in the $\log g$–$\log T$ diagram, where g and T denote respectively the surface gravity and the temperature in the atmosphere. If the star rotates
rapidly, it can easily enter into the region of instability, because rotation further reduces
the effective gravity. Thus, it appears that the phenomenon of gaseous ring around
Be stars is due to a right combination of rotation and absorption in resonance lines

Lucy and Solomon have followed Parker's theory of the solar wind in their development of mass flow from the star. They can therefore explain the high velocities
reaching several thousand km s^{-1} as observed. However, in the case of Be stars the
problem is slightly different. In the first place there is the rapid rotation of the star.
While the stellar wind of a rapidly rotating star has been studied by Limber (1967) it

may not be applicable to the present case, because what we observe here are rotating gaseous rings but not stellar winds. Whether superimposed on the ring structure there is also a stellar wind from the star is at present not clear.

3.2. MODE OF GASEOUS FLOW

A simple consideration indicates that the specific angular momentum of a particle rotating in a circular orbit under the gravitational force of a central mass is proportional to the square root of the orbital radius. Consequently even at the point where the star is rotationally unstable at its equator, there is not enough angular momentum to lift up the mass at the equator in order to form a detached rotating ring. Consequently in the absence of magnetic field or other agencies that transport angular momentum, only a fraction of matter that is lifted up from the equator can go into the rotating ring. If the radius of the star at the equator is r_0, and the radius of the ring is r_1, the fraction of the matter that goes into the ring is $(r_0/r_1)^{1/2}$, if the rest either falls back into, or escapes from, the star, carrying away no net angular momentum. Actually the fallen and escaped gas may carry away some amount of angular momentum, hence the actual fraction of ejected matter that goes into the rotating ring is likely less than $(r_0/r_1)^{1/2}$.

Thus we see that during the formation of the gaseous ring we expect to observe some mass either falling into or escaping from the stars. Perhaps the two simultaneous events – the gaseous ring and the falling matter – might have led Hutchings (1970, also Crampton and Hutchings, 1972) to suggest the idea of falling rings because observationally the two separate events of gaseous rings and falling matter on the one hand and the single event of falling rings on the other are not greatly different. In general we may also expect to see some matter escaping from the star. Whether the escaping matter takes the form of the stellar wind, or follows the process of evaporation cannot at present be answered either by observations or by theory. Also the question of what fraction of ejected matter is going to become a gaseous ring, what fraction is falling and what fraction is escaping is not known. Suffices it to say one of the controlling factors remains the radial and tangential components of the ejection at the stellar surface.

Thus we may conclude that the mixed mode of mass ejection from a rotating star in the absence of stellar magnetic field leads in general to several substrata of mass around the star in the equatorial region. There are (1) the ring or disk structure and (2) the falling and/or expanding substratum and finally (3) the general envelope supported by random motion. Away from the equatorial region, matter is ejected if at all at a much less rate than at the equator on account of smaller tangential velocities. Several possibilities exist for the ejected matter. It may escape into infinity, fall into the star or into the disk in the equatorial plane or stay in the stellar neighborhood to form an envelope. What course it actually follows obviously depends again upon the rate of rotation and degree of radial instability. Perhaps studies of pole-on stars such as one made by Togure (1969) may provide us with some empirical information.

3.3. INTERPRETATION OF EMISSION VARIATIONS

It may be argued that any new ejection of matter by the star disturbs the ring, because the newly ejected particles can knock the particles in the ring out of the orbits. This may explain the frequent fluctuation of the intensity, and disappearance and re-appearance of emission lines produced by the gaseous ring. However, the fact that emission rings are observed at all means that ejection of matter by the star is likely intermittent or if there is any continuous ejection, it must be exceedingly mild because each new violent ejection will likely destroy the existing ring. Once the ring has been destroyed the matter may or may not settle down into a new ring, which may or may not be identical to the previous one thereby creating a variety of possibilities that are to fascinate the observer. The observed result seems to show that the intervals between bursts are astronomically short and perhaps irregular. Whether the quiescent interval before, and the strength of, each burst follow some rule is a question that has not been extensively studied. Intuitively we may conjecture that the strength of each burst is likely to increase with the quiescent interval that precedes it.

The variation in the emission may be either irregular or roughly cyclic for a certain period of time and may take place either in the line profile only or in both the profile and the equivalent width. The cyclic variations themselves are complicated because periods can vary from decades to minutes or perhaps even seconds. Obviously we do not expect to find all variations due to the same cause. What we actually observe is of course the combined result arising from all causes. Thus our purpose is to disentangle and isolate various causes and treat them individually. In general, the variation in the profiles without the change of the total strength is most likely due to the effect of geometry, just like the effect of rotation on stellar absorption lines (e.g. Huang and Struve, 1960). Such a variation is comparatively easy to understand because it can be studied without resorting to the physical cause of the emission. It is this group of variations that will give us some information about gaseous motions around the star. The change in the profiles of this nature likely means the reshuffling of radial velocities of emitting atoms without any change in their total number.

We will now discuss several kinematic problems that can produce some kind of observed variations in emission. First, there is the well-known V/R variations with periods stretching from years to decades. While the relative intensity of two components of emission varies, their total strength seems to be roughly constant. Theories for this kind of variation were last reviewed by McLaughlin (1961), who has devoted his time and energy in studying this kind of variations more than anyone else. He reviewed three models, namely (1) the rotating-pulsating model (2) the intermittently expanding-rotating two-layer model and (3) the eccentric rotating model.

The artificiality of the first model is obvious. McLaughlin has pointed out several defects from the observational point of view. Theoretically it is difficult to see pulsations of such long periods taking place in the neighborhood of stars on or close to the main sequence. Also, there is no conceivable mechanism to drive the pulsation of a tenuous detached ring which, being maintained simply by the angular momentum, possesses no elasticity of its own.

McLaughlin's second model assumes two layers – the inner emitting and the outer absorbing layer. By considering the cyclic change of the motion in both layers it is possible to obtain the V/R variation. McLaughlin has found that the observed facts are incompatible to this model. Theoretically it is hard to understand how two layers can maintain their separate existence cycle by cycle.

In the eccentric rotating model the V/R variation is attributed to the apsidal motion of a rotating elliptical ring of emitting gases. This possibility was first mentioned by Struve (1931) but it was McLaughlin who seriously embraced it and made some preliminary explorations. Indeed both V/R variations and velocity shifts of emission and absorption components based on this model seem to be consistent with observed results. Also Johnson (1958) examined the adequacy of the star's oblateness as the cause of the apsidal motion of the elliptical ring and found the rate of this motion indeed adequate for explaining the observed periods of V/R variations. We have made a quantitative study of the change in emission profiles based on this model. Since the observed V/R variations are related to the eccentricity of the elliptical ring, the latter can therefore be determined empirically. It appears that in most cases a small eccentricity is enough to account for the observed amplitude of V/R variations (Huang, unpublished). Also according to this model the absorption line remains always in the center of two emission edges as observed.

Rapid variations of emission profiles in Be stars have been found recently by many investigators like Hutchings and associates here in Victoria, Bahng in Northwestern University and a large group in Meudon. The emission may vary either in the profile only or in both the profile and the strength. Obviously no single cause can be assigned to such a heterogenous group of variations.

One of the possibilities may be the non-uniform distribution of emitting gas in the ring (Huang, 1972). It can be easily seen that even in the circular ring, such a distribution produces in the emission line not only asymmetry but also time-dependence. The variation is due to the fact that the emitting particles revolve with different Keplerian velocities on circles of different radii. If the gas is tenuous, it produces rapid variation in profiles but no change in the total intensity of the line. Such a variation may or may not show any periodicity. Even if it should indicate some periodicity it does not repeat itself exactly in every cycle. On the other hand it has a definite tendency of being damped due to the smoothing effect on the uneven density distribution in the ring by both differential and random motions. The time scale of damping depends upon the width and the size of the ring as well as the mass of the central star.

Another possibility of having asymmetric profiles and time variations may result from the multi-substratum nature of the medium around the Be star mentioned previously. The size of the rotating ring which is fixed by its angular momentum cannot change easily without angular momentum transfer. However, there are other substrata of falling, escaping and/or amorphous gas. The emission arising from these substrata will modify the emission due to the ring. As a result the resultant emission will be asymmetric. As has been mentioned before the phenomena that have been interpreted by Hutchings (1970) and by Crampton and Hutchings (1972) as due to falling or expanding rings result, according to the present model, from the multi-substratum nature of the gaseous medium.

3.4. TIME SEQUENCE OF RING FORMATION

Why is the ring elliptical in some cases and circular in others? Why is the density in the ring not uniformly distributed? Why does there exist the other substrata in addition to the ring? All these questions lead to the fundamental problem of mass ejection from the star. Observations show that the ejection must be intermittent. Each new violent ejection destroys the previous ring structure if it exists and makes the medium chaotic and full of large turbulent eddies. In this initial phase there is no structure in the medium even in the equatorial plane. The emission is likely strong and complex. Not only the profile but also the total strength may vary greatly because of the violent nature, such as collision of eddies, shocks, etc., in the medium. However, because of the angular momentum, some of the matter may gradually settle down to form a ring while the rest may fall into the star and/or escape into space. This is the multi-substratum phase we have emphasized before. In this second phase the emission can also have various kinds of variations. Falling and escaping components in the substratum will be dissipated first. Eventually even the amorphous component will be dissipated, leaving in the final phase only the ring itself until it is destroyed by the next violent ejection from the star or by gradual dissipation. During the final phase variations in the emission are expected only in the profile.

According to this scenario it is easy to see that the density in the ring cannot be uniform when it is formed out of the chaotic medium. Also some small ellipticity could be expected in the ring. If, however, the ring should be left alone for a long time without any disturbance, it will eventually reach a circular state, as are Saturn's rings. But around the Be star, the ring is never left alone for any long period of time. That is why we observe so many variations in the emission lines. Conversely these variations indicated how active Be stars are.

The characteristic feature of our model that differs from other ones is its developing nature. It does not advance any single supporting mechanism for the envelope. Indeed from the comprehensive study by Limber and Marlborough (1968), it is quite evident that no single mechanism can satisfactorily explain how the envelope can be maintained. Although they have concluded that the support is centrifugal, as most previous investigators since Struve had done, they have found difficulty in finding how such a support can be achieved through thermal and radiative viscosity and have had to resort to turbulent and/or magnetic viscosity. At the same time they have further pointed out that if the turbulent viscosity is responsible for transporting the angular momentum it has to be supersonic. Whether such turbulence can develop and is consistent with the observations is not clear. Therefore they have concluded that the "magnetic field remains a prime candidate for the required viscous agent even though the case for its importance is largely circumstantial".

The idea of magnetic transport of angular momentum has also been taken by Hazlehurst (1967) in his theory of ring formation. He follows the concept of magnetic braking mechanism advanced by Lüst and Schlüter (1955; also Mestel, 1959). While the idea is attractive we must point out that all magnetic theories for ring formation would predict that the ring should be formed around not only rapidly rotating stars, but also moderately rotating stars. This is because magnetic field is a very efficient agency for transporting angular momentum from the star to the envelope. But the observation shows evidence that only rapidly rotating stars possess rings. For this reason we are uncertain whether the magnetic field plays a role in the ring formation around Be stars.

In our picture we consider the medium that evolves with time. During the first two stages the envelope is not in the hydrostatic equilibrium. Consequently the questions of supporting mechanisms does not arise. In this first stage the medium is dominated by the ejected matter from the star. In the second stage, the envelope adjusts itself after the disturbance suffered in the first stage. Some particles are supposed indeed to fall into the star, while others remain in the envelope. Only in the final stage, we assume the stage of the centrifugal support. Because of what happens in the second stage, we have achieved the centrifugal support in the final stage, without resorting to the magnetic or other kind of viscosity.

What makes the ejection of matter from the star intermittent instead of continuous which has been seen in the observation and accepted in our model, is perhaps due to the nature of marginal instability. Once instability is relieved by ejection of matter, it may take some time to rebuild the potential for the next instability. For example, after a large amount of angular momentum is lost in an outburst, the equatorial matter may become stable for a while until its rotation is accelerated by a little more of the contraction of the core of the star (Crampin and Hoyle, 1960) and the transport of angular momentum outward (Limber, 1969). Perhaps a small change in the rotational velocity at the equator will tip the balance of stability and instability, because the change affects not only the tangential velocity but also the effective gravity that controls the vertical instability. Long-period variations of emission sometimes can last a few cycles indicating that the ring can persist without much disturbance several tens of years. Other events such as disappearance and reappearance of rings seem to confirm that the interval between two consecutive violent eruptions is also of this same order of magnitude. Tentatively we adopt this value as the average time scale for building up the instability. Obviously the actual time scale varies from star to star.

In this discussion we have emphasized only the qualitative nature of the evolution of the gaseous medium around Be stars. It is very difficult to treat the first two phases because of the time-dependent nature of the entire envelope. The envelope is at first chaotic, its motion being dominated by the ejection mode which can not at present be quantitatively described. In the process of ring formation, it dissipated matter as well as energy belonging to the random motion. It is unlikely that one can express this situation in any unique way. Perhaps a large number of plausible models may be built for different stages of development in this second phase. One of such models has been given by Limber (1969).

4. Rings and Disks in Binary Systems

4.1. MATHEMATICAL CONSIDERATIONS

What has been said so far from single stars may not all be applied to stars in binary systems. Because of the orbital motion, it becomes meaningless to consider the pure radial or pure tangential mode of ejection. For the same reason the gaseous flow in the binary system is more complicated than that around the single star. By considering the topographical structure of the equipotential surface in the binary system, Kuiper (1941) pointed out the importance of gaseous flow through the Lagrangian points, a concept that proves to be extremely useful in our interpretation of the observed phenomena in binary stars. Following Kuiper, who treated the gaseous flow within the framework of the restricted three-body problem, many investigators have computed numerous orbits that a particle can move in a binary system. However, the value of such calculation has been questioned because the three-body calculation necessarily neglects collisions among the gaseous particles. If one should take a look at some single trajectory with its many loops and cusps, it is quite clear that the neglect of collisions is most unrealistic. Also there are an infinite number of trajectories one can compute. How could they be interpreted in terms of actual flow? For these reasons Prendergast (1960) has tried to study this problem from the hydrodynamical consideration. He has written down the appropriate hydrodynamic equations, but made no attempt to discuss how suitable boundary conditions may be imposed to the differential equation of flow.

Thus we see that the approach based purely on the restricted three-body problem suffers the difficulty of neglecting collisions and the hydrodynamic approach encounters the unsurmountable problem of how to impose suitable boundary conditions that are consistent with the physical reality. By neglecting the pressure and by making some approximation Prendergast succeeded in obtaining a particular solution depicting circulatory flows around each component and around the entire system. However, it does not tell you how such a flow can be achieved. Consequently it does not provide the answer to the question of when and why a gaseous ring is expected around the more massive component, the less massive component, or each of both components, in a particular system. We may also ask what leads to the flow around the entire system. It requires an extremely large value of specific angular momentum in order to have a circulatory flow around the entire system. How can this angular momentum be imparted to the matter or what is the ejection mechanism? Actually under the ordinary circumstance of ejection the motion of the envelope surrounding the entire system possesses likely a radial component. All these difficulties arise from the fact that we do not know how to impose the realistic boundary condition to the hydrodynamic equations. It follows that even though the hydrodynamic approach is a legitimate one, it has not so far provided us with any new insight to the complicated problem of gaseous flow in the binary system.

For these reason Huang (1965a) has proposed a statistical approach based on the three-body problem. By considering the three-body problem we can take proper care of the boundary condition such as the ejection at the Lagrangian points, etc. By considering statistics of Jacobian constants of particles in the stream the collision among them are accounted for. However, so far the result obtained from this approach is no more than the particular solution that Prendergast has derived from the hydrodynamic considerations. But we still hope that something observationally significant could be obtained later by this approach.

What makes the trajectories derived from the restricted three-body problem meaningless are collisions among particles. Now it may be argued that we can find trajectories in which collisions do not play an important role. If so, such trajectories may represent some realistic case of motion of gaseous streams. What are then the conditions for such trajectories? Obviously each must be devoid of loops or cusps so that it does not intersect itself, because any intersection only means collisions. However, a single trajectory without loops and cusps is not sufficient to justify a gaseous stream that follows the trajectory. It must be stable and be surrounded by a family of similar non-intersecting trajectories such that the two adjacent ones differ only infinitesimally in all aspects. In this way we identify a gaseous stream to a family of non-intersecting trajectories. Such a condition obviously is satisfied in the case of a family of non-intersecting periodic orbits, although this is by no means the sole possibility.

Poincaré has shown that periodic orbits exist around each of the two finite bodies in the restricted three-body problem. We have developed a perturbation method for finding such orbits (Huang, 1967a). Figure 1 illustrates a few such orbits in a family around the more massive component in the orbital plane together with the innermost contact surface, corresponding to a system of the mass ratio of 4 to 1. Indeed these orbits form a family of similar and non-intersecting trajectories in the plane of orbit. We suggest that the existence of such a family of stable periodic orbits makes gaseous rings around the component star in a binary system possible, just as the existence of periodic orbits makes rings around single stars possible. In a sense the existence of periodic orbits provides the particles with some kind of invisible tracks. Those particles that move with velocities quite different from these corresponding to the periodic orbits simply depart from the tracks. But those that move with velocities close to that corresponding to the periodic orbits are trapped in tracks and become a part of the ring system. Because small perturbations from the stable periodic orbits will remain small, particles do not move away easily once being captured in the ring. They will stay there for a long time until some drastic disturbance kicks them out of the periodic orbits and thereby destroys the ring structure. In this way we link the observed ring with the existence of a family of stable periodic orbits around the star. Perhaps most particles in Saturn's rings are moving in periodic orbits because of the long time of their existence. However the particles in the gaseous rings in Be stars are likely moving in nearly periodic orbits, as their motions have not been left undisturbed for any long period of time.

The existence of periodic orbits circling around each component shows nothing

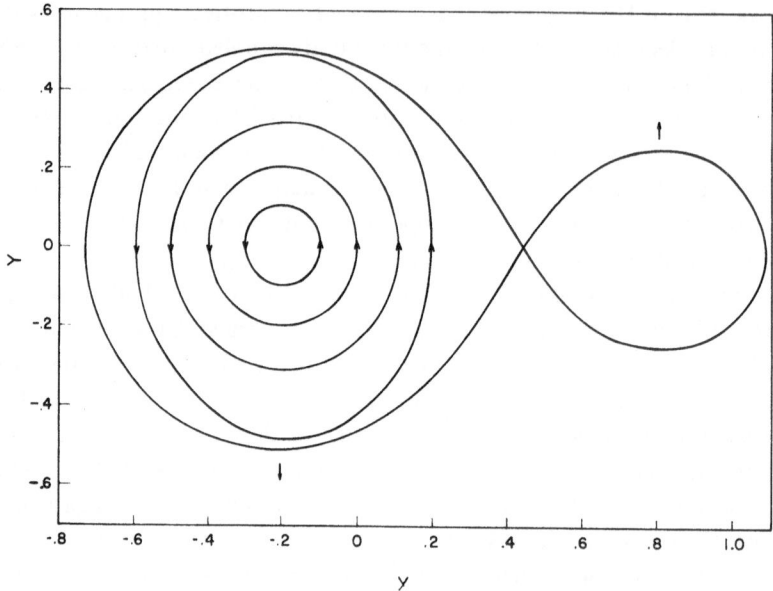

Fig. 1. The family of periodic orbits around the more massive component in a binary system. According to our interpretation the existence of such a family makes the ring formation possible. Note the increasing departure of these orbits from the circular form as their size increases.

more than the possibility of the gaseous ring. It does not explain its actual appearance. The fact that it revolves around the component star indicates the large amount of angular momentum it possesses. The question is: why does the matter possess such a large amount of angular momentum? Here we encounter a physical problem in contradistinction to the mathematical problem of periodic orbits just described.

4.2. PHYSICAL CONSIDERATIONS

Several ways that rotating rings and disks may be formed in binary systems can be visualized (Huang, 1971).

(i) Rotational instability – just as the emission rings are formed around Be stars as the result of a combination of rapid rotation at the equator and radial instability, rings can be similarly formed around the components of binaries if the star is rotating rapidly. Obviously it would be the case if at least one of the components in a binary is a Be star (Hutchings and Wright, 1971). However, many emission rings found in binaries do not appear to be formed in this way, because the component stars in the center of rings are not rotating at such a rapid rate as to become rotationally unstable and most of their spectral types are too late for radial instability according to Lucy and Solomon's mechanism.

(ii) Angular momentum transport through magnetic braking – we will discuss the magnetic braking theories later. In the process the stellar angular momentum is being transferred to the surrounding medium which collapses into a rotating disk when enough angular momentum has been acquired. This may happen whether the star is single or a component of a binary system.

The angular momentum transport through magnetic braking takes place, most likely, in the pre-main-sequence stage of stellar evolution (see references in Huang, 1969). According to Hall (1971a) the secondary component of BM Orionis is a case of disk formation in the pre-main-sequence stage in a binary system. While its primary B star may have already reached the main sequence, the secondary component could very well be still in the stage of gravitational contraction leading to the main sequence. Also, the cooler components in RS CVn binaries may be in pre-main-sequence contraction too (Catalano and Rodonò, 1967; Hall, 1972), although this suggestion has not been unanimously accepted (Koch, 1970). Hall (1972) further suggested that the cool star in RS CVn itself may be a T Tauri-like star. This suggestion has been obtained from a careful analysis of the light curve but does not come from spectroscopic data.

(iii) Cataclysmic ejection – when a star explodes, it leaves some debris around. If the star originally possesses a large amount of angular momentum, such a debris will easily collapse into a disk. Furthermore, if the explosion is of the supernova nature, this debris will be rich in heavy elements which may condense into solid particles. We suspect that the secondary component of ε Aur may possess a disk just like this (Huang, 1965b; Kopal, 1971).

(iv) Ejection as a result of stellar evolution reaching the innermost contact surface – For the star in a binary system there is the instability caused by, on the nuclear time scale, the evolutionary expansion which brings its radius into direct contact with the Roche limit. The mass flow due to this cause can be considerable. Therefore once matter has started to flow out, the loss of mass itself may induce further instabilities not only on the nuclear time scale but also on short time scales, such as Kelvin and pulsational time scales (Morton, 1960). Considering the fact that the binary elements and the mass ratio are themselves related to the mass loss and/or gain of component stars, we are confronted with a very complicated problem, in which the time scale of dynamical evolution of the binary itself must also be taken into account. In other words, stellar evolution of component stars and dynamical evolution of the binary system are coupled together (Smak, 1962). The coupling is provided of course by the ejected matter. That makes the modes of ejection and flow exceedingly important not only for those who are interested in stellar spectroscopy by also those in stellar evolution (e.g. Paczński, 1971; Plavec, 1968).

Ejection in this catagory is expected to take place intensively, though not exclusively, at the Lagrangian points (Kuiper, 1941) and provides a cause for formation of rotating rings and disks. Because of symmetry, it is reasonable to expect that the ejection taking place at the Lagrangian points produces a medium confined to the orbital plane and its immediate neighborhood. In other words, whether around the companion of the ejecting star or around the entire system, the ejected gas will be distributed more like a disk than a sphere. The structure of the disk could be either amorphous or rotating, depending upon the amount of angular momentum the medium possesses.

Observationally we have found cases of the rings revolving around the more massive component while the less massive component fills the Roche surface (e.g. Huang and Struve, 1956). That happens in most Algol-type binaries. It may be argued that the angular momentum per unit mass of the less massive component is greater than that of the more massive component. Thus when the mass ejected by the less massive component reaches the neighborhood of the more massive component, it finds itself possessing an extra angular momentum. Such a gaseous medium around a central field (in this case, produced by the massive component) will simply collapse into a rotating ring or disk (Huang, 1957).

The angular momentum of a particle varies as it moves in the gravitational field of two revolving stars. However, once it has entered into a periodic orbit, its average value remains constant. Consequently a definite amount of angular momentum can be assigned to each ring of gases. Thus, in order to form the ring, this amount of angular momentum must be available to the ejected matter. The latter's angular momentum is supplied by the ejecting star at the time of ejection and further supplemented by acquisition during its flight from the point of ejection to the place where the ring is formed. This is because the ejected matter being acted on by the Coriolis force, moves in such a way its angular momentum increases in general after its ejection at the inner Lagrangian point. However, if the angular momentum of the ejected matter is not sufficiently large, it may fall directly to the massive component, forming as in the case of U Cephei what Batten and Plavec (1971) term a luminous bridge. When the matter falls into the primary component, its angular momentum is transported to the atmosphere and makes the latter rotate faster than otherwise would be the case. In this way we have explained the non-synchronized rotation of the primary component of this system (Huang, 1966b). In such cases, one may visualize the gaseous ring as coalescing into the atmosphere. Such a view seems to be supported by observational results obtained by Batten (1969) who has found that the velocity from the emission lines in the spectra of U Cep is virtually identical to the rotational velocity of the primary star. Thus the pictures presented by Batten, Plavec and myself are consistent.

There remains the question raised by Batten (1970) whether the two events – formation of a gaseous ring around the massive component and the driving of the atmosphere into rapid rotation by the falling gas – are coexisting or not. We agree with him in that there is no reason for the two events to be mutually exclusive. Whether the two events can actually occur side by side depends upon the mode of ejection as well as the mode of dissipation of the ring. If the ejection velocities cover a wide range, some particles will fall into the star while others form the ring. In such a case the two events occur simultaneously. During the dissipation of the ring the constituent particles may be swept into the atmosphere of the star to drive the atmospheric rotation instead of being driven away into space. In such a case the one event becomes consequential to the other. However under some different modes of ejection and dissipation, the two events could become mutually exclusive.

We have attributed formation of rotating rings mainly to the disparity of masses of two component stars. As an educated guess, rings may be formed around more massive components in binary systems whose mass ratio, M_1/M_2, is between 2 and 8 (Huang and Struve, 1956). What would happen if the mass ratio becomes greater than, say 10? The specific angular momentum of the matter ejected by the less massive component will be very high in such systems. But too high a value of angular momentum is not necessarily conducive to formation of the ring around the component star because the range of permissible rings is limited there. In such cases rotating rings or disks may more readily form around the entire system. Scarcity of observable systems in this range of mass ratios makes the discussion academic.

If there is not enough angular momentum present, the medium around the component star can not collapse into a regularly rotating ring or disk. Such a disk disperses more readily than the rotating one unless it is constantly replenished with new matter by ejection. Kraft (1958, 1959, 1962) has suggested that disks exist around some eruptive components of binaries such as DQ Her, T Cor B, U Gem, etc. According to him the masses of two components in each of such systems are nearly equal. Often the disk is associated with the slightly less massive eruptive component while the slightly more massive component fills the innermost contact surface and is losing mass. Whether such disks are completely supported by the centrifugal force is at present not clear.

4.3. OBSERVATIONAL MANIFESTATIONS

The rotating rings and disks around component stars in close binary systems will reveal their existence to us in several ways, depending upon (1) the nature of the central star, (2) the nature of the gaseous ring and disk itself. When the star is hot and luminous, and the matter in the ring is rare, emission lines of hydrogen and other elements perhaps will be the dominant feature. Indeed the gaseous emission rings around the more massive components of many eclipsing binaries have been discovered in this way. While Joy (1942, 1947) first proposed the existence of a gaseous ring that produces emission lines in RW Tauri, it was Struve (e.g. Sahade, 1960) who made a large number of discoveries of this kind of objects. With the use of rapid scanning techniques, such as effectively employed by Hutchings and others, the gaseous rings could be quantitatively studied. The relation between the rotational velocity of the ring and the period of the binary especially can be studied.

Let M_1 and M_2 respectively be masses of two component stars, $(M_1 \geqslant M_2)$, a, the semi-major axis of the relative orbit, and ϱ the radius of the gaseous ring around M_1. In the two-body approximation we may assume that the ring is circular making ϱ constant in all phases. However, in the approximation based on periodic orbits in the restricted three-body problem, the ring cannot be circular, as is evident in Figure 3.1. If ϱ/a is to be determined by the time of eclipse of the emission line, ϱ must be the radius in the direction perpendicular to the line joining two component stars. Similarly the rotational velocity of the emission ring should be constant in the two-body approximation but is variable in the three-body approximation. Therefore we specify the rotational velocity, V, to be at the point where ϱ/a is measured. It has been shown that based on the periodic orbits in the restricted three-body problem, V, in the c.g.s. unit is given by

$$V = \left(\frac{GM_1}{\varrho}\right)^{1/2} f\left(\mu, \frac{\varrho}{a}\right) \sin i, \tag{1}$$

where

$$\mu = \frac{M_2}{M_1 + M_2}, \tag{2}$$

and G is the gravitational constant, and i is the orbital inclination while $f(\mu, \varrho/a)$ is a function which has been given in a tabulated form (Huang, 1967a). Needless to say we have assumed that the plane of the ring coincides with the orbital plane. For small μ and small ϱ/a

$$f\left(\mu, \frac{\varrho}{a}\right) \to 1, \tag{3}$$

Equation (1) reduces itself to the case of the two-body approximation. Since V and ϱ/a are measurable, we have from Equation (1) a relation between total mass, the mass ratio and the separation. If K_1 denotes the semi-amplitude of the velocity curve of M_1, and e the eccentricity of the binary orbit, the relation between V and K_1 is given by

$$\left(\frac{V}{K_1}\right)^2 = \frac{M_1(M_1 + M_2)}{M_2^2} (1 - e^2) \left(\frac{a}{\varrho}\right) f^2\left(\mu, \frac{\varrho}{a}\right), \tag{4}$$

which determines the mass ratio of the system even when we do not have the radial velocity curve of M_2

Thus the motion of the gaseous particles in the ring gives us some information about the mass ratio as good as the motion of M_2 itself. However since $f(\mu, \varrho/a)$ has been determined on the assumption of the restricted three-body problem we can apply this method for determining the mass ratio only for systems with small eccentricities. For binaries of large eccentricities one can make similar calculations as we did for circular orbits. In such cases the function $f(\mu, \varrho/a)$ should be replaced by another one containing e as an additional variable. In reality there cannot be many binaries of high eccentricities which possess rings because of their large disturbing effect on the stability of the ring.

Struve (1950) has found that the rotational velocity of the gaseous ring is related to the orbital period, P, of the binary in the following way

$$V^3 \propto \frac{1}{P}. \tag{5}$$

Within the restricted three-body approximation such a relation would mean that

$$\frac{\varrho}{a} \propto M_1^{2/3} \left(\frac{M_1}{M_1 + M_2}\right)^{1/3} f^2\left(\mu, \frac{\varrho}{a}\right). \tag{6}$$

Batten (1972) has pointed out that most of the observational data available to

Struve concerned Algol-type systems, in which indeed M_1 and $M_1/(M_1+M_2)$ do not vary greatly. With data now available for other kinds of binaries he has found that they all fit a somewhat different relation namely

$$V^{4.6} \propto \frac{1}{P}. \tag{7}$$

On the basis of the restricted three-body approximation, it can be easily shown that this relation indicates that

$$\frac{\varrho}{a} \propto \frac{1}{a^{0.35}} \left(\frac{M_1}{M_1+M_2}\right)^{0.22} M_1^{0.78} f^2 \left(\mu, \frac{\varrho}{a}\right). \tag{8}$$

Hence ϱ/a seems to be related not only to the mass and the mass ratio as in relation (6) but also on the size of the binary. Since small masses often occur in small systems, the effect of M_1 and a may cancel out and leave ϱ/a not greatly different from very small to very large systems. Thus, the relation given by relation (7) and the narrow range of ϱ/a in systems of greatly different sizes both pointed out by Batten in this symposium, represent the two facts that are closely related. Both show that the mass of the star around which the gaseous ring revolves increases with the separation of the two components in the binary.

Finally it must be said that the present analysis is tentative at best. In the first place the ring cannot be geometrically thin as assumed in the calculation. Secondly, there is also the question of whether the matter in the disk is completely supported by the centrifugal force. However, the suggestion that the statistical relation obtained by Batten indicated a mass-separation relation among those binaries included in his statistical study has the intuitive appeal.

It is reasonable to expect that almost all kinds of variation in emission found in Be stars can happen in binaries with rings. However, in most binaries the emission is too weak to be observed outside eclipse. Consequently such variations have not been recognized because they are overshadowed by the intensity variation caused by eclipse itself.

If the matter in the disk increases, we would expect that the continuous opacity, say due to electron scattering, will be considerable and the disk may become semi-transparent. Thus instead of emission, the disk will reveal itself by its obscuration. It was this argument that led us to suggest that the secondary component of β Lyrae system may possess just such a structure (Huang, 1963a). Indeed, it seems that all essential features of this peculiar system can be understood in terms of this structure. More recently, Hall (1971a) has shown that the secondary of BM Orionis may also possess a semi-transparent disk of this kind.

We may further argue that the rotating gaseous disk exists in such a state that it both produces emission lines and causes obscuration of the companion star. This situation can arise if the density of gas is rare at some places (likely near the edges) which produces emission, but dense at other places (likely in the middle) where obscuration occurs. Hall (1971b) suggested that RS Cephei may represent an example for such an intermediate case. Hall and Garrison (1972) have reported SW Cygni as another example.

If the star is not too hot or the gaseous ring is far away from the star, the matter in the disk may condense into solid particles or planetesimals. In such a case, the presence of the disk could be detected by its emission in infrared radiation as well as its obscuration of the secondary component. The secondary component of ε Aur may possess such an opaque disk. If so, the disk would be in many respects like the solar nebula (Huang, 1965b; Kopal, 1971) and may even develop into a planetary system.

However the interpretations of the light curve of ε Aur based on a semi-transparent disk that is composed of solid particles has its difficulty. We have mentioned infrared emission as one of the characteristics of such a disk. However, Low and Mitchell (1965) found no infrared excess in this system. Therefore at present we can only take the suggested disk composed of solid particles in ε Aur as tentative. As we have emphasized elsewhere (Huang, 1971), a critical test of this hypothesis will be detection of eclipse in infrared when the primary component, which may be deficient in infrared radiation and thereby obliterate the infrared emission of the disk, obscures the latter from the observer. We have found from the disk model that the next infrared eclipses would occur in June 1974–June 1976 according to the spectroscopic orbital elements given by Kuiper *et al.* (1937) or in October 1971–October 1973 according to those given by Morris (1962). Since this is a very critical test, we hope that some infrared astronomers will observe this star in the next few years.

Apart from the circulatory motions that are associated with the ring and disk structure and the jet motion starting from the Lagrangian points, there must be other kinds of flow taking place in the binary system. We have already mentioned those possibilities in Section 2.2. At distances large compared with the separation between the two components, perhaps the flow pattern may not be greatly different from that around the rotating star. The matter which must have a high concentration in the orbital plane and decrease rapidly as we move away from it is likely in a state of expansion as has been observed in β Lyrae (e.g. Struve, 1958). The tangential velocity would depend upon how effective the angular-momentum transfer is from the binary system to the medium.

An intriguing question that is related to the gaseous flow in the binary system is the change in the orbital period. As soon as there is gaseous flow in the system the orbital period will change. Hence the change in orbital period in close binaries is the general rule instead of the exception because some kind of flow is inevitable in close binaries. In general such a change is small and may not be observationally detected. In those cases where the variation in period has been detected, it is also difficult to interpret them physically because the variation depends on the total mass involved in the flow, the mode of ejection and flow pattern and many other factors not connected with mass flow (e.g., Kruszewski, 1966). Huang (1963b) has expressed the change of period in terms of specific energy and specific angular momentum – a practice that has been followed by Piotrowski (1964). By doing so the period changes in a few simple, but nevertheless realistic, cases have been treated. On the other hand Van 't Veer's (1972) recent study of period change of contact binaries has imposed some different assumptions on the modes of mass ejection and flow. In any case the mode of ejection and flow must be assumed before the period change can be calculated. As we have not solved the flow problem in a binary system it is premature to undertake a general treatment of period change, although a serious attempt in this direction has been made by Piotrowski (1964; also Kruszewski, 1966). Therefore we may conclude that since there is no way at present to ascertain the exact mode of ejection and flow pattern that actually takes place in any particular system, the rate of mass change and/or loss derived from the rate of change in period is tentative.

5. Consequences of Ejection by Stars with Magnetic Fields

The mass of gaseous streams around stars is usually very small compared with the star itself. But observationally they play a critical role since the streams, being outside the photosphere, are exactly what will imprint their marks on the light we observe. Thus, gas streams with their low densities are the cause of many peculiarities found both spectroscopically and photometrically, and a complete understanding of stellar spectra, especially those of peculiar ones, must include the behavior of the gaseous stream around the stars, whether they are single or in binary systems.

In spite of the clear marks that are imprinted on stellar spectra, the ejected matter plays in general an insignificant role in shaping the over-all structure of the ejecting star itself. However, when the star also possesses a magnetic field, the ejection process could modify the nature of the star considerably. We shall discuss the following three cases that belong to this category.

5.1. BRAKING OF STELLAR ROTATION

One of the most puzzling phenomena in stellar astronomy is the discontinuity of the rotational behavior of the single main-sequences stars, discovered by Struve (1930). While rotation is common among early-type stars, stars later than spectral division F5 have never been found to rotate in any measurable way. That puts the equatorial velocities of the late type stars less than $5-10$ km s^{-1} at the present accuracy of determination.

It is difficult to understand this peculiar behavior of stellar rotation when one considers the fact that whatever is the mechanism of star formation, it is always accompanied by acquisition of some angular momentum. It is hard to find a plausible reason for such a discontinuity in stellar angular momenta at the time of star formation. Several theories (See Kraft, 1970, for a general review) have been proposed in order to explain this strange phenomenon. In general it has been thought that this behavior is likely a result of braking of stellar rotation after the star has been formed. If so, there are three questions that we must answer. First, what is the braking mechanism? Second, what is the observational evidence for such a mechanism? And finally, why does it brake only the late-type stars? When we reviewed stellar rotation (Huang and Struve, 1960), none of the three questions could be satisfactorily answered.

But in the past decade new developments in different fields all lead us to believe that magnetic activities occur in the early stage of stellar evolution. General reviews of such developments have been given elsewhere (Huang, 1969; 1973). Therefore it is reasonable now to say that the braking of stellar rotation is of electro-magnetic nature. However, there are many ways that braking of stellar rotation can be effected by the magnetic field. We have mentioned already one by Lüst and Schlüter (1955) in connection with Hazlehurst's theory of ring formation. It would be out of the scope of this paper to present theories of magnetic braking so far advanced, all of which concern the motion of gases in the stellar neighborhood. Suffice it to say here that all theories are in principle attractive as far as the braking is concerned. But most require some assumptions which will lead to results contrary to what has actually been observed. It is our judgement that Schatzman's (1962) theory of flare-like activities seems to have the benefit of observational support. According to him when a jet of ionized gas is ejected during flares from the stellar surface, the matter is forced to turn with the stellar magnetic field up to a distance until the matter breaks away. He has applied this idea to the early stage of stellar evolution when the star is dominantly convective and therefore it is reasonable to expect magnetic activities on the surface just as the solar magnetic activity is attributed to the convection and/or differential rotation of the Sun.

This theory is consistent with the evidence derived from observation. In the first place it has been known that T Tauri and T Tauri-like stars which are in the pre-main-sequence stage are indeed ejecting matter (e.g. Herbig, 1962; Kuhi, 1964, 1966). Flares are often observed in some of these stars. According to Herbig (1952), T Tauri stars are indeed rotating with appreciable velocities, indicating that the magnetic braking process is not yet completed.

Schatzman's theory was proposed without the benefit of Hayashi's (1961) theory of pre-main-sequence evolution that has now been generally accepted. In the light of Hayashi's theory, Schatzman's argument for the differentiation of rotational behavior at F5 can no longer be applied. However his basic assumption that convection is the source of flares is still attractive. Therefore we may slightly modify his theory to assume that the braking of rotation takes place in the Hayashi phase of stellar evolution (Huang, 1967b). In massive stars the Hayashi phase is short. Consequently the rotation has not been greatly braked. While in less massive stars the Hayashi phase is long. Consequently the rotation has been effectively braked. We have proposed this explanation more on intuitive grounds than on actual calculation. But results of recent calculations by Larson (1969, 1972) have fully confirmed our intuitive suggestion because he has found that stars of masses greater than $1.5 \odot$ never undergo the Hayashi phase. Instead they first appear on the radiative pre-main-sequence track. His results provide a strong support to our hypothesis that braking of rotation occurs in the Hayashi phase, because the limiting mass of $1.5 \odot$ he has derived is very close to the $1.3 \odot$ that corresponds to the mass of stars of spectral type F5, where the sudden change of rotational behavior occurs. The small difference could be the result of uncertainty either in the calibration of stellar masses or in the computation.

5.2. FORMATION OF PLANETARY SYSTEMS IN THE UNIVERSE

Schatzman's (1962) paper is devoted solely to the braking of stellar rotation through mass ejection during flare-like activities. He assumes that the ejected matter escapes from the star, carrying away the angular momentum to infinity. Actually one may argue (Huang, 1967b; Nakano, 1970) that when the star is formed from the interstellar medium, there must be left a remnant or residue matter in its neighborbood. If so, those ejected particles will mix up with or stir the residue medium and impart their angular momentum to it. Once the medium has acquired the angular momentum, it naturally collapses into a rotating disk in which the gases will condense to form grains and grains will accret to become planetesimals and finally planets themselves. In this way we arrive quite naturally at a system that has all the major features of our own planetary system, such as the near-circular and near co-planar nature and the disparity in the angular momentum distribution between the Sun and the planets. Since, according to our reasoning, the disk is formed as a result of braking, it follows that planetary systems probably accompany most main-sequence stars of spectral type later than F5. According to this theory planetary systems are being formed at the pre-main-sequence stage.

5.3. FORMATION OF CLOSE BINARIES OF THE W URSAE MAJORIS TYPE

If the rotational angular momentum can be reduced by magnetic braking, the orbital angular momentum of a binary system can be similarly dissipated. Consequently as long as the braking mechanism continues, the separation between two components of the system will gradually decrease until the two become physically in contact (Huang, 1966a; Mestel, 1967). Such a process may be the cause for forming W U Ma stars. This theory has the attraction in that W U Ma systems are mainly composed of stars of spectral type F and later, just in the spectral range where rotation is effectively braked. Hence we are able to invoke the same mechanism for explaining the very slow rotation of stars later than F5 and the high abundance of W U Ma systems.

What would happen if the contacting binary is further braked? It can be easily seen that mass will flow from the less to the more massive component. As an extreme case the less massive component could be completely absorbed, by what may be called a fusion process, in the more massive component. In such a case the less massive component turns inside out. The matter in its core becomes the atmosphere of the merged star. If thermonuclear reactions of converting hydrogen into helium have already started in the interior of the less massive component, such a mode of mass transfer will result in an overabundance of helium in the atmosphere of the final star (Huang, 1966a).

This theory of the origin of contact binaries has its weakness in that while contact binaries are indeed composed mainly of stars of spectral type F and later, there are contact binaries with components of earlier spectral types. For this latter group of binaries, it is difficult to apply the magnetic braking idea. Consequently we may have to conclude that magnetic braking is not the sole agency for making contact binaries, although it may play a dominant role in shaping up W U Ma systems.

Acknowledgments

I sincerely thank Dr. Batten for sending me beforehand the draft of his paper to be presented in this syposium. As has been mentioned in the text, Dr. Batten's paper induced me to discuss the velocity-period relation of rings in binaries which would not otherwise be discussed here. Also I take pleasure in thanking Dr. Hall who sent me the reprint of a paper on S W Cygni and Dr. Van 't Veer who sent me the reprint of a paper on period changes in contact systems, before their publication. Preparation of this article has been supported by a grant from the National Aeronautics and Space Administration.

References

Batten, A. H.: 1969, *Publ. Astron. Soc. Pacific* **81**, 904.
Batten, A. H.: 1970, *Publ. Astron. Soc. Pacific* **82**, 574.
Batten, A. H.: 1972, presented at this symposium (p. 1).
Batten, A. H. and Plavec, M.: 1971, *Sky Telesc.* **42**, 147 and 213.
Buerger, P.: 1969, *Astrophys. J.* **158**, 1151.
Carruthers, G. R.: 1968, *Astrophys. J.* **151**, 269.
Catalano, S. and Rodono, M.: 1967, *Mem. Soc. Astron. Ital.* **38**, 395.
Crampin, J. and Hoyle, F.: 1960, *Monthly Notices Roy. Astron. Soc.* **120**, 33.
Crampton, D. and Hutchings, J. B.: 1972, *Nature* **237**, 92.
Curtiss, R. H.: 1926, *J. Roy. Astron. Soc. Can.* **20**, 19.
Hall, D. S.: 1971a, *Veröffentl. Remeis-Sternw. Bamberg* **9**, 217.
Hall, D. S.: 1971b, *IAU Colloquium No. 16* (in press).
Hall, D. S.: 1972, *Publ. Astron. Soc. Pacific* **84**, 323.
Hall, D. S. and Garrison, L. M., Jr.: 1972, *Publ. Astron. Soc. Pacific* **84**, 552.
Hardorp, J. and Strittmatter, P. A.: 1968, *Astrophys. J.* **153** 465.
Hayashi, C.: 1961, *Publ. Astron. Soc. Japan* **13**, 450.
Hazlehurst, J.: 1967, *Z. Astrophys.* **65**, 311.
Herbig, G. H.: 1952, *J. Roy. Astron. Soc. Can.* **46**, 222.
Herbig, G. H.: 1962, *Adv. Astron. Astrophys.* **1**, 47.
Huang, S.-S.: 1957, *J. Roy. Astron. Soc. Can.* **51**, 19.
Huang, S.-S.: 1963a, *Astrophys. J.* **138**, 342.
Huang, S.-S.: 1963b, *Astrophys. J.* **138**, 471.
Huang, S.-S.: 1965a, *Astrophys. J.* **141**, 201.
Huang, S.-S.: 1965b, *Astrophys. J.* **141**, 976.
Huang, S.-S.: 1966a, *Ann. Astrophys.* **29**, 331.
Huang, S.-S.: 1966b, *Ann. Rev. Astron. Astrophys.* **4**, 35.
Huang, S.-S.: 1967a, *Astrophys. J.* **148**, 793.
Huang, S.-S.: 1967b, *Astrophys. J.* **150**, 229.
Huang, S.-S.: 1969, *Vistas Astron.* **11**, 217.
Huang, S.-S.: 1971, *IAU Colloquium No. 16* (in press).
Huang, S.-S.: 1972, *Astrophys. J.* **171**, 549.
Huang, S.-S.: 1973, to be published.
Huang, S.-S. and Struve, O.: 1956, *Astron. J.* **61**, 300.
Huang, S.-S. and Struve, O.: 1960, in J. L. Greenstein (ed.), *Stellar Atmospheres*, Chicago Univ. Press, Ch. 8.
Hutchings, J. B.: 1970, *Monthly Notices Roy. Astron. Soc.* **150**, 55.
Hutchings, J. B. and Wright, K. O.: 1971, *Monthly Notices Roy. Astron. Soc.* **155**, 203.

Jeans, J. H.: 1925, *The Dynamical Theory of Gases*, Cambridge Univ. Press, Cambridge, Ch. 15.
Johnson, M.: 1958, in 'Etoiles à raies d'émission', *Mem. Soc. Roy. Sci. Liège, 4ème Ser.* **20**, p. 219.
Joy, A. H.: 1942, *Publ. Astron. Soc. Pacific* **54**, 35.
Joy, A. H.: 1947, *Publ. Astron. Soc. Pacific* **59**, 171.
Koch, R. H.: 1970, in K. Gyldenkerne and R. M. West (eds.), 'Mass Loss and Evolution in Close Binaries', *Proc. IAU Colloquium* No. 6, Copenhagen Univ. Copenhagen, p. 65.
Kopal, Z.: 1971, *Astrophys. Space Sci.* **10**, 332.
Kraft, R. P.: 1958, *Astrophys. J.* **127**, 625.
Kraft, R. P.: 1959, *Astrophys. J.* **130**, 110.
Kraft, R. P.: 1962, *Astrophys. J.* **135**, 408.
Kraft, R. P.: 1970, in G. H. Herbig (ed.), *Spectroscopic Astrophysics*, Univ. of California Press, Berkeley, p. 385.
Kruszewski, A.: 1966, *Adv. Astron. Astrophys.* **4**, 233.
Kuhi, L. V.: 1964, *Astrophys. J.* **140**, 1409.
Kuhi, L. V.: 1966, *Astrophys. J.* **143**, 991.
Kuiper, G. P.: 1941, *Astrophys. J.* **93**, 133.
Kuiper, G. P., Struve, O., and Strömgren, B.: 1937, *Astrophys. J.* **86**, 570.
Larson, R. B.: 1969, *Monthly Notices Roy. Astron. Soc.* **145**, 271.
Larson, R. B.: 1972, *Monthly Notices Roy. Astron. Soc.* **157**, 121.
Limber, D. N.: 1967, *Astrophys. J.* **148**, 141.
Limber, D. N.: 1969, *Astrophys. J.* **157**, 785.
Limber, D. N.: 1970, in A. Slettebak (ed.), *Stellar Rotation*, Gordon and Breach, New York, p. 274.
Limber, D. N. and Marlborough, J. M.: 1968, *Astrophys. J.* **152**, 181.
Low, F. J. and Mitchell, R. I.: 1965, *Astrophys. J.* **141**, 327.
Lucy, L. B. and Solomon, P. M.: 1970, *Astrophys. J.* **159**, 879.
Lüst, R. and Schlüter, A.: 1955, *Z. Astrophys.* **38**, 190.
McLaughlin, D. B.: 1943, *Publ. Michigan Obs.* **8**, 149.
McLaughlin, D. B.: 1950, *Pop. Astron.* **58**, 50.
McLaughlin, D. B.: 1961, *J. Roy. Astron. Soc. Can.* **55**, 13 and 73.
Merrill, P. W. and Burwell, C. G.: 1933, *Astrophys. J.* **78**, 87.
Mestel, L.: 1959, *Monthly Notices Roy. Astron. Soc.* **119**, 249.
Mestel, L.: 1967, in 'Instabilité Gravitationelle et Formation des Étoiles, des Galaxies, et leurs Structures Caractéristiques', *Mem Soc. Roy. Sci. Liège, 5ème Ser.* **15**, p. 351.
Morris, S. C.: 1962, *J. Roy. Astron. Soc. Can.* **56**, 210.
Morton, D. C.: 1960, *Astrophys. J.* **132**, 146.
Nakano, T.: 1970, *Prog. Theor. Phys.* **44**, 77.
Öpik, E. J.: 1963, *Geophys. J. Roy. Astron. Soc.* **7**, 490.
Paczński, B.: 1971, *Ann. Rev. Astron. Astrophys.* **9**, 183.
Parker, E. N.: 1963, *Interplanetary Dynamical Processes*, Interscience Publ., New York.
Piotrowski, S. L.: 1964, *Acta Astron.* **14**, 251.
Plavec, M.: 1968, *Adv. Astron. Astrophys.* **6**, 201.
Pottasch, S.: 1959, *Ann. Astrophys.* **22**, 297.
Prendergast, K. H.: 1960, *Astrophys. J.* **132**, 162.
Sahade, J.: 1960 in J. L. Greenstein (ed.), *Stellar Atmospheres*, Univ. of Chicago Press, Chicago, Ch. 12.
Schatzman, E.: 1962, *Ann. Astrophys.* **25**, 18.
Slettebak, A.: 1966, *Astrophys. J.* **145**, 126.
Smak, J.: 1962, *Acta Astron.* **12**, 28.
Spitzer, L., Jr.: 1952, in G. P. Kuiper (ed.), *The Atmospheres of the Earth and Planets*, Univ. of Chicago Press, Chicago, Ch. 7.
Struve, O.: 1930, *Astrophys. J.* **72**, 1.
Struve, O.: 1931, *Astrophys. J.* **73**, 94.
Struve, O.: 1941, *Astrophys. J.* **93**, 104.
Struve, O.: 1950, *Stellar Evolution*, Princeton University Press, Princeton.
Struve, O.: 1958, *Publ. Astron. Soc. Pacific* **70**, 5.
Togure, T.: 1969, *Astron. Astrophys* **1**, 253.
Van 't Veer, F.: 1972, preprint.

FIRST DISCUSSION SESSION

(Wednesday afternoon; 6 September, 1972)

(following review papers by Batten and Huang)

Chairman: A. D. THACKERAY

Thackeray: First we should express our thanks to Dr. Batten and Dr. Huang for the papers they have just presented. I've been asked to encourage as much freedom as possible in the discussion of these papers and not to limit ourselves to the order on the programme. Would anyone like to comment on any point?

Sahade: Could we ask Dr. Hall to present his ideas on SW Cyg?

Hall: Dr. Batten, in his review paper, talked about streams and disks, and clouds. In my analysis of the light curve of SW Cyg (Hall and Garrison, 1972), I found the need to hypothesize a different kind of feature, which I will call a lump. Perhaps I should not call it 'lump', but that is easier to say than 'protuberance' – the term used in the paper. Let me explain exactly why I found this idea necessary. Perhaps someone can suggest another way to explain the peculiar feature.

The binary SW Cyg is a rather typical Algol-like eclipsing binary, with an A2e primary and a K0 subgiant secondary which fills its Roche lobe. Primary eclipse is total, with totality lasting about 2.5 h. The peculiarity which bothered me most was that totality was not exactly flat. This was true in the visual and in the blue and also in the ultra-violet, but there is something else wrong with the ultraviolet light-curve which I will discuss later. The light rises gradually, as you go from mid-eclipse to third contact, by something like $0^{m}.05$. There is admittedly some scatter in my 1968 observations, because SW Cyg is quite faint at minimum, but this same effect shows up very nicely in Walter's 1964 observations and also in Wendell's 1900 visual obser-servations. So there can be little doubt that this effect is real and at least semi-perma-nent. This same rise occurs in blue and ultra-violet, and the amount of the rise corresponds to the same percentage of the light of the hotter star. In other words, this increase has the colour of the hotter star, which makes me think that the cause is something that is actually part of the hotter star.

This is the way the lump model works. The lump effectively extends the photosphere of the hotter star by about 10% on its leading hemisphere. As the hotter star moves from mid-primary eclipse towards third contact, the lump will be the first part of it to emerge from behind the cooler star. Since the lump has about 10% of the 'radius' of the hotter star, it will have about 1% of the projected area. Thus, with the same surface brightness, it will account for about 1% of the light. This light, relative to the light of the cooler star visible throughout totality, can produce the extra bright-ness observed around third contact.

Several other peculiarities can be accounted for by this lump. First, Dr. Walter noticed in his light curve of SW Cyg that the descending branch was fainter than the ascending branch, at least the upper part of it was. This could be the eclipse of the

Batten (ed.), Extended Atmospheres and Circumstellar Matter in Spectroscopic Binary Systems, 48–60.

lump. Second, Dr. Walter noticed dips in his light curve outside eclipse, where there should not have been any. In particular I notice the two depressions around phase $0.^P2$ and phase $0.^P8$. These two dips, if they were separated by exactly 180°, could be explained very nicely as the phases at which we see the hotter star end-on, (if the star-plus-lump combination is thought of as an elongated star). The effect could then be analogous to the familiar ellipticity effect caused by tidal elongation, except that the elongation is not in the direction of the line of centers.

Whenever the descending branch is fainter than the ascending branch, as in the light curves of U Cep and RZ Sct, it is traditional to consider the faintness as an effect of absorption by the stream. But a stream cannot easily explain the other features. Between mid-primary eclipse and third contact the stream is behind the cooler star and cannot be used to explain peculiarities. Furthermore, a stream cannot conveniently explain the dips around $0.^P2$ and $0.^P8$. Another problem is that the stream cannot easily explain a peculiarity which has the colour of the hotter star unless it absorbs non-selectively. Electron scattering is non-selective but, although this source of opacity can be operating in streams near B stars such as those in U Cep and RZ Sct, can there be electron scattering in the stream near the A2 star in SW Cyg?

The essential characteristic of the lump, I want to emphasize, is that its density is sufficiently large that it is optically thick in the continuum observed in the V bandpass. Therefore the surface of the hotter star, as defined by its photosphere, is not spherical but actually has a bulge in it. Why the lump is there is another question. That is not my problem!

Another peculiarity shows up in the ultraviolet when we look at second contact. Second contact is quite well defined in V and B, but in U it is rounded, so much so as to give the impression that the eclipse is partial in U. At the phase of second contact, the light in U is still about $0.^m3$ above the level it eventually reaches at minimum. This light almost certainly comes from the stream or else from the region where the stream encounters the hotter star on its trailing side. Emission from the stream should come primarily in the Balmer continuum and therefore mostly in the U but almost not at all in the B or V.

McNamara: Your B curve also exhibits a rounded edge at second contact, although it is not as pronounced as in the U curve.

Hall: Dr. McNamara noticed in Figure 1 of my paper (Hall and Garrison, 1972) that a suggestion of the rounding around second contact appears in the blue, although not at all in the visual. Perhaps this might be the effect of including several of the Balmer lines in the B bandpass, or perhaps the short wavelength tail of the B bandpass reaches down far enough to include a little bit of the Balmer continuum.

Thackeray: Would Dr. Walter like to comment on this?

Walter: Let me first consider the problems of SW Cyg from a general point of view. In an Algol-type system with a period of about five days, like SW Cyg, the gas stream, which leaves the subgiant near the Lagrangian point, L_1 can be made to intersect different parts of the surface of the bright star by quite small changes in its direction. For phases outside eclipse, one must take into account the effects on the light curve of absorption by the gas stream, and emission from regions where the particles in the stream strike the bright component. If the particles meet the surface of the bright star almost tangentially, as in AD Her (Korsch and Walter 1969), the photometric effects of the stream can only be seen in the part of the light curve between secondary and primary eclipses. The absorption is particularly strong immediately after secondary eclipse, when we are looking along the stream. If the particles arrive somewhere on the part of the following side of the bright component that faces the subgiant, this region of the surface will be brighter and can be seen without absorption immediately after primary eclipse. At this phase, the rectified light curves of many Algol-type systems show a hump. The light curve of U Cep shows a large hump – smaller ones are seen in the light curves of SW Cyg and TT Lyr (Walter, 1971a). The light curve of SW Cyg shows this asymmetric gas-stream effect clearly (Walter, 1971b).

My observations of SW Cyg, obtained in 1964 in Sicily, were not numerous enough to conclude any more from the light curve outside eclipses. The observations of the eclipse phases, made in B and V, were good and seemed to be suitable for a determination of the photometric elements. For a long time, however, although I tried hard, I could not obtain a satisfactory solution in both colours with the same values for the ratio of the radii, k, and the orbital inclination, i; but, after many vain attempts, I found that solutions with quite usual limb-darkening coefficients fitted the light curves very well, in the neighbourhood of total eclipse.

I then abandoned the usual Russell-Merrill method, in which the form of the eclipse curve is used to derive the photometric elements. If a light curve is suspected to be distorted by luminous regions (on the surface of one of the stars), the geometric parameters must be obtained from data which are independent of the intensity distribution over the stellar surface. For SW Cyg, the phase of second contact, and the depths of the minima were used. Because the primary eclipse is asymmetrical, I assumed an extinction by the gas stream of two per cent in the descending branch.

In this way, I obtained good representations of the central parts of primary eclipse, using exactly the same values of k and i in both colours. In the outer parts of the the eclipse, however, a residual excess luminosity of about two per cent remained, in both colours. This can be interpreted as additional light from hot regions which happen to be eclipsed within the primary eclipse. The known geometry of the eclipse of the stars allows us to locate these hot regions on the bright component. In the case of SW Cyg, the surprising result was that these regions cannot be located near the equator of the bright star; they must be situated in high latitudes.

The old, very good observations of SW Cyg made with the polarizing photometer by Wendell can be represented in a quite similar manner, if exactly the same geometrical elements are used as were obtained from my V observations. Probably there was rather more additional light in 1900. The asymmetry of the primary eclipse was also larger then, and I adopted an extinction of three per cent.

The fact that the luminous regions are at high latitudes suggests the action of magnetic forces which compel the ionized particles of the gas stream to move out of the orbital plane. If there are magnetic fields, either on the bright component, or between the components, where the gas stream flows, then such a binary may be considered to work like a big mass spectrograph. If there are magnetic fields, some particles are strongly influenced by them, follow the lines of force, and arrive at high latitudes. Other particles may be more strongly influenced by gravitational forces and arrive near the equator of the bright star. In my report at the I.A.U. Colloquium No. 16 (Walter, 1971a) I distinguished between g (gravitational) and m (magnetic) regions for hot spots. So I think my observations and Dr. Hall's do not contradict each other, but concern complementary aspects of hot-spot phenomena.

Biermann: I would like to comment just very briefly on this. What you expect from the theoretical calculations is a gas stream leaving the Lagrangian point and expanding a bit (depending on the mass ratio of the system and the angular momentum per unit mass). You get some sort of hot spot that is a standing shock where the stream meets the star, and another standing shock where the part of the stream that goes around the star meets the directly expanding stream. Are Dr. Hall's observations consistent with the hypothesis that his 'lump' is a kind of weak shock between the two streams?

Hall: I think of the lump more as a significant accumulation of matter of sufficiently high density to be optically thick in the continuum. The colour of the lump seems to be different from that of the stream. The stream emits almost entirely in the ultraviolet, whereas the lump has the colour of the hotter star.

Biermann: I can't propose any detailed model for the shock. I would expect the stream by itself to have very different characteristics from a shock....

Hall: What would you predict the colour of the shock to be?

Biermann: Oh, I wouldn't predict any colour.

Plavec: I think we might try to get at some prediction of the colour. It's surprising that the colour should be the same as that of an A-type star because when the stream falls on the star and creates a hot spot, if its kinetic energy is very rapidly converted into energy of thermal motion, we can write that the maximum possible temperature would be $1.5 \times 10^7 \, m/R$ (where m and R are the mass and radius of the star, respectively, in solar units). Now, for an A3 star, m/R is approximately unity, and you have a very high temperature. Even X-rays could be produced. Of course, we are not considering the problem of intensity, but certainly the spot would be really hot. When the material falls at a high speed onto the star, even if you assume that the efficiency of the process is very low (say one per cent) you still get a very high temperature – certainly nothing like the surface temperature of an A-type star. So the bulge in Dr. Hall's picture is, first, very difficult to explain dynamically – in the direction he wants it – and second, its colour index indicating (if it is measured accurately enough) so low a temperature is also difficult to explain. Perhaps Dr. Biermann's suggestions might save the picture proposed by Dr. Hall, if two streams are meeting – but they may meet rather far from the surface of the star, and the velocities will be so terribly high.

Underhill: This brings up the problem of detecting a gas stream around a hot star. To detect the gas stream you must observe at a wavelength in which the stream is emitting strongly. If we have a spot with a temperature in the range of one to ten million degrees, the spot will radiate very strongly at wavelengths shorter than 2000 Å. This means you must observe from above the Earth's atmosphere using either a photometer behind a small telescope or a spectrometer. However, at present there is no instrument up in space capable of observing faint stars. OAO-2 does have photometers but they can be used only on stars brighter than about 6th magnitude. I don't know how faint SW Cyg is, but I suspect it is too faint for OAO-2. There is a lot of work to be done in photometry in the far UV but first of all we have to get aloft an instrument capable of doing photometry on faint objects. A second goal is to launch an instrument capable of doing spectrographic work. At present there is one instrument planned for purely spectrographic work in addition to the two OAO's which are now flying. If the gas in the streams or rings Batten was telling us about has a temperature between 10000° and 30000°, it may be detected by observing the Balmer lines of hydrogen, which can be done from the ground. If the temperature is in the range 5000° to 10000°, the emission in the Balmer lines will be weak and the gas may be visible in the resonance lines of Ca II or better still in the resonance lines of Mg II at 2800 Å. To observe these requires a satellite-borne instrument. Thus we see that space astronomy has much to offer to the study of binary stars and of the gas streams in binary systems, for it will permit the detection of very hot streams and spots as well as cool streams of gas.

Smak: The restricted three-body approach, used to compute the particles trajectories, was surprisingly successful in explaining several properties of the circumstellar matter. And, while it is obvious that there are regions of higher density where this approach cannot be correct, I believe that – for reasons of simplicity – we should go with it as far as possible trying to interpret at least some of those many problems which are not yet explained.

Before I shall discuss one such problem, let me mention once again a few trivial facts. Suppose the secondary component loses matter through the inner Lagrangian point. The stream of matter goes toward the primary component and – if there is no circumstellar matter – it lands on the surface of the star (Figure 1a). If the amount of matter is small, it can be accumulated by the star. If its amount is large, however, and – particularly – if the amount of momentum it carries is large, then a rotating disk must be formed. The mean radius of such a disk can be estimated from the angular momentum considerations and this is usually done under the assumption that other momentum transfer processes can be neglected. With the disk being present, the stream of matter from the secondary component must collide with its outer portions and a hot spot must be formed at the place of collision (Figure 1b). This is what is typically observed in systems of the nova and U Gem type. It may be added that usually the amount of momentum carried by the stream is so large that the radius of the disk must also be large and – as a rule – the spot must be formed in a relatively narrow range of phase angles roughly in a position indicated in our schematic picture.

Let us assume, however, that the outer dimensions of the disk are not always the same. Indeed, we know already examples of variable dimensions of the disk, like

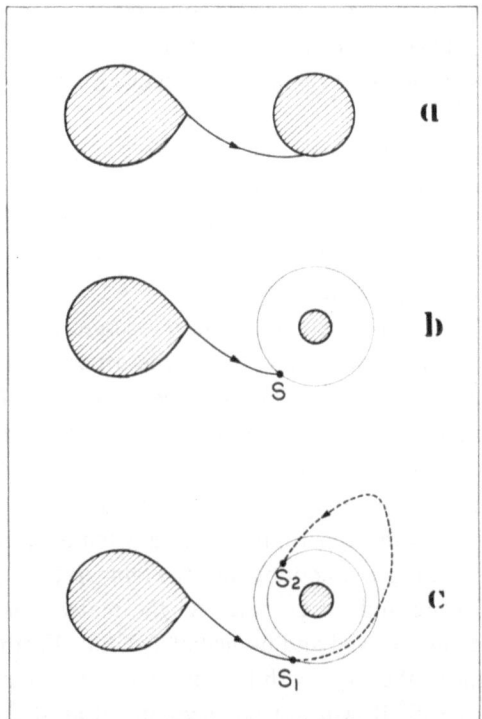

Fig. 1.   Trajectories in U Gem systems.

U Gem or RW Per. Such phenomena are connected presumably with the variable rate of the momentum transfer from the disk to the primary component or to the orbital momentum and/or with some other instabilities occurring within the disk. Let us further assume that – at one time – the disk can become much smaller. Then the stream of matter can go around the primary component and collide with the disk at an almost opposite phase angle. Figure 1c shows two situations corresponding to the two different radii of the disk with the two resulting locations of the hot spot, S1 and S2. If we let the radius of the disk vary continuously, we get two ranges for S1 and S2, respectively. Detailed calculations would probably show that – for any given system – both these ranges are rather narrow. However, with the stream particles travelling so close to the edge of the disk, we should consider here the effect of slowing them down by the lower density medium extending far beyond the conventional limits of the disk. With this effect taken into account we could probably obtain nearly any location for the hot spot although it is clear that such a spot may no longer be a distinct, point-like feature.

These considerations, if correct, may offer an explanation for the so-called periastron, or O'Connell effect including the cases where this effect is variable. Indeed, a hot spot located at an arbitrary phase angle can produce an arbitrary distortion of the photometric curve. Its variable location and variable intensity of its radiation can account for the variable light curves. Of course, as long as we do not know too much about the behaviour of the disks, we cannot insist on applying this explanation to all cases but at least in some of them this mechanism may be more plausible than the others.

Finally, I wish to point out that the situation presented in Figure 1c may, in fact, occur in certain Algol-type systems. I refer to the extensive distortions in the radial velocity curves which have been successfully interpreted in terms of the streams of gas circulating nearly around the primary component; in contrast to the disks, they produce absorption lines and are less regular.

R. E. Wilson: I'd like to comment that the temperature Dr. Plavec suggested is probably too large by considerably more than one order of magnitude. By way of illustration, if one does a similar calculation for very compact objects – white dwarfs or neutron stars – one finds temperatures of 10^9 to 10^{12} degrees, but the observed temperatures of typical X-ray sources are, of course, only about 10^7 degrees. The temperature obtained from the formula is up to five orders of magnitude too high, and it is, indeed, a very extreme upper limit. Normally temperature calculations involve a balance between heating and cooling, but the temperature quoted by Dr. Plavec corresponds to all heating and no cooling. As the material is heated it begins to reradiate the energy it has received and immediately cools. Furthermore, the material coming in shares its thermal energy with the material of the star, so I think temperatures of the order of 3×10^7 K will not be encountered at all in this kind of situation.

Plavec: No doubt this is an upper limit, but I doubt if it is off by, say, three orders of magnitude. We will probably have discussion on X-ray sources later. I wanted to comment actually on Smak's picture. It's true that in certain cases (mostly in those in which he is interested) the primary component is so small that the ring forms and then the hot spot is situated mainly on the outside of the ring. But, unfortunately, in the Algol systems, which are of relatively short period, the radius of the primary component is about 0.2 (expressed in units of the orbital radius) which means, if you perform the calculations Smak referred to, that the stream will hit this component directly on its first approach. It is very difficult to escape from this fact, and I think Dr. Batten will agree that our idea now is, that in the case of U Cep and U Sge, the ring or disk we observe may actually be due to some scattering of particles by the primary, after impact. In any case, the ring will probably not be dense. Hot spots on the ring can form in binaries where the primary component is very much smaller – smaller, for example, than 0.1. This easily happens in systems with periods of about 200 days, or in systems containing degenerate stars (where the radius of the star itself is quite small). Then you can look for a possible hot spot on the ring. But generally in Algol systems like SW Cyg, I think we must assume that most of the material falls on the star just because of its larger size.

Huang: Most of you talk about the hot spot on the ring. If you have a gas that hits on an extended medium, you produce a hot spot – but the ring is tenuous. Why do you talk about a hot spot on the ring? If you have a ring maintained only by angular momentum with no elasticity, the stream of gas will destroy the ring instead of having a shock or anything.

Smak: Even with densities in the disk of the order of 10^{12}–10^{13} particles cm^{-3}, which I believe is an underestimate, one finds that the mean free path of the stream particles is one or two orders of magnitude smaller than the disk's radius. So, the collision must effectively take place in the outermost part of the disk.

Underhill: I am concerned about the estimates of density in these rings or streams. The average particle density in a late B type star or an A type atmosphere is 10^{13} to 10^{14} particles per cm^{-3}. In supergiant atmospheres the density is about 10^{12}. If you have coronal conditions the density is about 10^7; when forbidden lines appear the density is 10^4 to 10^7. Now some of the densities Batten quoted were 10^{16} to 10^{17} particles cm^{-3} and I heard Smak say he would like them an order of magnitude greater. I don't understand it. You cannot have such large densities and obtain the spectroscopic observations which you have. Even in a dense main-sequence star like the Sun the density in the atmosphere is not much greater than 10^{15} particles per cubic cm. It cannot be an order of magnitude greater in a gas stream from a K type giant.

Batten: I think that I have quoted densities as high as 10^{16} or 10^{17} particles cm^{-3} only in order to criticize them. I don't think they are that high myself. I do think they may be as high as 10^{13} particles cm^{-3} because, after all, at least in the case of Algol systems, the streams originate in stellar atmospheres of about that density. Forbidden lines, of course, must come from regions of lower density – as I pointed out in my review. You must distinguish between regions of higher density (which I call streams and disks) and regions of low density (which I call clouds). Even if you don't like my terminology, I have made the distinction.

McNamara: When Batten discovered the emission lines in the spectrum of U Cep, I think he mentioned that their detection depended very critically on when the plates were exposed (in relation to the contacts of the eclipse). Might it not be that the line intensities are changing quite dramatically because of real changes in the disk?

Batten: You've quoted me correctly but I've changed my mind about this. I still think that timing of exposures may be important, but, just about a year ago, Mr. Baldwin and I made simultaneous photometric and spectroscopic observations during eclipses of U Cep. We know that we began and ended our exposures within a few minutes of second or third contacts, and we are reasonably confident that if emission had been detectable we would have detected it again. We did not. Therefore we believe that either the disk itself, or the state of excitation of matter within the disk really does change.

Hall: I would like to ask Dr. Batten if he remembers a relatively old paper on U Cep by Miczaika (1953), in which he hypothesized a lump in U Cep to explain the same sort of photometric peculiarities in his light curve which caused me to hypothesize a lump in SW Cyg. Let me point out that Miczaika's model referred to his yellow light curve, which should have been relatively unaffected by emission lines *per se*, and furthermore that his lump had the same orientation: on the leading side of the hotter star. In both U Cep and SW Cyg the first part of the descending branch of primary eclipse is relatively too faint (caused, I would say, by the eclipse of the lump by the cooler star). In both systems there is an increase in light during totality as you go from mid-eclipse to third contact (caused, I would say, by the emergence of the lump from behind the cooler star before the hotter star itself emerges). And there are two dips in the light curve outside eclipse approximately 180° apart from each other (caused, I think, by the fact that at these phases we see the elongated star-plus-lump end-on). I wanted to ask Dr. Batten to what extent he considered Miczaika's model believable, at least in the case of U Cep.

Batten: Yes, I remember that paper. I believe the light curve of U Cep that Dr. Walter showed us was Miczaika's. My own feeling is that the way in which Miczaika made up his normal points concealed the true nature of the variation outside eclipse. As I mentioned in my review, most light curves of U Cep show a sudden dip of about $0.^m1$ somewhere between $0.^P8$ and $0.^P9$ in phase. This feature is seen very clearly in light curves of different colours, even an infrared light curve obtained by Khozov and Minaev (1969) at 8100 Å shows it quite clearly. I think your light curve, Dr. Catalano, shows it?

Catalano: Yes, the light curve we have in Catania shows a change near phase $0.^P8$ of about $0.^m1$. I should like to remark that the light change occurs just at the same phase as the radial-velocity curve shows a jump.

Batten: Miczaika's light curve is the only exception. I believe the distribution of his observations, and the way he combined them to form his normal points, have produced a misleading picture of the out-of-eclipse light variation, and for this reason I am a little skeptical of his model.

Hall: Miczaika found that the duration of totality was variable from cycle to cycle. At third contact he found that the exact phase at which the light suddenly increased was significantly different from one cycle to the next, as if the photosphere of the leading edge of the hotter star was at a different place at different cycles.

Batten: There is evidence for real variation in the light of the secondary star, and different light curves show quite different sorts of variation during totality.

Thackeray: How sudden is the drop in brightness at phase $0.^P8$?

Batten: Quite sudden. At most it takes a few hundredths of a period, I believe.

Catalano: Yes, it is quite steep, but I don't remember now precisely how long it takes.

Underhill: I would like to ask Dr. Huang whether the decreasing percentage of stars with shells through types B and A shown in his table is real, or whether it is a result of the difficulty of detecting a shell at type A.

Huang: There are few A-type stars with rings. I believe it is an intrinsic deficiency. Since emission lines are found in the spectra of binaries with A-type primaries, it appears that A-type stars have temperatures high enough to excite the gas in the ring. Consequently the deficiency of gaseous rings about rapidly rotating A-type stars is intrinsic.

Underhill: If we had other criteria to rely on than the hydrogen line, for the detection of these shells, would you say that we could see more shells?

Huang: I am not sure of any other criterion that could be used to detect a rotating shell that is found by the hydrogen lines.

Underhill: There is nothing physically to prevent the cooler stars from having shells that produce emission lines. These cool shells would not emit in hydrogen, therefore we have not detected them that way, but is there any reason why there should not be extended atmospheres around late B-type and early A-type stars, similar to those around early B-type stars?

Huang: It does not appear physically that there should be a ring around A-type stars, and we don't know whether or not the extended atmospheres you have mentioned exist. But around stars with rings, there must also be another substratum, as I mentioned in my talk, because the ejected matter cannot all go into the ring. Some will be left there as an envelope that is distinct from the ring. I take the view that A-type stars have no envelope or ring, B-type stars have rings, and most O-type stars that show emission in their spectra do not possess rings.

Underhill: I don't think I agree with you....

Huang: Yes, of course. This is why at the beginning of my talk I emphasized that everything can happen. I used the phrase 'my present understanding' because I may change my mind tomorrow. Therefore I do not dispute your disagreement!

Biermann: I was thinking of the possibility of rings in single A-type stars and the question of those A-type stars in binaries, that show emission from their disks. How can you be sure that the emission is caused by the A-type star and not something else?

Huang: Because the other component star is of an even later spectral type and cannot be responsible for the excitation.

Biermann: But are you sure that the emission is not a feature of the gas stream itself? When you have an oblique shock, strong emission could be caused by it. Therefore, no emission observed in the spectra of single A-type stars does not rule out the possible existence of rings.

Huang: Oh, you can always have many possibilities, and in astrophysics there is no sure thing. You can make all kinds of assumptions. I wouldn't rule out any other possibility. I tried to describe the excitation problem purely from the observational point of view, and used the component stars with rings, in binaries, as a model for comparison. Of course, I completely agree that this comparison may be inadequate; but if you introduce other possibilities there would be no end to it. The excitation is not a settled problem.

Smak: I think the case of rings in binary systems is quite instructive. We observe emission in the spectra of very few B-type stars and from very many A-type stars – even of quite a number of F-type stars. (I refer to the spectral types of the primary components which are surrounded by disks.) We cannot blame the central A-type and F-type stars for the ionization of their disks. An even more convincing argument is provided by many novae, and U Gem type systems, in which the disks are optically thick so that the radiation from the very faint stars inside cannot penetrate the disk at all. So there must be another source of energy and I think we must agree that the ionization and excitation is of the collisional type with the kinetic energy being supplied by the stream coming from the secondary. Now, to return to the disks of single B-type and A-type stars, one can speculate that they are formed not only in the Be stars, but also in the A-type stars, except that the emission lines are not visible in the latter case because the ionization conditions are different. In addition to the different temperatures of stars, it may also be important to note that the amount of kinetic energy is different. If rotational break-up is, at least partly, responsible for disk formation, much larger velocities are observed in B-type stars. Therefore the ring of a B-type star can be bright and visible, and the ring of an A-type star may remain practically invisible, and the presence of absence of emission lines of hydrogen is not a definite test of the existence of a ring. We need another way of detecting rings.

Popper: I agree with Dr. Huang that it is not desirable to introduce unnecessary complications. But I also agree with Dr. Smak that the spectra of rings around A-type and F-type primaries are not different from those around hotter primaries in their

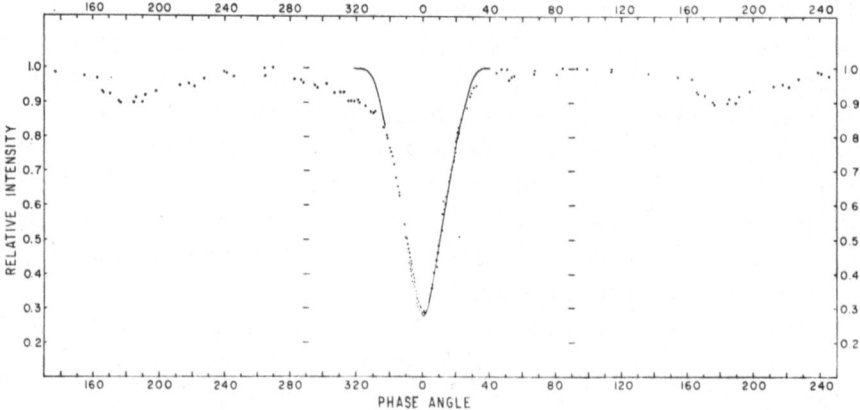

Fig. 2.   The *V* light curve of RZ Sct. The plotted data are normal points except for the smaller dots near the bottom of primary minimum which are individual observations. Redetermination of the period of the system, since this figure was drafted, has reduced the scatter of these latter points. The solid line represents a preliminary solution. A more definitive solution is in progress.

degree of excitation, which is higher than that of the photosphere of these later types. There is pretty strong evidence of non-thermal effects. The spectra of the rings around the F-type primaries of KU Cyg and RZ Oph, in addition to the hydrogen and calcium emission lines, show an ultra-violet continuum. This continuum is not a Balmer continuum since it extends strongly to longer wavelengths.

Hansen: First, I'd like to remind you of the light curve of U Cep which Dr. Batten showed earlier (Figure 2 on p. 6). Notice the break in the curve, which he has already pointed out, that is seen just before primary minimum when there is a definite decrease in light before the beginning of the actual eclipse. There is also a noticeable slope upward at the bottom of the primary eclipse between second and third contacts. But notice also that the secondary eclipse is not symmetrical, the ascending branch of this eclipse is much less steep than is the descending one. This was not pointed out earlier.

Dr. Hall has proposed a 'lump' model of the hotter star to explain the distortions in the primary eclipse, and I agree that such a model can explain these two features, but it does not explain the distortion of the secondary eclipse. Now look at the light curve of RZ Sct (Figure 2). Beginning at the left of the diagram, we notice that the secondary eclipse shows a relatively steep drop into eclipse and a much more gradual rise coming out. In the primary eclipse there is a long, gradual slope going into eclipse while the up-branch is steep all the way up, and the shoulder at the end of the eclipse is much more distinct.

I would like to ask how Dr. Hall would explain this feature of the secondary eclipse, the depressed ascending branch, in terms of his model? It seems clear to me that the system of RZ Sct has some sort of mass concentration, a stream or an eddy such as has been previously proposed, that is located on the side of the system corresponding to the advancing side of the secondary. Material in this position could cause an absorption during the approach to primary eclipse and also as the system is emerging from secondary eclipse. So the interesting thing to me is that RZ Sct has the same feature in the secondary minimum that is seen in U Cep, and I wonder how Dr. Hall's 'lump' model might explain this feature.

Hall: To explain it with the 'lump model', we would have to place the bright lump so that just after secondary eclipse it would be on the far side of the primary star, hidden from the observer. There should then be a minimum in the light curve because the lump is invisible (leaving the eclipses themselves out of account). This position for the lump is also the one needed if the loss of light before the beginning of primary eclipse is to be interpreted as an eclipse of the lump by the cooler star. If the lump is in this position, there should also be another minimum in the light curve, 180° in phase from the first, namely right after primary eclipse. At that phase the star-plus-lump (considered as an elongated star) would be seen 'end-on'. Your observed points do fall below your computed curve at fourth contact (as well as at first contact). You will have to tell me if these residuals can be interpreted as the result of such a minimum in the light curve.

There may be one difficulty with your interpretation of the dip after secondary eclipse as an effect of absorption by the stream and eddy. I think you said that the cooler star accounts for only ten per cent of the total light of the system, and the dip after secondary eclipse is about $0^{\rm m}.1$. If so, then the stream and eddy must absorb virtually all of the light of the cooler star.

Hansen: The cooler star does account for about 10% or 12% of the total light of the system, but the $0^{\rm m}.1$ that you refer to is the maximum depth of the secondary eclipse.

If in the secondary eclipse we compare the ascending branch, which is depressed, with the descending branch we find that the extra loss of light is, at most, about 35%, considering the amount that is lost at the midpoint of the secondary eclipse as 100%. A similar comparison in the primary eclipse shows that the stream absorption amounts to about 10%. It is true that to absorb about one third of the light of the cooler star requires a large, dense stream, but the spectroscopic evidence also indicates that RZ Sct does in fact possess an extensive stream and disk system of some sort.

One further comment might be of interest. The two halves of the light curve of RZ Sct look as if they might belong to two different systems. From primary minimum to secondary minimum it looks like the light curve of a typical Algol system, but from secondary minimum to primary minimum it looks more like the curve of a system such as β Lyr.

Batten: I have tried to explain the asymmetry of the secondary eclipse of U Cep in terms of a bright spot near the Lagrangian point of the secondary component, where the mass is actually ejected from the secondary star. I'm not very happy with the explanation but it does fit, qualitatively. Dr. Fracastoro has compiled an Atlas of light curves, and it is clear from this that asymmetric secondary eclipses are very common and we certainly must try to explain why this is so.

Devinney: I am surprised that Dr. Hall did not mention HS Her (B4+A4) in this connection. The rise to maximum following mid-secondary eclipse is too slow, in this system, giving the impression that the star recovers too slowly from the eclipse. In addition, the 'flat' secondary minimum slopes upwards. On the other hand, from the phases of contacts alone, we can obtain very good approximations of the fractional radii of this two-spectra system, and we find that both stars lie well within their Roche lobes. Thus it seems difficult to blame gas streaming for the peculiarity. I've found that a star spot covering ten per cent of the primary's disk, and having a temperature of 12000 K satisfies the light and colour curves.

Hall: I really should have thought to mention HS Her myself, but thank you for bringing it into the discussion. In this system we have a B4V primary and an A4V secondary. The major complication occurs right after secondary minimum. It is as if the system refuses to come out of secondary eclipse. Full light is not recovered until almost $0^P.1$ after where fourth contract should have been. This effect shows up most strongly in U and less strongly in B and V. I have described the system elsewhere (Hall and Hubbard 1971). The real puzzle is not the dip itself but that it occurs in a binary composed of two stars on the main sequence, both well within their Roche lobes and apparently perfectly normal. As I understand the situation, there is absolutely no good reason why there should be photometric complications in a system like HS Her.

Thackeray: It has been suggested that we adjourn this meeting until tomorrow morning.

References

Hall, D. S. and Garrison, L. M.: 1972, *Publ. Astron. Soc. Pacific* **84**, 552.
Hall, D. S. and Hubbard, G. S.: 1971, *Publ. Astron. Soc. Pacific* **83**, 459.
Khozov, G. V. and Minaev, N. A.: 1969, *Trudy Astr. Obs., Univ. Leningr. Gos.* **26**, 55.
Korsch, D. and Walter, K.: 1969, *Astron. Nachr.* **291**, 231.
Miczaika, G. R.: 1953, *Z. Astrophys.* **33**, 1.
Walter, K.: 1971a, *IAU Colloquium No. 16* (in press).
Walter, K.: 1971b, *Astron. Astrophys.* **13**, 249.

SECOND DISCUSSION SESSION

(Thursday morning; 7 September, 1972)

Chairman: F. B. WOOD

F. B. Wood: Yesterday we had talks about light curves of eclipsing stars, and people drew on the blackboard to illustrate. Today I have a copy of an Atlas of Light Curves of Eclipsing Variables compiled by M. G. Fracastoro. I can't very well pass it around, but if you wish to look at it during coffee break or afterwards, you will see that we have a beautiful series of various light curves; if any of them illustrate your particular paper, I'm sure you'd be free to borrow it. I think that when Fracastoro prepared this, he compared it with catalogues that gave solutions for L_1 and R_1 and so forth by saying that while you can read the vital statistics of bathing beauties (36-28-34 or whatever) it isn't quite the same thing as looking at pictures of the girls themselves. Now we have the 'pictures'. I have tried to arrange the first few talks to fit in with yesterday's discussion. First, will Dr. Catalano talk on photometric evidence of gaseous envelopes in close binaries?

Catalano: I should like to make a few remarks on what we can infer about gaseous envelopes in close binaries from the analysis of absorption effects in the light curve at phases just preceding and following the primary minimum. Such absorption effects are shown in Figure 1, where the V light curves around the primary minimum, of the systems V 548 Cyg (Rodonò, 1976), AR Lac (Blanco and Catalano, 1972), δ Lib (Koch, 1962) are given. Other systems could be added to these, such as RS Sgr (Baglow, 1948) and R CMa (Koch, 1960); U Oph (Huffer and Kopal, 1951) probably shows the same effect at the secondary minimum.

It is known that many systems show absorption effects in their light curves and atmospheric eclipses have been suggested in many cases, however what is interesting in the systems quoted is the duration and the depth of the disturbance.

The observed light curves could be explained assuming that the primary component is eclipsed by an extended semi-transparent envelope surrounding the secondary component. An estimate of the size of such envelopes can be made from the geometrical elements given in the literature and from the phases $\bar{\phi}_{abs}$ in the light curves of the beginning and the end of the absorption features. These data, together with the radii obtained for the envelopes, measured in units of the separation of the two components, are given in Table I. The fractional radii are also given. We note that the envelope is nearly in contact with the surface of the primary component ($\varrho + r_1 \approx 1$, ϱ fractional radius of the envelope and r_1 radius of the primary component). The values of the radii of the envelopes range from 0.6 for V 548 Cyg to 0.8 for AR Lac and are about twice the stellar radii, and 2.5 in the case of AR Lac. These figures are very close to the ones Dr. Batten quoted in his review for the size of the disk observed around the primary components of Algol-type systems. It is clear that these results

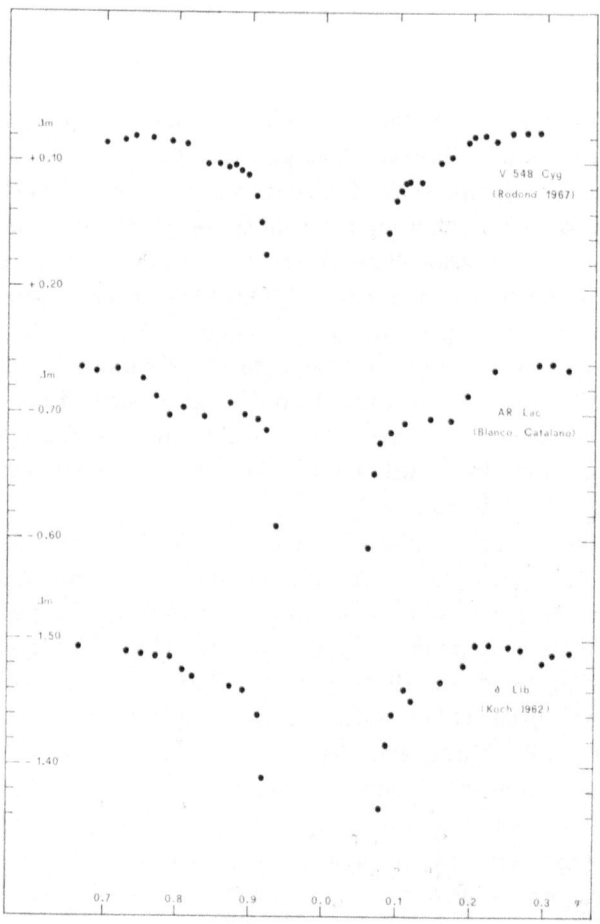

Fig. 1.   Light curves of three eclipsing binary systems showing effects
due to circumstellar matter.

TABLE I

Data for envelopes in five systems

System	Sp	i	r_1	r_2	$\overline{\phi}_{abs}$	ϱ_{env}	$\varrho - r_2$
R CMa	A9 + K5	77°	0.302	0.248	$0^P.18$	0.61	0.36
V 548 Cyg	A0 + F7	83	0.347	0.260	$0^P.20$	0.60	0.34
AR Lac	G2 + K0	90	0.194	0.319	$0^P.25$	0.80	0.49
δ Lib	A0 + G8	79	0.300	0.308	$0^P.22$	0.68	0.37
RS Sgr	B5 + A3	82	0.326	0.261	$0^P.21$	0.64	0.38

could be improved if a solution of the geometrical and light elements is carried on taking into account the atmospheric eclipse.

Considering now the plateau of the light curve during the absorption phases, we might think that the primary star is wholly eclipsed by the envelope for some time. However, if we compare the diameter of the eclipsed star, $2r_1$, and the thickness, $\varrho - r_2$, of the envelope, we find that this can happen only for AR Lac. To overcome this difficulty without changing drastically the radii of the two components, we could describe the absorption effect by a tongue of matter streaming near the primary component as in Fletcher's (1964) model of Algol. Then, assuming that (i) the absorption is due to electron scattering, (ii) at least one tenth of the stellar disk is obscured, and (iii) the stream extends no more than the separation of the star surfaces, we find values of the density of the stream of 1.6×10^{12} to 3.7×10^{12} (Table II). These values agree fairly well with those found in other systems at least as far as the stream hypothesis is concerned. For AR Lac the envelope assumption leads to a mean density of about 10^{11} electron cm^{-3} (Table II).

TABLE II

Electron densities

System	N_e cm^{-3} (envelope)	N_e cm^{-3} (stream)
R CMa		2.1×10^{12}
V 548 Cyg		2.8×10^{12}
AR Lac	$1.0 \quad 10^{11}$	3.1×10^{12}
δ Lib		1.6×10^{12}
RS Sgr		3.0×10^{12}

Oliver: I would like to comment on AR Lac which is apparently different from the other systems. The mass ratio of AR Lac is close to unity, which makes it a different kind of beast from the others. Kron (1947) observed quite substantial dips in the light curve (at one epoch, we can see three or four). If two of these just happened to be placed symmetrically on either side of primary minimum, they would produce the effect you are talking about. You estimate the envelope around the secondary star in AR Lac to extend to 0.8 of the separation, and that is well beyond the Roche limiting radius of 0.38. This means the envelope does not belong to the secondary: it belongs to the system. Finally, AR Lac is one of the R CVn systems. The secondary component is a late-type ($\sim$KO) subgiant and shows H and K emission lines in its spectrum. There are lots of photometric anomalies in systems of this sort that will probably be discussed later. I wonder if this system really belongs with the other four.

Catalano: The extension of the envelope is a problem, and also it is strange that we find an envelope around the secondary component. An alternative hypothesis is to suppose that there is a bright envelope around the primary component, but then there would be a problem about the density needed in the envelope to explain the loss of light we observed.

Oliver: Kron (1947) explained this phenomenon in the light curve of AR Lac by dark spots on the primary component.

Catalano: Yes, but I found that the light curve I obtained was rather different from Kron's. We showed some years ago that these variations in the light curve could be explained by variations in the light of the comparison star that Kron used. We have a stable light curve now, and the feature I showed repeats from cycle to cycle.

Oliver: Over how many cycles?

Catalano: We have observed it throughout a year, but Kron found that the spotted region on the primary star changed in about a month or less; and that is just the period we have found for the variations of the comparison star.

Oliver: I realize that, but, unfortunately, WY Cnc, SS Cam, and probably several others that are in other ways similar to AR Lac behave in the same way. For these stars it is certainly not variations in the comparison star that are responsible.

Van't Veer: Some of the systems in Table I do not seem to be quite possible dynamically. At least two or three systems are contact systems, or nearly so, as far as I can see from the radii. The width of the envelope ($\varrho - r_2$) as quoted by Catalano, is a real problem in these systems. The envelope extends far beyond the Lagrangian points into regions where the forces are directed outward from the system. It is not clear to me how you can explain the shoulders of the light curve if there is a homogeneous envelope around the whole system.

Catalano: Yes, this is true. I did not set out to explain everything in this one report. I am just exploring the consequences of assuming a gaseous envelope around the secondary component.

Smak: Is there any spectroscopic evidence supporting your interpretation?

Catalano: Perhaps, in R CMa, of which Dr. Kitamura could tell us. Sanford (1951) also remarked of the spectrum of AR Lac: "flat, shallow profiles of the rapidly rotating secondary characterize the intervals... preceding and following primary minimum". Struve found that the Ca II emission lines originate in all parts of the stellar surface, and not in a limited region like a spot.

Hall: Is your definition of ϱ the same as Dr. Batten's? Did he not find ϱ to be approximately 0.3?

Batten: Yes, but I was talking about disks around primary components.

Hall: Dr. Catalano mentioned that I found a similar effect in the light curve of RS Cep. Just before first contact, and just after fourth contact, there is loss of light. You have found light losses of about $0^m.05$, but in the light curve of RS Cep it is a little greater. Two other Algol-type systems, AQ Peg and VW Cyg have been investigated by Bob Tate, as part of his Master's thesis, and they show light losses of a few hundredths of a magnitude. My interpretation (Hall, 1972) is quite different from yours. I put the envelope or disk around the primary star. These systems are known to display the type of emission ascribed to a ring in RW Tau, and I think of my disk as the relatively dense, inner portion of the ring, sufficiently bright even when seen through broad-band filters like B and V to have some surface brightness of their own. So I think of the dips in the light curve as the eclipse of that luminous disk by the cool star. Then for RS Cep, I obtain $\varrho \approx 0.35$ – in good agreement with average value found by Batten. The eclipses of AQ Peg and VW Cyg must be total, but the light curves are not exactly flat during totality – the light curves look more like those of annular eclipses. My interpretation is that the surface brightness of the disk is concentrated towards the limb of the hotter star. The cooler star does totally eclipse the hot star itself, but it does not totally eclipse the disk. The eclipse of the luminous disk is 'annular'. If the disk is around the hot star, it is within the Roche lobe, whereas Dr. Catalano's envelopes are outside the Roche lobe of the cooler star.

Fracastoro: I support Dr. Oliver's view on AR Lac; it looks to me, too, to be quite different from the other four systems. The spectrum of the primary star and the orbital inclination are different. Most of all, the value of r_1 is much smaller than in the other systems. I have another comment on what Dr. Hall says. I would be careful before speaking about dark clouds or bright clouds. Any solar observer can tell you that prominences look bright against the background of the sky, but dark against the surface of the Sun.

Hall: The loss of light just before first contact involves that part of the envelope or disk which is projected against the sky, not against the star.

Fracastoro: It would add light – but we observe a depression in the light curve.

Hall: It has some light of its own, and when the cool star moves in front of the envelope, you have an eclipse of the envelope.

Catalano: But we know systems with bright rings around the primary components – U Cep and RW Tau – which do not show this feature in their light curves. I don't know if these rings are denser and their light greater.

Hall: Oh no. In systems like RS Cep, the relative radius of the primary is quite small. In Dr. Huang's original way of visualizing the problem, there is more empty space in which to put a ring and build up quite a dense disk. In U Cep there is relatively little empty space, so it is harder to build up a substantial disk.

F. B. Wood: Let's conclude that AR Lac is a subject for a symposium on its own, and move along. Dr. Catalano wishes to present another topic, and then Dr. Hutchings has a presentation on spectroscopic aspects. We'll discuss the two together.

Catalano: Dr. Fracastoro and I have compared the frequencies of elliptical orbits among all spectroscopic binaries, that are also eclipsing binaries as obtained from Batten's catalogue (1968). Only 19% of the observed spectroscopic eccentricities are confirmed by photometric observations. Of 39 systems that show spectroscopic eccentricities that are not confirmed by the light curves, 17 show asymmetric light curves – the maxima are of different heights. There is a correlation between the sense of the asymmetry of these light curves and the spectroscopically derived value of the longitude of periastron, ω. Systems for which $0° \leqslant \omega \leqslant 180°$ are fainter at the maximum following the primary minimum, and systems for which $180° \leqslant \omega \leqslant 360°$ are fainter at the maximum preceding the primary minimum.

Hutchings: The main part of what I want to say does not particularly tie in with the foregoing discussion. However, I would like to report on some computations which will bridge the gap as they apply to photometric observations of hot gas streams or envelopes in general. Figure 2 shows the effects on a light curve of an electron-scattering envelope centred on the hot component of a detached binary system. The density ($\sim 10^{12}$ particles cm^{-3}) and temperature ($\sim 10^4$ K) are of the order of these we have been discussing, and I have considered two geometrical con-figurations: an equatorial disk as drawn and a full sphere of the same outer radius. The effect is shown for two values of i (90° for total eclipse and 72° for a partial) and represents the deviation from the light-curve derived from the system without the envelope. Reasonable changes to the geometrical and other parameters changes absolute values by factors of up to ~ 4 but the general results are as illustrated. The main features are the drop in the light curve before first contact (and through it, in

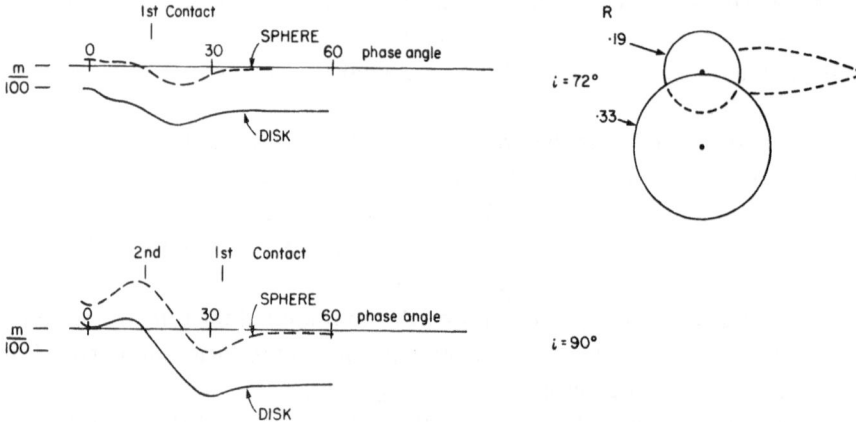

Fig. 2.   Effect of electron-scattering disk on light curve of an eclipsing system around primary eclipse. The disk is assumed to extend around the primary star to three times its radius to have a density of 10^{12} electrons cm^{-3}, and a temperature of about 10^4 K.

the total-eclipse case) of some one or two hundredths of a magnitude. Also, if the envelope is large enough, the eclipse depth is diminished. I would suggest, therefore, that this is an effect which may explain some of the pre-eclipse dips we have been looking at. I shall return to its relevance to my observations in a few minutes.

I should now like briefly to describe observations, both spectroscopic and photo-metric, of two emission-line stars which I think may be binary systems in the process of mass-exchange. They both have some extraordinary properties and I don't feel I understand them at this stage, so would welcome suggestions and discussion. The stars are HD 187 399 (B7e) and 173 219 (B0e). They are both shell stars of a sort and show dramatic, *periodic* spectrum variations.

In the Balmer and Ca II K lines in the spectrum of HD 187399 there are some points to notice: (i) the double absorption with variable separation; (ii) the strong increase in the blue-shifted absorption, accompanied by P Cyg emission at the periastron. I

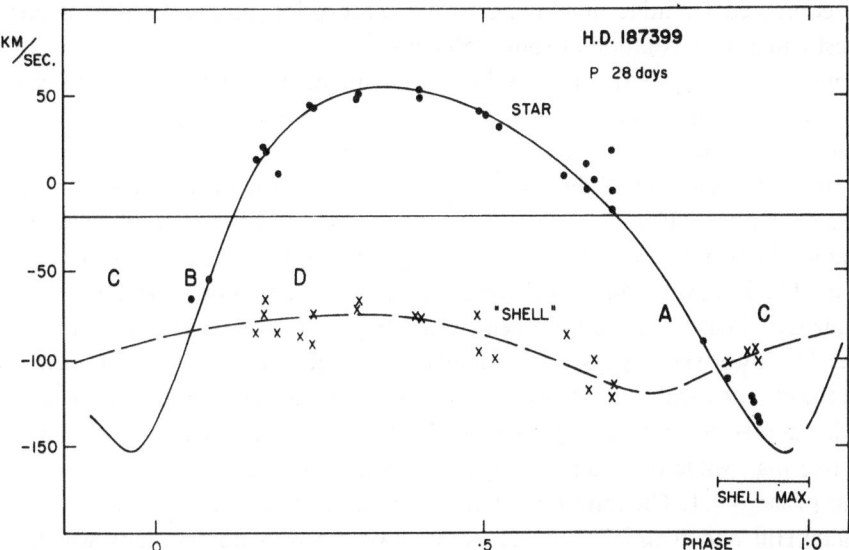

Fig. 3. Radial-velocity curve for HD 187399 from the absorption lines, through the 28d period. Dots refer to measures of all lines in the spectrum of the main component; crosses to shell components of Ca $_{II}$ K and Balmer lines.

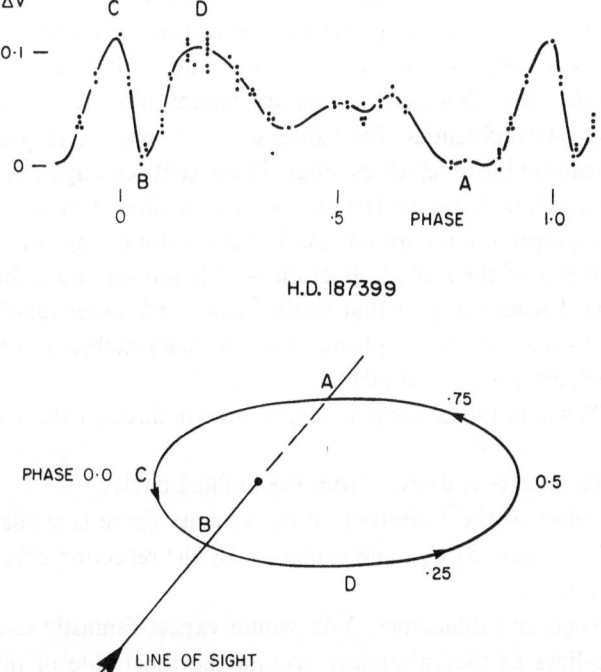

Fig. 4. *Upper:* Visual light-curve of HD 187399 from two seasons' observations. *Lower:* sketch of orbital plane, indicating expected phases of eclipse and visibility of gas streams.

have computed tentative underlying Balmer absorption profiles, which, if correct, suggest equator-on rotation at some 350 km s^{-1}.

Figure 4 shows firstly the radial-velocity measurements and derived orbital-velocity curve (the large-amplitude curve) and the velocities of the second absorption system, which has a nearly constant velocity, some 80 km s^{-1} negative of the main curve, but which may show small fluctuations about this with the orbital period, which are displaced in phase. The main curve is derived from Ti II, Fe II, He I, N II lines as well as Balmer lines; the lower one shows up only in the Balmer lines and the Ca II H and K lines. The 28-day period has held up for over 20 yrs so that one is inclined to believe the star is a binary. The orbit is eccentric and Figure 4 shows a sketch of the orbital plane. The lower velocity system is modulated so that it is directed towards and away from the observer with extreme values when the component stars are near line-of-sight. One is tempted to think of a gas stream towards the central star and in view of the tentative high value of e to postulate a slow eclipse of it, at phase $\sim 0^{P}.8$ and a faster one at phase $\sim 0^{P}.1$. The third item in the diagram is the light curve (!) as obtained by Graham Hill and Ron Hilditch. These are two season's observations which agree very well, so we are inclined to believe that the variations are real. The points to note are the long low centred on phase $\sim 0^{P}.8$ and the rapid one at $\sim 0^{P}.1$. Also the high values (phases $\sim 0^{P}.0$ and $0^{P}.25$) occur when the postulated gas stream is seen projected against the sky. The amplitude of the effect is $\sim 0^{m}.1$ which suggests, in conjunction with Figure 2 that the variations are caused by electron-scattering streams.

Figure 5 shows velocity curves for HD 173 219. This star has many peculiarities, chief of which are periodic velocity variations which differ in amplitude for different (absorption) lines, in the sense that the amplitude is higher for lines of lower excitation. The diagram shows this effect and the lines are orbital solutions for the lines grouped as shown by excitation potential. The values of ω, T_0, and e agree well – only the K's and (less significantly) the γ-velocities differ. The eccentricity again is high. There are further interesting data from the (Balmer) emission lines, which change a little in intensity. The absorption lines are not sharp and do not change in strength much. I have no photometry of the star (declination $-7°$ is too far south for us) but would like to see some. I would suggest that further work and understanding of these two stars may help us understand the phenomenon of mass-exchange, and have a list of other stars which are probably similar.

Fracastoro: What is the continuous curve drawn through the unfilled circles in Figure 5?

Hutchings: That curve is derived from the unfilled circles.

Scarfe: The effect of the excitation on the velocity curve is similar to that which Ovenden (1963) found in 57 Cyg and explained by the reflection effect. Could this be the explanation here?

Hutchings: There are difficulties. You would expect fantastic changes in the intensities of the lines as they disappear around the other side of the star, and this doesn't happen. I expect we are looking at some sort of extended envelope.

Batten: I don't think you can use Ovenden's explanation because no secondary

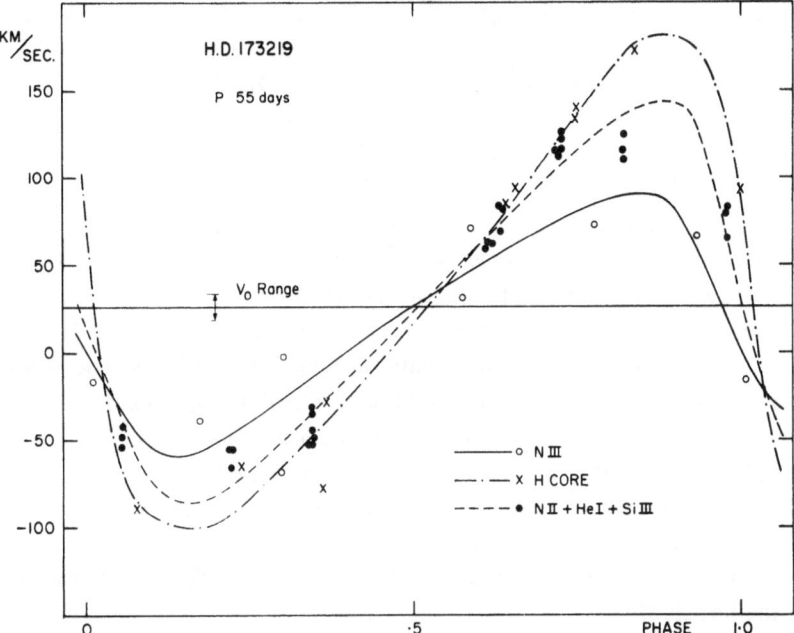

Fig. 5. Radial-velocity curves of HD 173219 from absorption lines of different ions, through the 55d period. Curves are orbital solutions based on velocity measurements as shown.

spectrum is seen. You couldn't have the primary star heated up by a secondary not bright enough to show a spectrum – unless that secondary was a very peculiar object.

Hall: I was very interested in your light curves of HD 187399 because a student at Vanderbilt (Tim Frazier) has recently got a somewhat incomplete light curve of the binary DN Cas – a pair of O-type stars. Originally the period was thought to be 1ᵈ15 but now we think it's twice that. With the new period, the light curve looks remarkably similar to the one you showed. The only difference is that we have no evidence for unequal durations of the eclipses – as if the orbit of our system is circular. The two eclipses are nearly the same depth. After secondary eclipse the light curve rises quite high, comes down to a plateau, and then the primary eclipse begins. Both eclipses are about 0ᵐ1 below the plateau, and the rise is also about 0ᵐ1 above the plateau. Since both components of the binary are O-type stars, there might well be an electron-scattering envelope. You did say you could get effects as big as 0ᵐ1 by electron-scattering?

Hutchings: Yes, that's the sort of change we think could be caused by a hot gas stream.

Hall: Good; I can't think of any other way to explain our light curve of DN Cas.

Devinney: I'd like to mention that C. R. Lynds (1959) observed some B-type stars that showed very strange light curves. Maybe they should be considered in conjunction with what Dr. Hutchings has observed.

Oliver: Do you have the light curve in several colours? That would give some information on electron-scattering.

Hutchings: Yes, we have the light curve in four colours, and it looks pretty much the same in all four.

Hall: We get the same picture in *U*, *B*, and *V* for DN Cas.

F. B. Wood: I think Dr. McNamara's contribution might fit in very well here.

McNamara: Tippets and I have secured observations of the primary eclipse of U Sge over two cycles in the *uvby β* photometric system. A careful examination of the data indicates that the duration of total eclipse is wavelength dependent in the sense that the duration is greater at longer wavelengths than at shorter ones. This implies either that the B-type star (eclipsed star) is larger at shorter wavelengths than at longer wavelengths or that the opacity of the gas surrounding the G2 IV–III (eclipsing) star is wavelength dependent, being greater at longer wavelengths. The latter possibility seems probably the more likely, and the source of opacity might be the negative hydrogen ion. If this is so, the length of totality should increase for $\lambda > 8000$ Å since the absorption coefficient of the negative hydrogen ion reaches a maximum at 8000 Å.

Hall: You mentioned two possible explanations. In one you assumed the trouble is caused by the cool star, and in the other you assumed it is caused by the hot star. What sort of effect would you predict if there were a slightly extended atmosphere around the hot star?

McNamara: I think that absorption by the negative hydrogen ion is the most likely explanation, because of the behaviour of the c_1 index during eclipse. If the B-type star being eclipsed was simply a little more extended in the ultraviolet than it was in the yellow, I don't think you would expect the c_1 index to go down and back up, as it does. So I believe the most likely explanation is in terms of some peculiar absorption effects around the G-type star.

Smak: Are the colour-index values at the bottom of the eclipse of some help to the interpretation?

McNamara: They don't help out very much, although if you look at the m_1 index and take it literally as indicating metal abundance, you find that the secondary component of U Sge does turn out to be slightly metal deficient – as I believe is pointed out by Miner (1966). Someone is to discuss the ultraviolet excesses of these stars later. I might say that when you look at the hydrogen lines in these spectra, they seem quite normal, but from the values of the m_1 index you would have to conclude that the secondary components are slightly metal poor.

F. B. Wood: Dr. Smak's comments on radiation from circumstellar matter would fit in here.

Smak: I wish to make a few comments and ask a few questions concerning the radiation of the circumstellar matter. First, there is a great need for quantitative measurements of the emission lines in terms of their absolute intensities. Even from what we know now qualitatively, it appears that there is a large diversity among different types of systems. Yet we do not know whether the absolute intensity of the emission lines coming from disks in the Algol-type systems is systematically greater than in the case of novae and U Gem type systems. Another question is concerned with the visibility of disks in Algol-type systems with primaries of different spectral types. Namely, why is it that in nearly all systems with F-type primaries, where emission is present, it is visible throughout the cycle? Is it due to the fainter background of the F-star radiation (as compared with A's and B's), or is it because the emission is indeed stronger?

Passing to the continuous radiation, we may note that it appears to be – relatively – much stronger in the case of novae and U Gem type systems, where it often accounts for nearly 100% of the total radiation of the system. An important observable parameter is then the line-to-continuum ratio. Even in the case of novae and U Gem type systems we find widely different ratios, from cases of rather strong line emission to those where the lines are practically absent. The best eaxmple here is VV Pup, in which the hot spot contributes about 50% of the total continuum radiation but is practically invisible in the Balmer lines. Before we ask theoreticians to compute meaningful models of disks, spots, etc., we really need more quantitative measurements of radiation coming from these objects.

Underhill: About the real problem of getting absolute intensities, I agree with Dr. Smak. Unless you have absolute energy units, or energy units relative to some standard star, you cannot decide whether the emission is more visible because it is referred to a weaker background, or just because it is stronger itself.

F. B. Wood: Dr. Bolton would like to speak on correlated optical and radio observations of Algol.

Bolton: Algol was first detected as a radio source in the fall of 1971 by Wade and Hjellming (1972). In January of this year Hughes and Woodsworth (1972) and Hjellming *et al.* (1972) discovered strong radio flares from Algol that sometimes reached intensities in excess of 0.3 flux units. Since the flares were first detected, I have monitored Algol optically to search for spectroscopic changes that might be related to the radio flaring.

The characteristics of the radio flares have been summarized recently by Hjellming (1972). The radio emission from Algol appears to be thermal bremsstrahlung during quiescent periods. During the flare the radio spectral index decreases as though the plasma is becoming partially optically thick. We now have approximately 100 photographic spectra at a dispersion of 12 Å mm^{-1} that have been taken simultaneously with radio observations at either NRAO or ARO. Most of these show no abnormalities.

However, whenever the radio spectral index is less than about -0.3 we see weak emission in the red wing of the Ca II K line. The strength of this emission seems to be correlated with the spectral index in the sense of the emission strength increasing with decreasing spectral index. There is also a suggestion of activity in the Mg II λ 4481 line along with that in the K line.

In view of these possible correlations, I wondered if there might be evidence on on older Algol plates of past radio activity. I was able to inspect over 100 plates of Algol taken by Dr. MacLaughlin with the University of Michigan telescope during the 1927–28 observing season. He usually took three plates per night at a prism dispersion of 17 Å mm^{-1} at Hδ. About a dozen of his plates show some K-line emission and one series shows the emission increasing in strength over about a one-hour period. Thus I'm inclined to think that the radio activity is not a new phenomenon and that it is also not connected with the sudden period changes that are occasionally seen in Algol.

In addition to these possible correlations, we have found that the strengths and line profiles of most of the lines vary throughout the orbit and not just during the partial eclipse of the primary. In particular, we find that the half-widths of the hydrogen lines vary by as much as 75% and that the residual intensities of the hydrogen lines change by at least 15% and probably much more. These changes occur in exactly the sense necessary to explain Andrews' photometry of Hα that Dr. Batten discussed. However, the line profiles suggest to me that we are seeing additional absorption outside the eclipses rather than emission during the eclipses.

F. B. Wood: I'm almost afraid to open a discussion on Algol – discovered in 1680 by Montanari, and still there are new findings.

Batten: I'd like to make three comments. First, those results from Hjellming's paper indicate, as I said yesterday, that the radio flare is very probably coming from something other than the stream, if there is a stream in Algol, and it seems to me, therefore, that one should not necessarily expect a close correlation between the radio observations and the kind of emission feature that you have seen at the K-line, which presumably comes from regions where matter is considerably denser. Second, I was very interested in your slide of the hydrogen lines because something much like this happens in the spectrum of U Cep. Most of the time the hydrogen-line profile is basically a very shallow, broad feature. But on a number of plates, which seem to be concentrated around phase $0^P.5$ or $0^P.6$ after the primary eclipse, you get a definitely deeper profile. The equivalent width is different, the actual width is different, and the depth is different. These two profiles seem to change one into the other quite quickly. I have no explanation for it. On the other hand, there is also definitely at some phases a filling up of either one profile or the other, by something which I think can only be emission from the stream, and it has an irregular profile. So I think there are both types of variation of the hydrogen line present. There is a real change in profile which probably matches what you've shown for Algol, but it isn't quite so dramatic. There is also, I think, a modification of these two basic profiles which is probably emission from the stream. My final comment, which appears in my review paper is that I wish we could find another word than 'flare' to describe this sort of thing. It is a confusing term, and if anyone comes up with another term while we're discussing here, I think we should all agree to use it.

Bolton: I agree that these flares may very well not be associated with the gas streaming although it is hard for me to see how flares of this magnitude could take place without moving some fairly substantial amounts of matter around. I have been intrigued for some time by the so-called B flare stars. There are at least five early-type stars that are reported to show flares of $0^m.5–0^m.8$. These include a Be star, a β CMa variable, and several ordinary main-sequence stars. As near as I have been able to determine, these are all ordinary stars for their type — except for the reported flares.

I have looked at three of these stars and have good evidence that all three are binaries, and I wonder if these are related in some way to what is going on in Algol. Finally, I think that it should be emphasized that the correlation I have found is very tentative. We have additional simultaneous observations scheduled for the next month so that we may perhaps firm up our conclusions.

Van't Veer: If you wish to correlate a flare, or let us say outburst, with a period change, you have to wait a very long time. With a period change of one second, you have to wait for the accumulation of many seconds before you can see it – one or two years.

Bolton: What I am saying is that if the correlation between the K-line emission and the radio flaring is real, then the emission on the 1927 plates indicates that the radio activity is not correlated with a period change because there was no period change in 1927.

Plavec: How does an outburst in Algol compare with a solar outburst?

Bolton: The radio outburst is much stronger. If Hjellming is right about Algol being a transient X-ray source near the outburst peak, the energy output would be $\sim 10^{35}$ ergs s^{-1}.

Plavec: The variation in hydrogen-line profiles both in depth and width is observed not only in the spectra of Algol and U Cep, but also, for example, in those of RW Tau and U Sge. So one is tempted to regard it as typical for primary components of Algol systems. But all these stars are of spectral types very close to B8 or B9, and it might be interesting to see if this variation does not occur in single, normal stars of type B8 or B9.

Hill: Somebody had seen K-line emission on plates of Algol taken a number of years ago at the Dominion Astrophysical Observatory, but nothing was ever published. I'd also like to comment that one must be careful in interpreting the spectrum of Algol in primary eclipse, because the line profiles can be distorted by the spectrum of the third component.

Popper: What is the size of the radio error-box from which this variation occurs?

Bolton: Four-tenths of a second of arc.

Whelan: Are there any polarization observations?

Bolton: No, not yet. We hope some may be made on the next run.

Smak: I have a question for Dr. Popper. At first your question seemed quite funny, but did you mean to imply that you suspect the third body to be responsible for the radiation – or any other field star located nearby?

Popper: I just wanted to inquire if there was any possibility of a coincidence. It's quite unlikely, I suspect, but we should ask if the radio emission really does come from Algol.

Smak: Could the third body be responsible?

Bolton: I think the radio errors are such that that would be permissible. I think that the argument that has been made against the third body being involved is that in the one case (Antares) where the radio emission in a binary system can clearly be assigned to one of the component stars the emission was from the B-type star and not the companion, which was losing mass. This is perhaps not too strong an argument, but it was what led Hjellming to look at other B-type binaries where there was mass exchange. There is the additional argument that the size of the plasma cloud involved in the flaring seems to be about the size of the Roche lobes about the close pair in the Algol system.

Fracastoro: I wouldn't dare to ask you to go on observing this radio emission in order to see whether there is any correlation between the radio-outburst frequency and the orbital period of the third body, since you have already found that there is no correlation with the $2\overset{d}{.}86$ photometric period. It would be very interesting to find something like the 11-yr solar cycle.

F. B. Wood: We are deliberately keeping time for discussion. Anyone who has signified a wish to speak will get an opportunity. We are altering the order of speakers a little. A group on period changes should be kept together at some session. Dr. Walter will now talk to us on Magnetic Fields and Hot Spots.

Walter: I mentioned this problem yesterday, in connection with SW Cyg. A young colleague and I have a paper on TW And in press (Ammann and Walter, 1972). This system has period of four days, and we believe that the existence of hot spots on the surface of its primary component can be clearly demonstrated.

The light curve of this system seemed to be without irregularities, but because of the possibility of gas streams within the systems we tried to base the rectification only on the second quarter of the light curve, which, in many systems, seems to be free of the effects of gas streams. We call this method 'short-region rectification' in contrast to the usual 'long-region rectification' in which the whole light curve outside eclipses is used. The short-region rectification showed the presence of additional light before and after primary eclipse. We interpreted this as the light of hot spots. The interpretation seems to be supported by the observed intensities in B and V during the primary eclipse when the additional light is also totally eclipsed – very similarly to the situation in SW Cyg. Again, in TW And, the hot spots seem to be located at high latitudes and thus hint of the existence of magnetic fields. The short-region rectification not only gave a better representation of the normal points outside eclipses than the usual rectification, but the sum of the squares of the residuals of the intensities within the primary eclipse was also much reduced by the new method of solution.

In Algol systems, we have to distinguish between stable geometrical elements and variable properties which arise from gas streams and/or magnetic fields. This could be shown, for example, by a comparison of Dugan's photometric observations of RV Oph in 1906–14, with my own obtained in 1964. Another system which may show variations of the light curve is TV Cas (P = $1\overset{d}{.}8$) In 1971 its light curve was of the kind to be expected from systems containing gas streams. For such systems it may be advisable to wait some time until the system exhibits a light curve that can be explained by some known model. Catalano's (1966) rectification of the light curve of u Her provides an example of the great differences which may result from different methods of rectification. Short-region rectification of this light curve gives results that strongly suggest hot-spot activity and UV-radiation from the spots.

Smak: Where does the evidence of magnetic fields actually come from?

Walter: From discussion of the observations of the systems SW Cyg and TW And. The geometry of the eclipses is well known, and it can be shown that the luminous regions are situated on the surface of the primary component at about latitudes $\pm 60°$ out of the orbital plane. The additional light of the hot spot is seen outside eclipse for about one half the orbital period. It is fully seen at the beginning and the end of the primary eclipse. Within this stellar eclipse there is a second one – a total eclipse of this additional light – and it is possible to determine the region in which this light must be situated as is shown in my paper on SW Cyg (Walter, 1971).

Smak: I have two comments. First, the lines you drew actually correspond to the beginning and end of the eclipse of the spot. If the spot is slightly elongated, it can be on the equator and show the same effect. Second, all the numbers that enter, namely the radii, inclination, etc. are based on your specific solution. You use only parts of the light curve, and this determines the answers. Since you admit that there is a hot spot, or detached material, it becomes a matter of judgment or arbitrary decision which part of the light curve you use.

Walter: We have made a photometric solution of a rather deep total-eclipse curve, and the distortions of the light curve, which have been influenced by the method of rectification, concern only a few per cent of the intensities. The photometric elements cannot be changed very much and still give a reasonable solution. I think we have indeed obtained a geometrical solution that is close to the truth.

Kitamura: Do you think that the existence of a magnetic field provides a unique solution for the residuals?

Walter: It seemed to me the simplest hypothesis and I did not try any others.

Kitamura: Your solution for the light curve is strongly based on 'short-region recti-fication'. I think that any theory of rectification is approximate. In order to treat such a peculiar light curve you must check the rectified light curve in various ways. In par-ticular, did you test whether your rectified light curve, outside eclipse, can be repre-sented well by the eclipse function?

Hall: To put the same question another way: if you turn the problem around and solve the primary eclipse, would that solution represent the light curve out of eclipse?

Walter: I see. First, we tried to rectify the whole light curve outside eclipse. The comparison of the result with the observed curve was not satisfactory. We could obtain a satisfactory comparison from the short-region rectification, but once we had made this, it was difficult to return to the old method. The original light curve seemed quite symmetric, normal, and free from the effects of gas streams.

Kitamura: Even so, in my opinion, with such a peculiar system we should be very careful with rectification and we should check that the light curve within eclipse can be well represented by the eclipse function only.

Milone: You mentioned the need to look at these light curves over many years, and I would like to emphasize this very strongly. O'Connell (1951) published a list of systems whose light curves show unequal maxima. We have observed a few of these again, and found that the sign of the difference between maxima has changed. A couple of these cases have been very well studied, RS CVn by Catalano and Rodonò (1967) and RT Lac by Doug Hall and myself. We do in fact find some sort of long-term cyclic variation and we try to take it out of the light curve before we make a solution. It seems to me to make sense to look at long-term changes in the light curves as well as to examine those parts of the light curves that show short-term changes. We can remove the time-dependent part of the light curve and try to rectify in the usual manner.

Walter: Yes, we considered variations that might have happened during the years of observation. There were some small variations just before and after primary eclipse, but they were so small that we could neglect them.

Linnell: It's surprising that you were able to obtain acceptable values of rectification coefficients with such a short phase range. I think that Dr. Kitamura's point is very pertinent. I wonder whether the nature of the residuals would be different if you had chosen short phase-range segments in different phase regions, rather than just on either side of the secondary minimum. In other words, how sensitive is the nature of the residuals to the particular phase range chosen for the determination of the rectification coefficients?

Walter: I do not have the exact figures here, but, of course, for both rectifications we used the method of least squares. The coefficients A_1 and A_2 of the $\cos\theta$ and $\sin\theta$ terms (θ is the phase angle) had very small mean errors as determined from the long region. The errors of these terms in the short-region rectification are, of course, larger, but the values of A_1 and A_2 are significant and may be used for rectification. I don't believe our solution is exceptional. I think TW And is a quite normal Algol-type system.

Biermann: I think your geometric argument is based on the assumption that you see the same part of the hot spot as it goes into and comes out of eclipse. As Smak said, it might be an elongated region of which you see different parts at the two phases, and then you can't use your argument. Even assuming that your picture is correct, magnetic fields would have to be rather high. Have you looked for any evidence of that in the spectrum?

Walter: I have no instrument for it, but I think that fields of 100 G would suffice. In your own dissertation, you mentioned that fields of this order of magnitude would follow from estimates of the kinetic energy and magnetic energy of particles. Magnetic fields of this order of magnitude could affect the gas streams but could not be detected from the spectra.

Hilditch: You mentioned light curves of u Her. There is a red curve by Catalano and a blue one by Ruiz (Merrill, 1963). I have made a preliminary analysis of these light curves, using the synthesis programme of Graham Hill and John Hutchings, and I can fit the light curves very well with the normal model. All I need to fit the light curves completely are improved values of the mass ratio and of the temperature of the primary component. It is not necessary to introduce any gas streams.

Walter: It is still unexplained, however, that the long-region rectification of the excellent light curves by Shapley and Calder (1935) and Catalano are not satisfactory at all.

Cowley: What are the spectral types of the components of TW And and of the other systems in which you find hot spots?

Walter: The components of TW And are F0 V and G6 IV. The brighter component of SW Cyg is an A-type star.

Plavec: What are the radii of the two stars, for example in TW And?

Walter: One is about 0.16 and the other 0.25 of the distance between the centres of the stars. The larger star fills its Roche lobe.

Plavec: We should realize that normally the subgiant is relatively large in comparison with the other star. Therefore, the nozzle (or whatever it is) through which the material flows from the late-type star is not very small – maybe larger than, or of the same size as, the radius of the other star. We are discussing whether the hot spot is very near the equator or near the pole. In either case we are assuming that the stream is very nicely focussed from this relatively large nozzle into a relatively small spot where it hits the primary star. I can't say this assumption is wrong, but we should bear in mind that it requires very strong focussing of the flowing material.

Hutchings: I'd like to refer back to Dr. Hilditch's remarks, and make a comment on behalf of those in the room – and there are several – who don't believe in rectification. It seems to me that rectification has been controversial, even among those who do believe in it. I would like to see this sort of system analyzed by the synthesis method. If we can produce the same sort of residuals by a totally different approach, we might well be much farther along the way to establishing the existence or otherwise, and the positions of these hot spots. It seems that we are talking about magnetic fields and such like things a little bit prematurely.

Van't Veer: I agree with Hutchings that we can do better now with the method of synthetic light curves. You should try to reproduce the light curve artificially and compare your synthetic light curve with your observations. It is a method of trial and error, but it avoids the use of unreasonable physical hypotheses.

Walter: You can use the method of artificial light curves if there are no gas streams present. If there are no gas streams, the short-region rectification should give the same result as the long-region rectification – but, in fact the two give different results.

Smak: I am one of those who has never done any rectification, but it is my understanding that rectification, in the meaning of the conventional Russell method, is also making use of a systhetic light curve based on very specific assumptions; namely, there are two stars, one or both of them being possibly elliptical and possibly affected by a reflection effect that can be described by a cosine term. If we have something else in the system, either from the very beginning, or as a result of our rectification, then the Russell rectification procedure should not be applied blindly.

F. B. Wood: Blindly! Blindly, no. I'm tempted to say more, but I'd better not! We spent three days on this subject in Philadelphia, a year ago. I will just say that we are moving into an area in which there is intense disagreement amongst those working in it.

Oliver: If you do believe that there might be spots, either hot spots or dark spots, on a star, neither traditional rectification nor synthetic light curves are really going to show you them, because they are not included in either of those models. Either way, all you will see is a slightly greater scatter – or, if you are lucky, some systematic trends – in the residuals. If you are not willing to accept the possibility of some other model and try it, for example by not fitting your synthetic light curve to points in the region where you suspect some anomaly, you are not going to find the anomaly. Even in the synthetic-light-curve method, you should at least try 'short-region rectification' by leaving out the region on either side of primary minimum when fitting your synthetic light curve and then looking to see if the same trend is found in the residuals. You cannot judge, I think, whether or not a hot spot is present from the general scatter of the residuals. Who's to say how much scatter there should be? The scatters in the two plots we saw did not differ by more than a factor of two. Who can judge what the real scatter should be, *a priori*?

Chen: I agree with John Oliver's argument and I think that at the IAU Colloquium No. 16 in Philadelphia we showed, at least the invited speakers showed, that the solutions for an Algol-type light curve, using the synthetic-light-curve technique and the rectification technique separately were the same. The differences of elements were in the third or fourth decimal. This, I think, is beyond the accuracy of the observations.

R. E. Wilson: I don't believe we should complain about someone failing to use a light-curve-synthesis program when he does not have access to one. Such programs do not exist everywhere in the world. I would like, however, to comment on the free-hand light curve Dr. Walter drew through his normal points of TW And. I think it is not a good practice to publish hand-drawn curves through the observations. One can publish only the observational points and see what the observations are like, or one can publish the observations with a theoretical curve through them, which then shows the comparison of theory and observation. But a hand-drawn curve through the observational points only misleads the eye, and, as far as I can see, serves no purpose whatsoever, although many people publish them. I would like to see this practice ended.

Walter: I can answer you directly: this curve will not be published! We are publishing only the normal points and the observations. Nothing has been deduced from the curve – it's only to help your imagination.

R. E. Wilson: My point is that it hampers the imagination. With a hand-drawn curve through the points, it is difficult to judge the run of the observations.

Walter: Then, please, think away the line, and look only at the points! The line should only show that the light curve, which was composed of some hundred observations from about eighty nights, is smooth. The observations don't show much scatter. The line was not meant to make one think of hidden problems of gas streams.

Linnell: It seems to me that in general the attempt to detect photometric anomalies in systems such as this where you believe that there may be gas streams present, has to depend on residuals from an accurately calculated light curve whether you are going to do it by synthesis technique or whatever. The detection of trends which are statistically significant must depend on extensive observational residuals. I think that we should try to get away from the use of normal points. In view of the discussion yesterday concerning the transient nature of some of these phenomena, the assembly of individual points into normal points from several different cycles may, in fact, tend to hide some of the photometric effects that you hope to investigate. When you construct a synthetic light curve, and then calculate residuals, it's very important to look at the individual residuals on each night, and compare those residuals with those of the extinction star from the extinction line, so that you can determine whether or not there is genuine statistical significance to postulated trends in the residuals of the variable.

Walter: For just this reason, my method was to avoid giving single nights too much weight or preference. I make only three or four observations a night in each colour – especially outside eclipse – and combine the results of many nights. I think, when this method is used, it makes no difference whether we use normal points or single observations. Normal points can be helpful, and often help us to see more from the observations. I doubt if we would have arrived at what we think is a solid interpretation of an interesting property of TW And if we had used only single observations. We would have given all the observations to a computer and would have taken the results and published them. We would have no need to consider further, because we would have considered the problem solved.

F. B. Wood: We are well over schedule, but the participants have been interested. Let's have one last comment from Dr. Leung, and then we must continue either privately or at a later session. This communication illustrates well that any good light curve creates a lot more problems than it solves!

Leung: A very good example of a variable light curve is provided by the W UMa system VW Cep. If you realize that some phenomena will not be very stable so that there are likely to be cycle-to-cycle variations, you can always pick a season of a few weeks' time and solve the light curve from a few particular cycles. You can account for all sorts of distortions by hot spots, cold spots, or gas streams and then obtain the orbital elements. This will not be very meaningful bacause the normal points from the next few months will define a light curve that will give you a completely different set of geometrical elements. It's rather dangerous to introduce too many complications into the solution of a light curve obtained over only a short period of time. I don't use normal points from, say, two weeks, but I try to use all the individual observations to define a mean light curve and look for periodical fluctuations from it. In this way, I try to separate whatever is periodic, transient, or non-secular. Then I have a light curve I can analyze.

References

Amman, M. and Walter, K.: 1972, *Astron. Astrophys.* (in press).

Baglow, R. L.: 1948, *Monthly Notices Roy. Astron. Soc.* **108**, 343.

Batten, A. H.: 1968, *Publ. Dominion Astrophys. Obs.* **13**, 119.

Blanco, C. and Catalano, S.: 1972, unpublished.

Catalano, S.: 1966, *Publ. Catania Oss.*, No. 97.

Catalano, S. and Rodonò, M.: 1967, *Mem. Soc. Astron. Ital.* **38**, 395.

Fletcher, E. S.: 1964, *Astron. J.* **69**, 357.

Hall, D. S.: 1972, *Proc. IAU Colloquium* No. 16 (in press).

Hjellming, R. M.: 1972, *Nature Phys. Sci.* **238**, 52.

Hjellming, R. M., Wade, C. M., and Webster, E.: 1972, *Nature Phys. Sci.* **236**, 43.

Hughes, V. A. and Woodsworth, A.: 1972, *Nature Phys. Sci.* **236**, 42.

Huffer, C. M. and Kopal, Z.: 1951, *Astrophys. J.* **114**, 297.

Koch, R. H.: 1960, *Astron. J.* **65**, 326.

Koch, R. H.: 1962, *Astron. J.* **67**, 130.

Kron, G. E.: 1947, *Publ. Astron. Soc. Pacific* **59**, 261.

Lynds, C. R.: 1959, *Astron. J.* **130**, 577.

Merrill, J. E.: 1963, in F. B. Wood (ed.), *Photoelectric Astronomy for Amateurs*, MacMillan, New York, p. 165.

Miner, E. D.: 1966, *Astrophys. J.* **144**, 1101.

O'Connell, D. J. K.: 1951, *Publ. Riverview College Obs.* **2**, 85.

Ovenden, M. W.: 1963, *Monthly Notices Roy. Astron. Soc.* **126**, 77.

Rodonò, M.: 1967, *Mem. Soc. Astron. Ital.* **38**, 465.

Sanford, R. F.: 1951, *Astrophys. J.* **113**, 299.

Shapley, H. and Calder, W. A.: 1935, *Harvard Circ.* No. 398.

Wade, C. M. and Hjellming, R. M.: 1972, *Nature* **235**, 270.

Walter, K.: 1971, *Astron. Astrophys.* **13**, 249.

EXPANDING ENVELOPES OF STARS

A. A. BOYARCHUK*

Krymskaya Astrofizicheskaya Observatoria, U.S.S.R.

Abstract. Observations of Be stars are summarized and discussed, with a view to establishing dimensions of, and physical conditions within, their envelopes. Shell stars and Be stars are considered together, the observed differences between them being considered as due to the size of the envelope and the angle that the rotational axis makes with the line of sight. Rotational ejection of matter is insufficient to account for the formation of the envelopes. Theories of the structure of envelopes are also discussed. The envelopes of Wolf-Rayet stars are considered in a similar way. The interpretation of these is complicated by the known binary nature of many Wolf-Rayet stars. Other groups of stars possessing envelope-like structures – U Gem stars and old novae, symbiotic stars, and T Tauri stars – are briefly mentioned. The problems of stellar envelopes, first listed by Struve thirty years ago, are still unsolved.

1. Introduction

A rather large number of stars have in their spectra some characteristics which indicate the existence of a great mass of gas above their photospheres. In general, in these cases, one says that the stars have extended atmospheres or envelopes.

The distinction between an envelope of a star and an extended atmosphere of a star is somewhat indefinite. Often, some astronomers consider a particular star to have an envelope while other astronomers consider the same star to have an extended atmosphere. One can consider from a general point of view that the density of the gas in envelopes decreases outwards slower than that in extended atmospheres. The density in an envelope can even begin to increase a little at a certain distance from the stellar photosphere. Unfortunately, not many investigations of the density of a gas in stellar envelopes have been made. Therefore, in order to define to which group a given star is related we will make this distinction: if in a stellar spectrum different groups of lines characterize different densities of gas (broad hydrogen lines with strong Stark wings and narrow deep cores; rotationally broadened helium lines and very narrow ionized metallic lines) then we will consider that this star has an envelope. On the contrary, if the spectral characteristics give evidence that the conditions in the outer levels of the atmosphere change smoothly, then we will say that this star has an extended atmosphere.

Thirty years ago, Otto Struve (1942) who gave very much attention to investigations of envelope stars, published a well-known review of problems of gas envelopes in which he formulated the basic problems:

(1) Why do some stars possess tenuous outer atmospheres or shells, while other stars, apparently of identical physical characteristics, do not have such shells?

(2) What is the origin of a shell and how is it supported, in apparent violation of the laws of mechanics?

* Presented at Parksville by J. Smak.

Batten (ed.), Extended Atmospheres and Circumstellar Matter in Spectroscopic Binary Systems, 81–94.

(3) How can we account for the remarkable tendency of nearly all shells to vary either periodically or, more often, in an irregular manner?

(4) Why do some shells expand, while others are stationary?

In spite of many papers on stellar envelopes being published during the last thirty years, the problems mentioned above have not been solved. Now we will give a brief review of our knowledge on envelopes of stars. We will begin with Be stars, then we will consider other types of envelopes.

2. Be Stars

These stars are the most typical representatives of stars having an envelope. The existence of rather narrow emission lines belonging to hydrogen, to helium, and, more rarely, to ionized metals, in addition to the normal B-type spectra indicate the existence of an envelope. Besides the Be stars, there are the stars whose spectra have narrow absorption lines of hydrogen, helium, and once-ionized metals. According to a suggestion of Struve, such stars are called now 'shell-stars'.

Figures 1–3 show the line profiles observed in spectra of different Be stars. Normal

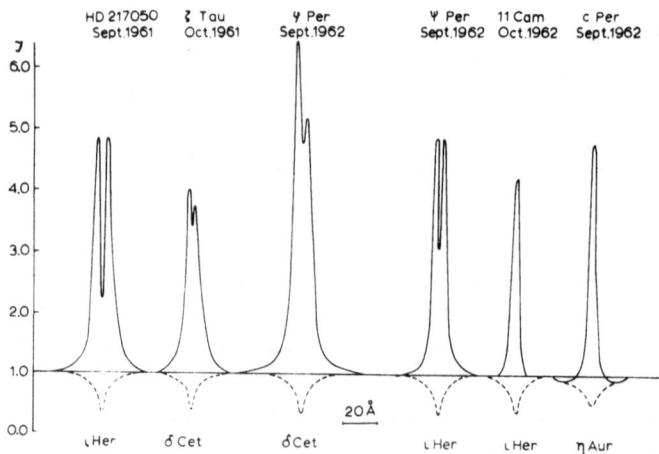

Fig. 1.   The profiles of Hα line in the spectra of some Be and B stars (Boyarchuk and Pronik, 1964)

Be stars and shell stars do not form two different groups. The observed differences between Be and shell spectra is due rather to the dimensions of the envelope and to the angle of inclination of the rotational axis to the line of sight. Figure 4, which is based on a figure given by Hack and Struve (1971), shows schematically the formation of a line in the envelope. If the observer is at O_1, the line of sight is parallel to the rotational axis, the profile of the spectral lines will be characterized by stellar absorption on which a central emission due to the extended equatorial parts of the envelope is superimposed. The envelope absorption is absent, because there is no absorbing gas between the stellar surface and the observer. This type of Be star is called 'pole-on-stars'. When the observer is at O_3, the line of sight is perpendicular to the rotational

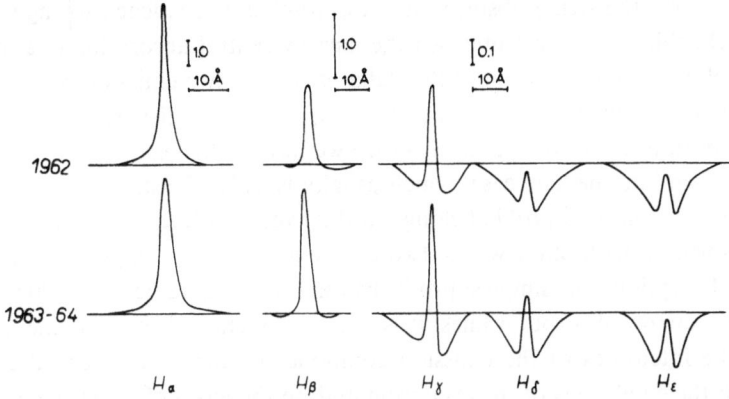

Fig. 2.   The profiles of Hα to Hε in the spectra of χ Oph (Boyarchuk and Pronik, 1965).

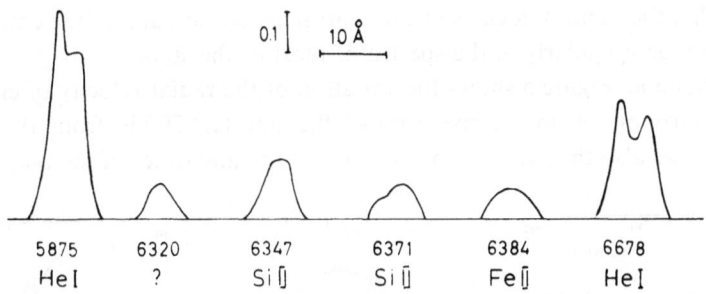

Fig. 3.   The profiles of some emission lines in the spectrum of X Per. (Boyarchuk and Pronik, 1965.)

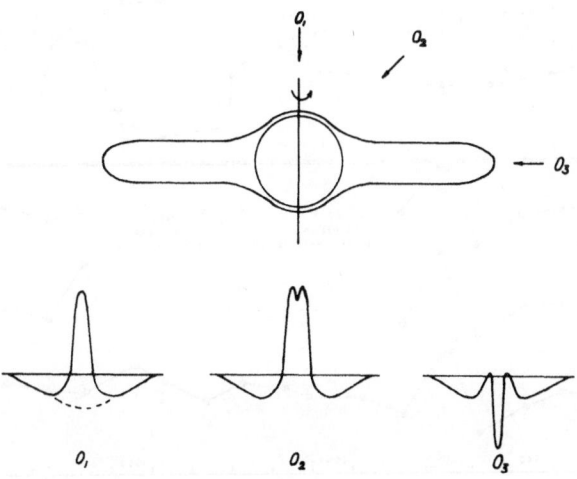

Fig. 4.   Sketch of a Be star envelope.

axis. We observe the stellar absorption lines, which are broadened strongly by stellar
rotation. On this line is superimposed the narrow central absorption due to the part
of the envelope which is in front of the stellar disk, and two emission components due
to the rest of the envelope. The emission components are displaced from the centre
by the rotation of the envelope. In this case we have a shell star. If the observer is at
O_2 we have an intermediate case, which usually is called Be stars.

The most complicated profile belongs to the hydrogen lines which have the strong
stellar absorption with Stark wings, two envelope emission components, and central
envelope absorption. The simplest profile belongs to the metallic lines, which often are
only rather narrow absorption lines. It is easy to understand that the line profile, the
ratio of the intensities of the emission components, and of the central absorption
depend on the conditions of the excitation and on the gas motion in the envelope. A
great amount of data about the displacement of spectral lines formed in the stellar
envelope has been accumulated. McLaughlin, Struve, Swings and Merrill were main
contributors in this type of work. The best review of results of measurements of line
displacements was given by McLaughlin (1961). The main conclusion which can be
drawn is that the radial velocities of the central absorption and of the emission components change irregularly in the spectra of most of the stars.

As an example. Figure 5 shows the variation of the radial velocity of emission, E,
and of absorption, A, in the spectrum of the star HD 20336 from 1915 to 1957.
Figure 5 shows also the variation of the ratio of the intensities of the blue and of the

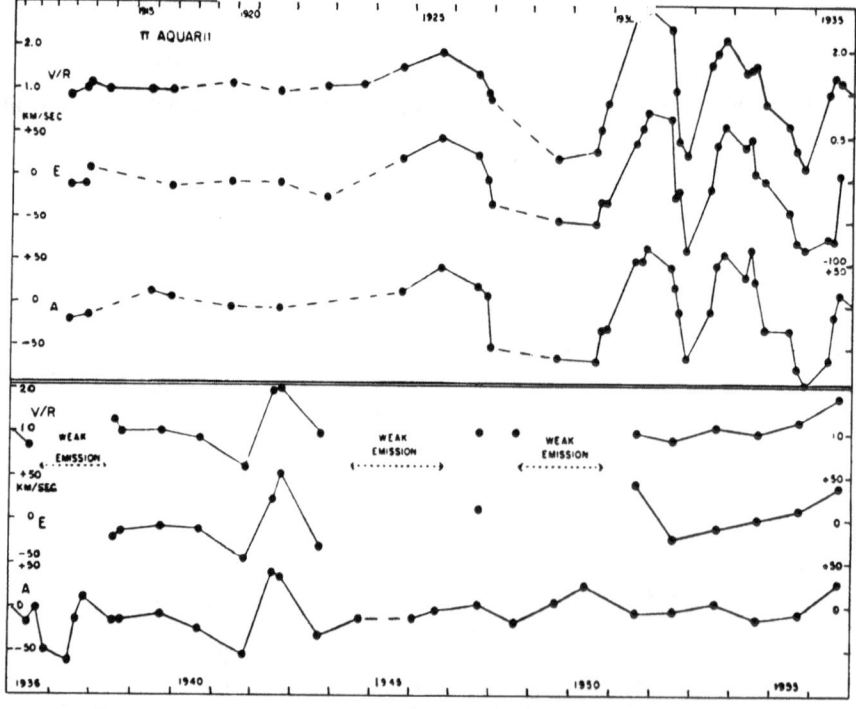

Fig. 5.   Velocity and V/R variations of HD 20336 (McLaughlin, 1961).

red emission components, V/R. The quasi-periodic variations were well observed from 1915 to 1935 while from 1940 there are only small non-periodic variations. Most Be stars show similar variations. A few stars show periodic variation of their radial velocities for a long time. For example, ζ Tau and ϕ Per have variations of the radial velocities with periods $132^{d}91$ and $126^{d}67$ respectively. These variations have been interpreted as due to the orbital motions in binary systems.

Besides the variations of the radial velocities and of the relative intensities of emission components, irregular variations of the total intensity of emission, E/C, and of the intensity of the shell spectra have been observed. Sometimes those variations were so large that they lead to the appearance or disappearance of the features of the envelope. McLaughlin (1961) pointed out that sometimes the spectrum of π Aqu has no emission. On the other hand, we have the well-known star Pleione. From 1905 on until 1938 it was a normal B5 star with absorption lines greatly broadened by rotation. However, quite suddenly, in 1938, McLaughlin and Mohler discovered that the spectrum of Pleione had hydrogen emission lines together with the shell spectra of ionized metals. Great variations also took place in the spectrum of γ Cas.

Thus, the observed radial velocities of envelopes of Be stars as a rule change irregularly. This means that huge masses of gas move towards and away from the star changing irregularly in velocity and direction. Moreover, as we will see below, the motions involve rather large volumes of matter.

An interesting fact was discovered by Pringle and McNamara (1962). They found that the radial velocities of ζ Tau determined by absorption lines redward of the Balmer discontinuity are systematically more negative by about 40 km s^{-1} than those determined from lines shortward of the Balmer discontinuity. This fact can be easily explained if the expansional velocity of the envelope decreases outwards. The irregularity of processes which occur in the envelopes of Be stars cause many difficulties for our understanding the reasons of the origin of envelopes.

Let us consider now the character of the local motions of gas. Information about this type of motion can be obtained from line profiles or from the curve of growth (i.e. the ratio between the value of the equivalent width of line and the number of absorption atoms which form it). By analogy with the analysis of spectra of stellar atmospheres we will call the motions found from the line profiles 'macroturbulence' and the motions found from the curve of growth 'microturbulence'.

As far as all Be stars have very large rotational velocities (Sletteback, 1949) the profiles of both the stellar and the envelope lines are broadened by rotation. In such cases, we can use the profiles for the analysis of turbulent motions if the profile width exceeds the rotational width. It was found that the width of Hα emission is about two times larger than that of Hβ emission in the spectra of shell stars and some other Be stars (Underhill, 1953; Ringulet-Koswalder, 1963). Underhill (1961) suggested that the motions with the velocities above 1000 km s^{-1} exist in the shell stars. Later Boyarchuk and Pronik (1964) studied the profile of the Hα line in the spectra of some Be stars and shown that the broad wings are due to radiation damping and not to the Doppler effect. Figure 6 shows the analysis of the Hα profile. The observed points

A. A. BOYARCHUK

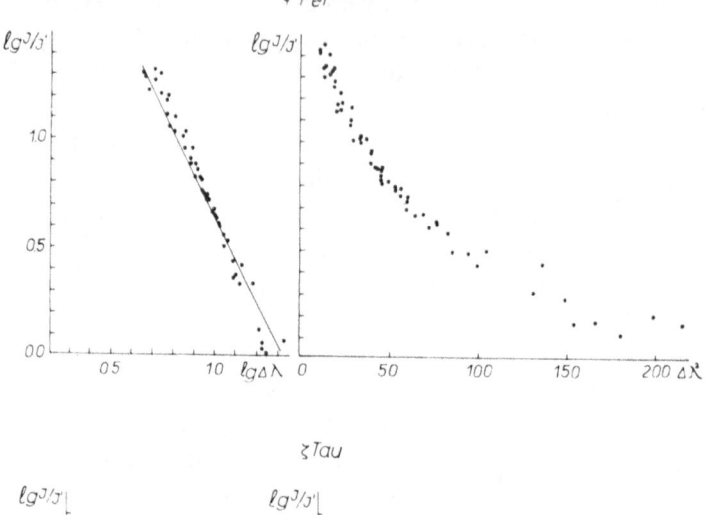

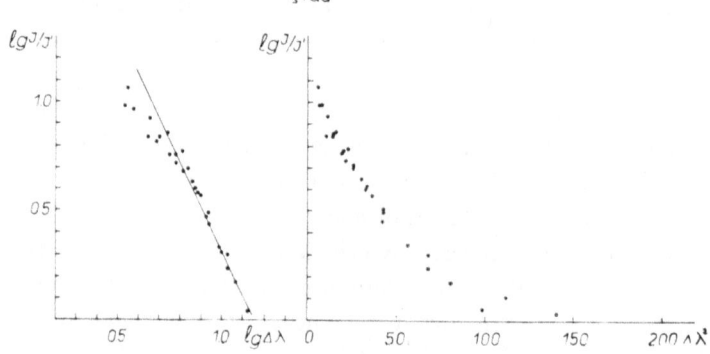

Fig. 6.   Analysis of Hα profile.

give the straight line with a slope of about two in the plot of lg I against lg $\Delta\lambda$. This slope is the one expected for broadening by radiation damping. There is a significant deviation from a straight line, however, which corresponds to Doppler broadening. Thus the rotation of envelopes leave us a small possibility to use emission line profiles for the analysis of macroturbulence.

The analysis of the equivalent widths gives us another possibility to study motions in the envelopes. In this case we investigate the motions of gas volumes which are smaller than the thickness of the envelope. The value of turbulent velocity is determined by the vertical shift of the observed curve of growth. Therefore this method can be used only for shell stars with strong absorption metallic spectra, which can be observed with a high dispersion. This is the main reason why curves of growth were constructed only for a few stars (Boyarchuk and Pronik, 1963, 1964, 1967; Ozemre, 1967). These investigators found that the value of the turbulent velocity should be changed from ~ 10 km s^{-1} to ~ 20 km s^{-1}. The value of the turbulent velocity in the envelope of Be stars is larger than the average value found for the atmospheres of supergiants.

The width of emission and absorption lines that arise in envelopes can give some information about the dimensions of the envelope. Struve and Wurm (1938) showed that the envelopes of Be stars are rotating in accordance with the conservation of the angular momentum:

$$v \cdot r = \text{const}, \tag{1}$$

where v is the rotational velocity at the distance r from the axis of rotation. Struve (1942) was the first to determine the dimensions of the envelope of a Be star. He found that the radius of envelope is equal to about two radii of the star and therefore the dilution factor is about 0.1. Later investigations confirmed these values. Boyarchuk (1958) found from the study of 12 Be stars that the mean radius of envelope is equal 1.6 R_{star}, if the emission lines are used and 2–3 R_{star}, if the absorption lines are used.

Let us consider the conditions of the excitation in the envelope of a Be star. There is a little doubt that the source of the excitation of emission lines in the envelopes is the radiation of a central star. The same mechanism which takes place in the planetary nebulae acts in Be-star envelopes: the stellar radiation ionizes the atoms in an envelope which then undergo recombinations and cascade transitions. The observed emission lines are the result. However, the conditions in the envelopes of Be stars differ significantly from those in planetary nebulae. The envelopes are optically thick for lines of subordinate series. The envelopes have a dilution factor about 10^{-1}–10^{-2} while planetary nebulae have a dilution factor much smaller $\sim 10^{-12}$. It is necessary to keep in mind these differences when we calculate the Balmer decrement. First, we have to take into account the ionization from excited levels. Second, we have to consider simultaneously a set of non-linear integral-differential equations. The calculation of Balmer decrements in Be-star envelopes met many mathematical difficulties. There are two main approaches: the moving-envelope theory and the static-atmosphere theory.

These theories may give the two extreme cases with respect to the state of motion inside the envelope. The moving-envelope theory was developed by Sobolev (1946, 1962). He supposed that if the envelope is in a state of motion with varying velocity, photons emitted in the interior region have a probability, β_{ik}, to leave the envelope without any absorption because of the Doppler effect and to contribute to the formation of emission lines. The introduction of this probability of photons leaving makes it possible to change a system of complicated equations of radiative transfer into a system of rather simple algebraic equations. Many astronomers have calculated the Balmer decrement by using Sobolev's theory.

The most extensive calculations were made by Boyarchuk (1966), by Hirata and Uesugi (1967) and by Ilmas (1971). Boyarchuk gave the Balmer decrement for a wide range of values of the electron temperature, the stellar temperature, the dilution factor and the probability of L_{α}-photon exit. The first thirty levels were considered in the solution, a correction being applied for the remaining levels. Ilmas (1971) has taken into account transition by electronic collisions.

The static-atmosphere theory has been developed by Miyamoto (1949, 1952) and by Kogure (1959, 1961, 1967). The envelope is supposed to be static and opaque for the Lyman and Balmer radiations so that the Balmer lines are formed in the outermost part of the envelope. The mathematical difficulties of the solution of the system of transfer equations permit to consider only seven energy levels (discrete + continuum). Figure 7 shows the relation between $H\gamma/H\beta$ and $H\delta/H\beta$. The observed values were published by Rojas and Herman (1958) for B0–B4 stars (open circles) and for B5–B9 stars (dots) and by Burbridge and Burbridge (1953) for B0–B4 stars (squares). One theoretical curve corresponds to the solution case VII of the static-atmosphere theory (Kogure, 1967), another theoretical curve corresponds to a solution of the moving-envelope theory (Boyarchuk, 1966).

Sobolev's theory gives a good agreement with observations. The static-atmosphere

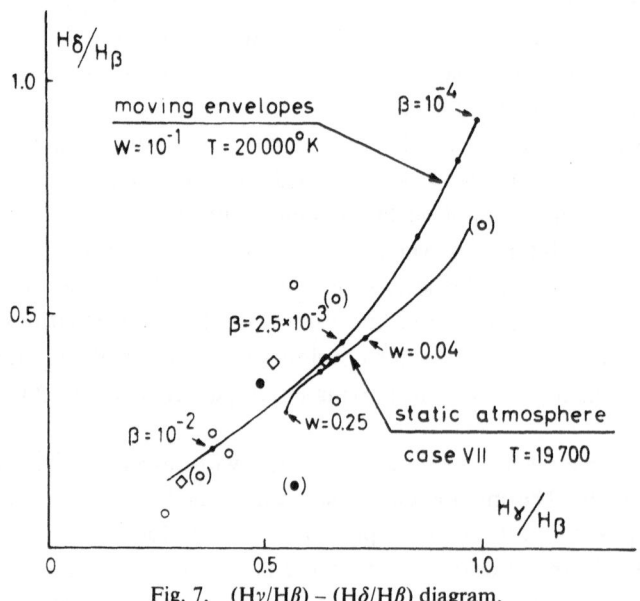

Fig. 7. $(H\gamma/H\beta) - (H\delta/H\beta)$ diagram.

theory agrees with observations when the dilution factor becomes larger than 0.25. The latter contradicts the values of $W = 0.1, 0.01$, which are deduced by the law of conservation of angular momentum.

It should be noted that the observed Balmer decrement depends on the manner of drawing in the underlying background. If an emission line is located above a pure continuum, it is easy to draw a background. But if an emission line is superimposed on strong stellar absorption lines the problem of drawing a background becomes more complicated. The opacity of the envelope for line radiations introduces some additional difficulties in the determination of the intensities of the background stellar radiation. In practice astronomers draw usually the underlying background by an arbitrary interpolation, and this can give a rather large error for weak emission such as $H\gamma$, $H\delta$, etc.

Several astronomers have investigated emission profiles (Rublev, 1964; Kogure, 1969; Marlborough, 1969). All of them explained a general shape of emission lines, – the differences in details due apparently to an inhomogeneity of the envelopes.

Only a few investigations of relative intensities of lines of other elements than hydrogen have been made. It is necessary to point out the calculations in the relative intensities of neutral helium lines by Struve and Würm (1938) and by Wellman (1952). They have shown that the metastable levels are overpopulated under conditions of dilute radiation, and absorption lines arising from levels such as $\lambda\,3889$ and $\lambda\,3965$ are strengthened. This is in qualitative agreement with observations.

If we have some information about physical conditions in the envelopes, we can apply methods which are used usually in the analysis of the stellar atmospheres. We can use the Inglis-Teller formula

$$\lg n_e = 23.26 - 7.5 \lg n,\tag{2}$$

(where n is the upper level quantum number of the last Balmer line resolved) for the estimation of the electron density. Boyarchuk and Pronik (1965) have found the value $n_e \sim 2 \times 10^{11}$ cm^{-3} for the envelope of ζ Tau in 1964. Searle (1958), from a study of three shell stars, concluded that the electron densities of these shells are about 5×10^{11} cm^{-3}. We can find values n_e of the same order from tracings published by different astronomers. We conclude that the values $n_e = 10^{11}$–10^{12} cm^{-3} are found in large envelopes of Be stars. We cannot estimate the value of n_e for shell envelopes in which spectra the shell lines of envelopes disappear in the first Balmer lines. We can estimate the number of the hydrogen atoms, N_2, on the second energy level above 1 cm^2 of stellar surface by using the approximation of optically thin case (Unsöld, 1939):

$$W_\lambda = \frac{\pi^2 e^2 \lambda^2}{mc^2}\, f N_e,\tag{3}$$

where W_λ is the equivalent width of the shell absorption lines and f is the oscillator strength.

Boyarchuk (1958) has studied ten Be stars. He found that the values of $\lg N_2$ vary from about 16 for ζ Tau to 13 for 23 Tau. It is easy to calculate that such strong envelopes as the envelope of ζ Tau are opaque to the centers of the Balmer lines, but are transparent beyond the Balmer limit. The optical thicknesses of envelopes beyond the Lyman limit of all but ζ Tau stars are inferred to be $\tau_L \lesssim 1$. The envelope of ζ Tau has the value of $\tau_L \sim 10$. It means that the stellar radiation cannot ionize atoms in the envelope which have ionization potential higher than 13.6 eV. As a result, we observe in the spectrum of ζ Tau the numerous lines of the one-ionized metals. The shell metallic lines are not observed in the spectra of envelopes for which the value of $\lg N_2 < 15$. The problem of the origin of the envelope is still unsolved. It is known that Be stars are rotating very rapidly. Struve was the first to propose that the envelopes of Be stars formed through rotationally forced ejection. These stars are unstable at their equators and are losing mass through rotational break-up. But recently

Slettebak (1966) has found that the largest observed rotational velocities for stars O9.5–F0 V are always below the computed equatorial break-up. If a stable continuous ejection of gas caused by the stellar rotation takes place from stellar surface we should observe the stable expansion of the gas in the envelope. As we have seen earlier, on the average the envelopes do not expand. Thus, stellar rotation is not the main reason for the formation of envelopes; it creates favourable conditions for gas ejection, but an additional reason is needed. That reason, that mechanism does not have to work continually: its power changes with time.

Boyarchuk (1959) has proposed that forces similar to those producing solar activity act on the surface of Be stars. Limber and Marlborough (1968) and Henriksen (1969) assume that magnetic fields play an important role. There is a small group of hot stars which resemble Be stars in some respects. This is the group of P Cygni stars. The main features of the spectra of P Cygni stars are the emission lines accompanied by violetshifted absorptions. But in the case of a P Cyngi star we have an extended atmosphere rather than an expanding envelope and these stars will be discussed in other reports.

3. Envelopes of the Wolf-Rayet Stars

Wolf-Rayet stars are characterized by the presence in their spectra of numerous broad emission lines. Their spectra also contain a few absorption lines which arise from metastable energy levels, overpopulated because of dilution effects, and from relatively low levels.

Wolf-Rayet stars may be separated into two groups. The first group, indicated as WC, shows emission lines of carbon and oxygen, and the second group, WN, shows emission lines of nitrogen. Underhill (1958) suggested that the difference is due to a higher level of excitation in WN envelopes than in WC envelopes. But Kuhi (1968b) doubts this explanation, because N III and C III have similar ionization potentials.

The main process of excitation is ionization of the gas by short-wave stellar radiation and recombination. Several lines of He II, O III and N III are excited by monochromatic processes of Bowen's type.

Some authors have determined the temperature of the excitation radiation by Zanstra's method. Beals (1940) gives temperatures ranging from 59 000 K to 110 000 K for seven stars. Higher temperatures are found when lines from ions of low ionization potential are used. It can be interpreted that the level of ionization and excitation decreases outward quite significantly in Wolf-Rayet envelopes. Further investigations (Aller, 1943; Voronzov-Veliaminov, 1948; Aller and Faulkner, 1964) confirm Beals' results. However, it should be noted that conditions in envelopes of WR stars differ significantly from those of planetary nebula and Zanstra's method may lead to erroneous conclusions. Miyamoto (1952) points out that conditions in Wolf-Rayet envelopes probably resemble those in Be envelopes. He finds by using his own theory that the temperatures range between 30 000 K and 41 000 K – lower than those suggested by Beals. Rublev (1964) has determined the electron temperatures in envelopes of two Wolf-Rayet stars. He studied Pickering series of He II and found that the electron temperature was about 25000 K.

Underhill (1968) points out that the fact that no forbidden lines are observed in any spectra of WR stars indicates that their envelopes are not very rarefied, $n_e \gtrsim 10^{10}$ cm^{-3}. On the other hand $n_e \lesssim 10^{14}$ cm^{-3} because no significant Stark broadening of the He II lines is observed. Wallerstein (1968) estimates $n_e \sim 10^{11}$ from the X-ray intensity of a source which he relates to the WR star HD 211853.

Now let us consider motions in the envelopes of WR stars. The profiles of the emission lines give us important information. The emission lines which are observed in the spectra of WR stars are of two types: those with a gaussian profile (He II λ 4686) and those with a broad flat top (He I λ 5876). The profiles of the lines in the first group correspond to broadening by turbulent motions with velocities of the order of 500 km s^{-1} to 1000 km s^{-1}. The second group of profiles is characteristic of rapidly expanding spherical envelopes. The velocities of expansion are about 1000 km s^{-1}. (Beals, 1931; Sobolev, 1947).

The expansion of envelopes is confirmed by the presence of blue-shifted absorptions. The data of the measurements given by Hack and Struve (1971) show that the average of velocities of expansion is about 1200 km s^{-1}. The velocities are a little bigger for WC stars than for WN stars and in both cases they are larger in stars of lower temperatures. The observed velocities of expansion exceed the escape velocity ($V_{esc} \sim 600$ km s^{-1} for almost all stars.

Many WR stars are double stars. This makes it possible to determine the dimensions of the envelopes of WR stars, as has been done by Hiltner (1949), Kron and Gordon (1950) and Munch (1950). The most detailed investigation was made by Kuhi (1968). He observed several eclipses of the WR component of V 444 Cyg with a photoelectric spectrum scanner. He obtained a variety of shapes of eclipse curves for different lines. From the depth of the minimum we can compute the radius of the occulated WR envelope, assuming that the disk of the envelope has uniform surface brightness. Kuhi finds that the He II envelope has a radius of 16.9 $R_\odot$, while the N V envelope has a radius of 31 $R_\odot$. The WR component has a radius of 2 $R_\odot$ (Wilson, 1942). This means that the dilution coefficient in envelopes of WR stars is about 10^{-2}–10^{-3}, i.e. ten times smaller than in the envelopes of Be stars. Since lines of higher ionized ions have smaller width, the gas in envelopes of WR star must be radially accelerated. Kuhi points out also that there are lines whose intensities vary in a strange manner. They do not show any minimum or show a minimum early, before the minimum of continuous radiation. Kuhi explained these facts by inhomogeneity in the envelope caused by gas streams in the binary system. Kuhi's observations confirm the existence of a stratification of physical conditions in envelopes of WR stars. In general, the lines of higher ionized ions arise in deeper level than that of lower ionized ions.

The first model WR star was proposed by Beals (1930). He suggests that the broad emission lines should be produced in a radially expanding envelope surrounding the star. This model explains some features of WR stars, but there are two objections at least. First, not all emission lines have flat topped profiles accompanied by violet-displaced absorption components. Second, Wilson (1940) pointed out the absence of the expected phase shift between photometric and spectroscopic times of minima. In order to eliminate these difficulties, Underhill (1966) proposed more complicated model:

WR star has a moderately extended, moderately dense atmosphere in chaotic motion above a compact photosphere. This atmosphere is probably equivalent to the luminous disk of Kron and Gordon (1950). We must assume that the star is surrounded by a low density, rapidly expanding atmosphere (envelope) which is significantly opaque chiefly in lines which are strengthened under conditions of high temperature and moderate dilution.

It should be noted that although many papers have been published about WR stars, we do not know for sure the physical conditions (T_e, n_e) in their envelopes. We do not know what is the mechanism of formation of envelopes or what kind of forces work in the envelopes.

The Be stars and WR stars are groups of stars in which envelopes have been studied rather in detail. There are other objects that may have envelopes. But those envelopes have been studied only roughly. In some cases, we do not know whether we observe envelope or another phenomenon. Below we will consider several examples:

4. U Geminorum Stars and Old Novae

These stars are close binaries. The main component is surrounded by an envelope, which has the shape of a disk (Prendergast, 1960). The envelopes rotate with velocities of about 600 km s^{-1} (Krzeminski, 1965). The envelopes have the electron density of about 10^{13} cm^{-3} and the radius is about 10^{10} cm (Gorbatskij, 1970).

5. Symbiotic Stars

The term 'Symbiotic Stars' designates those astronomical objects whose spectra represent a combination of absorption features of a low temperature star with emission lines of high excitation. Swings and Struve (1941) have suggested that symbiotic stars are binaries: one of the components is a late-type giant, and the other a hot, small star which is the source of the excitation of an envelope surrounding both components. Boyarchuk (1970) has determined the size of an envelope $\sim 10^4 R_\odot$ and the electron density $\sim 10^7$ cm^{-3}. Thus the envelopes in which the emission lines originate are probably more closely related to planetary nebulae than to true stellar envelopes.

6. T Tauri Stars

This is a group of irregular variables of spectral class G which have emission lines in their spectra. Their peculiar spectral characteristics and other properties have been reviewed by Herbig (1962). The strongest emission lines (hydrogen and Ca II) often have shortward-displaced absorption component. Herbig has suggested also that the forbidden lines observed in the spectra of most T Tauri stars arise in a circumstellar envelope. Kuhi (1964) has interpreted the broad emission lines of H and Ca II in terms of an expanding envelope. A fit of computed line profiles to observed ones then gives the density at the surface of the star as $\sim 10^{10}$ atom cm^{-3} and expansion velocities of ~ 150 km s^{-1} as the average values for six stars. But some important characteristics of T Tauri stars are unknown still. For instance, we do not know what is the excitation source for the emission. It is possible that here we have a mechanism like that in the solar chromosphere. The T Tauri phenomenon is more closely related to stellar atmospheres than to envelopes. Recently, Walker (1969) has drawn attention to the fact that many T Tauri stars with the strong ultra-violet excess have redward-displaced absorption components of hydrogen and Ca II emission lines. These stars have an inverse P Cyg spectrum. Walker proposed that the inverse P Cyg spectrum in the T Tauri stars with ultra-violet excesses indicates actual infall of material. Thus I am not sure that in the case of T Tauri stars we are concerned with true envelopes. It seems to me the T Tauri phenomenon is more complicated.

In conclusion, it should be stressed that all of the problems of envelopes which were formulated by Otto Struve thirty years ago still require answers.

References

Aller, L. H.: 1943, *Astrophys. J.* **97**, 135.
Aller, L. H. and Faulkner, D. I.: 1964, *Astrophys. J.* **140**, 167.
Beals, C. S.: 1930, *Publ. Dominion Astrophys. Obs.* **4**, 228.
Beals, C. S.: 1931, *Monthly Notices Roy. Astron. Soc.* **91**, 966.
Beals, C. S.: 1940, *J. Roy. Astron. Soc. Can.* **34**, 169.
Boyarchuk, A. A.: 1958, *Mem. Soc. Roy. Sci. Liège, 4ème Série* **20**, 159.
Boyarchuk, A. A.: 1960, *Voprosi kosmogonyi* **7**, 231
Boyarchuk, A. A.: 1966, *Izv. Krymsk. Astrofiz. Obs.* **35**, 45.
Boyarchuk, A. A.: 1970, in A. A. Boyarchuk and R. E. Gershberg (eds.), *Eruptivnye Zvezdy*, Nauka, Moscow, p. 148.
Boyarchuk, A. A. and Pronik, I. I.: 1963, *Izv. Krymsk. Astrofiz. Obs.* **29**, 268.
Boyarchuk, A. A. and Pronik, I. I.: 1964, *Izv. Krymsk. Astrofiz. Obs.* **31**, 3.
Boyarchuk, A. A. and Pronik, I. I.: 1965, *Izv. Krymsk. Astrofiz. Obs.* **33**, 195.
Boyarchuk, A. A. and Pronik, I. I.: 1967, *Izv. Krymsk. Astrofiz. Obs.* **36**, 203.
Burbidge, G. R. and Burbidge E. M.: 1953, *Astrophys. J.* **118**, 252.
Gorbatsky, V. G.: 1970, in A. A. Boyarchuk and R. E. Gershberg (eds.), *Eruptivnye Zvezdy*, Nauka, Moscow, p. 63.
Hack, M. and Struve, O.: 1971, *Stellar Spectroscopy* **2**, Trieste, p. 14.
Henriksen, R. N.: 1969, *Astron. Astrophys.* **1**, 457.
Herbig, G. H.: 1962, *Adv. Astron. Astrophys.* **1**, 47.
Hiltner, W. A.: 1949, *Astrophys. J.* **110**, 95.
Hirata, R. and Uesugi, A.: 1967, *Contr. Kwasan Obs. Kyoto* No. 156.
Ilmas, M.: 1971, in *The Emission Lines in the Stellar Spectra*, Tartu, p. 47.
Kogure, T.: 1959, *Publ. Astron. Soc. Japan* **11**, 127, 278.
Kogure, T.: 1961, *Publ. Astron. Soc. Japan* **13**, 335.
Kogure, T.: 1967, *Publ. Astron. Soc. Japan* **19**, 30.
Kogure, T.: 1969, *Astron. Astrophys.* **1**, 253.
Kron, G. E. and Gordon, K. C.: 1950, *Astrophys. J.* **111**, 454.
Krzeminski, W.: 1965, *Astrophys. J.* **142**, 1051.
Kuhi, L. V.: 1964, *Astrophys. J.* **140**, 1409.
Kuhi, L. V.: 1968a, *Astrophys. J.* **152**, 101.
Kuhi, L. V.: 1968b, in K. B. Gebbie and R. N. Thomas (eds.), *Wolf-Rayet Stars*, Washington, p. 108.

Limber, D. N. and Marlborough, J. M.: 1968, *Astrophys. J.* **152**, 181.
McLaughlin, D. B.: 1961, *J. Roy. Astron. Soc. Can.* **55**, 76.
Marlborough, J. M.: 1969, *Astrophys. J.* **156**, 135.
Miyamoto, S.: 1949, *Jap. J. Astron.* **1**, 17.
Miyamoto, S.: 1952, *Publ. Astron. Soc. Japan* **4**, 1.
Münch, G.: 1950, *Astrophys. J.* **112**, 266.
Ozemre, K.: 1967, *Ann. Astrophys.* **30**, 495.
Prendergast, K.: 1960, *Astrophys. J.* **132**, 162.
Pringle, J. K. and McNamara, D. M.: 1962, *Publ. Astron. Soc. Pacific* **74**, 525.
Ringuelet-Kaswalder, A. E.: 1963, *Publ. Astron. Soc. Pacific* **75**, 323.
Rojas, H. and Herman, R.: 1958, *Mem. Soc. Roy. Sci. Liège, 4ème Série* **20**, 198.
Rublev, S. V.: 1964, *Astron. Zh.* **41**, 63 (*Soviet Astron.* **8**, 45).
Searle, L.: 1958, *Astrophys. J.* **128**, 61.
Slettebak, A.: 1949, *Astrophys. J.* **110**, 498.
Slettebak, A.: 1966, *Astrophys. J.* **145**, 126.
Sobolev, V. V.: 1947, *Moving Stellar Envelopes*, Leningrad.
Sobolev, V. V.: 1962, *Astron. Zh.* **39**, 632 (*Soviet Astron.* **6**, 531).
Struve, O.: 1942, *Astrophys. J.* **95**, 134.
Struve, O. and Würm, K.: 1938, *Astrophys. J.* **88**, 84.
Swings, P. and Struve, O.: 1941, *Astrophys. J.* **93**, 356.
Underhill, A. .: 1953, *Monthly Notices Roy. Astron. Soc.* **113**, 477.
Underhill, A. B.: 1958, *Mem. Soc. Roy. Sci. Liège, 4ème Série* **20**, 17.
Underhill, A. B.: 1961, *Publ. Dominion Astrophys. Obs.* **11**, 405.
Underhill, A. B.: 1968, in K. B. Gebbie and R. N. Thomas (eds.), *Wolf-Rayet Stars*, Washington, p. 195.
Unsöld, A.: 1938, *Physik der Sternatmosphären*, Springer Verlag, Berlin, p. 290.
Vorontsov-Velyaminov, B. A.: 1958, *Mem. Soc. Roy. Sci. Liège, 4ème Série* **20**, 55.
Walker, M.: 1969, in L. Detre (ed.), *Non-Periodic Phenomena in Variable Stars*, Academic Press, New York, p. 103.
Wallerstein, G.: 1968, *Astrophys. J. Letters* **151**, L121.
Wellman, P.: 1952, *Z. Astrophys.* **30**, 80.
Wilson, O. C.: 1940, *Astrophys. J.* **91**, 3, 79.
Wilson, O. C.: 1942, *Astrophys. J.* **95**, 402.

THIRD DISCUSSION SESSION

(Thursday Afternoon; 7 September, 1972)

(following the review paper by Boyarchuk)

Chairman: G. LARSSON-LEANDER

Larsson-Leander: Thank you, Dr. Smak, for giving us this account of Boyarchuk's paper. The paper is now open for discussion.

Underhill: The question with expanding envelopes is: how do you detect them? If you look at the resonance lines in the far ultra-violet spectra of O-type and early B-type supergiants, you see displaced absorption lines, corresponding to velocities of -1000 km s^{-1}, and we say that matter is leaving the star. No feature in the ordinary spectral region gives you that assurance – yet now we know that those stars have expanding envelopes. The question is not whether an expanding envelope is a rare thing, but when can we detect one? Perhaps all stars have expanding envelopes, in the sense that all are shedding material to some extent. I wonder, does this affect the interstellar matter? We say that stars are formed from the interstellar matter and that they deplete it. Are they creating as much as they are taking away? Again, as Huang asked yesterday: where does the energy come from? Following the arguments of Lucy and Solomon (1970), we think that radiation has a lot to do with it. Is there any other force that could be used to give a radial acceleration outward? This is just what Huang was talking about yesterday. How do you get an outward velocity for a shell, or for any material, whether it comes off in the plane of an orbit, or a plane of rotation?

Smak: If I understand properly, there is a sequence of questions one should answer. First, is the rotational velocity of Be stars sufficiently high to associate the formation of the disks with rotational break-up? If so, the second question will be: are those B-type stars without emission that are also rotating rapidly, rotating too slowly to form disks? The third question, if that is not the case, is: if the rotation is incapable of forming disks, is the Lucy-Solomon mechanism efficient enough in the case of the Be stars, but not in the case of the ordinary B-type stars? That question, I think, should be answered by both observers and theoreticians.

Underhill: I think that the Lucy-Solomon mechanism, radiation pressure driving the material off and giving it that needed outward acceleration, is the key factor. After all, some rapidly rotating B-type stars are known to show Be spectra for a while – then the emission disappears again. There seems to be no particular pattern in the visibility of the shell – it appears roughly every ten or fifteen years. I think this must be intimately connected with some extra source of radiation pressure, just below the surface of the star. Once a B-type star has begun to move off the main sequence, it will contract a bit and then start burning hydrogen in a shell. If the shell is near enough to the surface, there may be enough radiation to give that little extra

Batten (ed.), Extended Atmospheres and Circumstellar Matter in Spectroscopic Binary Systems, 95–116.

spurt of radiation pressure which emits a shell of gas for a while. This gradually drifts away, or falls back into the star, and then, ten or fifteen years later, the star gives itself a shake and gives another little burst. Mendoza (1958) has suggested that Be stars are just a bit above the main sequence, and that would fit with the idea that they are beginning shell-burning of hydrogen.

Huang: That's exactly the mechanism mentioned in my review paper, but I didn't have time to discuss it yesterday. Main-sequence stars are not exactly unstable, according to Lucy and Solomon. They become unstable only if they are rotating rapidly.

De Groot: But why, then, are there other rapidly rotating B-type stars which, as far as we know, are not going through an emission-line phase?

Underhill: They presumably have not advanced far enough off the main sequence to have shell-burning sufficiently near the surface that the radiation pressure gets out enough to give the needed outward acceleration.

Plavec: I'm afraid that this idea of having hydrogen shell-burning very close to the surface may be misleading because actually the hydrogen-burning shell is formed very deep inside the star, very far from its surface, and the hydrogen-burning extends over a period of the order of 10^5 yrs or something like that so it would be rather surprising that there would be such large changes in 10 or 15 yrs. As Dr. Huang mentioned, redistribution of angular momentum might be a better explanation than any serious change in the hydrogen-burning shell.

Hutchings: Did Dr. de Groot say that there are rapidly rotating stars that don't show emission lines? Are these rotating at break-up velocity? Is it not true to say that Be stars, rotational Be stars, are rotaing at maximum velocity?

De Groot: They are a little bit slower.

Underhill: Not all Be stars are rapidly rotating! There are so-called pole-on stars, and there are some stars that people believe have naturally sharp-lined spectra and show emissions.

Hutchings: I was just going to say that maybe this should be pointed out. Nevertheless, pole-on stars may still rotate. There are also B-type supergiants which are called Be but do not rotate rapidly. They are much more luminous than ordinary Be stars, and their surface gravity is low. Mass loss from them is a different matter altogether.

Underhill: It has even been suggested that the main-sequence Be stars with sharp-lined spectra can be divided into two groups (Schild, 1966). One group appears to contain intrinsically rapid rotators, while the other consists of slow rotators.

Hutchings: About Be stars lying off the main sequence, Roxburgh and Strittmatter (1965) said that this can be explained by an aspect effect. If you look at a rotating star equator-on, it is gravity-darkened and hence appears to lie off the main sequence. So the effect need not be an evolutionary one.

Bolton: There has been an idea around for some time that the Wolf-Rayet stars are remnants of mass-exchange in close binaries (Paczyński, 1967). There is a class of OB stars discovered and defined by Walborn (1971) in the spectra of which the absorption lines show carbon and nitrogen anomalies somewhat analogous to those found in the Wolf-Rayet spectra. Because a process like that suggested for the Wolf-Rayet stars might have operated on these stars, Mr. Lars Rogers and myself have begun to obtain spectra of those that are accessible from the Northern Hemisphere. We now have perhaps two dozen spectra of about eight of these stars and all but two of these can tentatively be identified as double-line binaries. The other two show indications of expanding atmospheres. One of these stars, HD 235679, is particularly interesting. It is a Be star with double emission with a V/R of 3 or 4 to 1. The emission is above the continuum at least down to λ 3889. The absorption lines are as sharp as any I've seen among early B-type stars, but on the Grant comparator the weak lines look clearly doubled. There is no indication of a shell spectrum or of any broad underlying absorptions.

Underhill: Not every stellar shell has a rapidly rotating star underlying.

Bolton: But there are definitely two components to the emission, a V and an R component. That's what is exceptional about this star. The Balmer-line emission is visible at least to λ 3889 on one plate and λ 3797 on another, so the Balmer decrement is quite small.

Underhill: This is an unusual object, and I don't think we should assign it to any class.

Bolton: I agree, but I think that its spectral anomalies suggest a possible relationship with the Wolf-Rayet stars, and at least suggest the possibility that mass exchange is going on, or has gone on.

Underhill: It can't be a Wolf-Rayet star; you never see double emission lines in the spectrum of a Wolf-Rayet star – they're just broad and rounded.

Bolton: I am *not* suggesting that the star is a Wolf-Rayet star. The carbon-nitrogen anomalies are in the absorption line spectrum. But the possible relationship to the Wolf-Rayet stars seems obvious to me.

Underhill: The fact that you have nitrogen lines excited rather than carbon is very probably correlated with the level of the temperature in the outer gas – for which you must consider non-LTE physics, radiation-dominated physics. It's very hard to make generalized statements, but if the temperature is above 100 000 K and the density is about 10^9 atoms cm^{-3}, the nitrogen ions will dominate. Below 100 000 K, but above 50 000 K, the carbon ions are dominant. I am only discussing this qualitatively, but keep in mind that you must not make generalizations based on LTE calculations.

De Groot: This suggestion by Paczyński, that a Wolf-Rayet star comes into being after mass exchange in a binary, is very attractive. I like it very much. But it does not give a complete answer to how you create the Wolf-Rayet phenomenon, because it does not explain how the Wolf-Rayet spectra of the nuclei of planetary nebulae appear. So, it may contribute to an understanding of Wolf-Rayet stars, but not to the more general phenomenon and there must be other reasons for this phenomenon as well.

Underhill: Well, you can follow that up by looking at the spectrum of Sco X-1, the optical spectrum of Sco X-1. If you read the descriptions given in the literature, it sounds very much like the spectrum of a Wolf-Rayet star, except the lines are not as broad as in the spectra of average population I Wolf-Rayet stars. I'm sure nobody is going to call Sco X-1 a Wolf-Rayet star. It just comes back to the fact that if you have a plasma of density around 10^9 to 10^{11} and an electron temperature around 10^5, you're going to get a Wolf-Rayet spectrum, and any combination of events that gives you those densities and those temperatures, gives you a Wolf-Rayet atmosphere.

Larsson-Leander: Thank you. Any more comments on this topic? Has Dr. Smak any comments on the U Gem stars which are perhaps always close binaries? Do any of us have any remarks on the symbiotic stars?

Underhill: So far as the symbiotic stars are concerned, I sometimes wonder whether it is completely beyond imagination that you're looking at the central part of a red giant that has been shedding matter and is getting down to the last remnants of its outer atmosphere and every once in a while you see through a little hole to the hot blue central core. The hole closes over and you see the atmosphere again. Is a symbiotic object really only one star – a long-period variable in the last gasps of emitting its outer atsmophere?

Hutchings: This would seem unlikely for AG Peg which has a very well-kept period of about 814 d. You wouldn't expect a star to open up, so that you can have a look inside it, fairly regularly like that.

Plavec: I'm going to talk on symbiotic stars in my review, so I don't want to spend too much time on them now. I think it is now very reasonably established that stars like AG Peg and Z And are binaries in which the blue component flares up (or, if you don't like that term, brightens up). In the case of Z And, a flare is rather short; in AG Peg it takes several years, but when it eventually decreases again, you can see very clearly the M-type spectrum of the other component. So it seems to me, at least, that the binary nature of this object is fairly well established.

Thackeray: AR Pav is a remarkable case of 'multiple symbiosis'. Its spectrum consists of a hot nebular spectrum with forbidden lines, indicating clouds of very low densities; a cooler nebular spectrum, a supergiant F shell absorption, varying in intensity, and TiO seen only near mid-eclipse. Mrs. Mayall's value for the period of 605 d (Mayall, 1937) has been maintained closely for some 50 cycles. There is an eclipse of some $2^{m}.5$, but there are large fluctuations of $\pm 1^{m}$ about a mean light-curve.

Sahade obtained some spectra in 1948 showing a P Cyg contour to the H lines with absorptions displaced -100 km s^{-1}. These contours have changed and Radcliffe spectra over 20 yrs show the rather sharp H lines cut into by an absorption. The model proposed in my paper of 1959 has had to be revised in one respect since radial velocities are now available.

The sharp H absorptions remain at more or less constant velocity, shifted by -22 km s^{-1} relative to the centre of gravity of the system, while the emissions oscillate back and forth in the period of 605 d. Thus, there is an expanding H cloud surrounding the whole system. The permitted He emissions vary with semi-amplitude $K = 13$ km s^{-1}, and yield a mass function 0.13 $M_{\odot}$. The supergiant F absorption, only occasionally seen, also varies cyclically and in phase with the He emission with large scatter and $K \sim 26$ km s^{-1}. Surprisingly, the forbidden emission velocities ([O III], [Ne III]) seem to vary in opposite phase as though associated at least statistically with the secondary.

However, new evidence that the secondary is in fact an M-type star, responsible for the TiO seen at mid-eclipse, has been provided by observations by Dr. Ian Glass at the recent 1972 eclipse. He found 'no change' in the infrared (J, H, K, L bands). It will be of interest to see if a secondary eclipse becomes visible in these infrared bands at phase $0^P.5$.

Leung: Are those large variations in brightness more or less semi-periodic or absolutely random?

Thackeray: Well, there has been no systematic photometry apart from Mrs. Mayall's light curve based on Harvard plates covering 40 yrs. We have a little photometry at Radcliffe, but this obviously should be followed up in detail with regular photometry at some place.

Larsson-Leander: Perhaps we could have a comment on the T Tau stars. I myself found Walker's (1972) interpretation of the ultra-violet excess and the blue continuum rather fascinating. Any more comments on Boyarchuk's paper?

De Groot: If I remember rightly, Dr. Smak said that in the paper it was mentioned that P Cyg stars show many similarities to the Be stars. I don't know exactly what Boyarchuk meant, but I don't see this clearly, and I'm afraid that this is a type of confusion about what stars we should call P Cyg stars. I do not think that all stars that have a P Cyg profile somewhere in the whole range of the spectrum should be called P Cyg stars, but this term should be reserved for those stars whose spectra show many more lines of various elements with this profile. Otherwise, we would mix into one pot various types of stars which really have quite different characteristics.

Larsson-Leander: We are free to leave 'expanding atmospheres' now, and to take topics left over from this morning. Mr. Bopp wants to talk about work by himself and Mr. Moffett on YY Gem.

Bopp: The eclipsing, double-lined spectroscopic binary YY Gem (Castor C) has been known to be a remarkable system since the work of Joy and Sanford (1926). The system consists of two late-type dwarfs, of spectral type dM1e, with the Balmer lines and Ca II H and K visible in emission. The period, from von Gent's (1931) photometric study is $0^d.8142822$. Joy and Sanford's initial study showed marked variation in the relative strengths of the two emission components, with the red emission component being consistently stronger than the blue. Struve's special interest in this system resulted in three separate spectroscopic investigations of this system. The first (Struve *et al.*, 1950) found equal intensities for the Balmer and Ca II components, with little, if any, evidence for variations. Two later investigations (Struve, 1952; Struve and Zebergs, 1959) found only small, irregular emission-line variations on a rapid time-scale.

As a further complication, the photoelectric investigations by Kron (1952) revealed the presence of secondary fluctuations in light that were periodic with the rotation, but temporary in duration. Kron attributed these variations to non-uniform illumination of the stellar surfaces ('star spots'). Most recently, YY Gem has been shown to be a flare star by Moffett and Bopp (1971).

Approximately forty coudé spectra at a dispersion of 18 Å mm^{-1} have been obtained with the McDonald Observatory 207-cm (82-inch) reflector during the periods 1971, February 6–17 and 1971, November 27 to 1972, February 1. The Balmer and Ca II emissions are generally of approximately equal strength, but plate No. 7154, taken on 1971, February 10, at phase 0^{p}715, is radically different. On this plate, the blue shifted component is markedly stronger than the red (Figure 1). Emission of Si I at λ3905 is prominent; this emission is not seen in the normal spectrum of YY Gem, but is known to be strongly enhanced in solar flares. Clearly a flare occurred, or was in progress during the hour-long exposure. However, on this spectrum the Balmer emission is not double, but triple (Figure 2). Midway between the two stellar

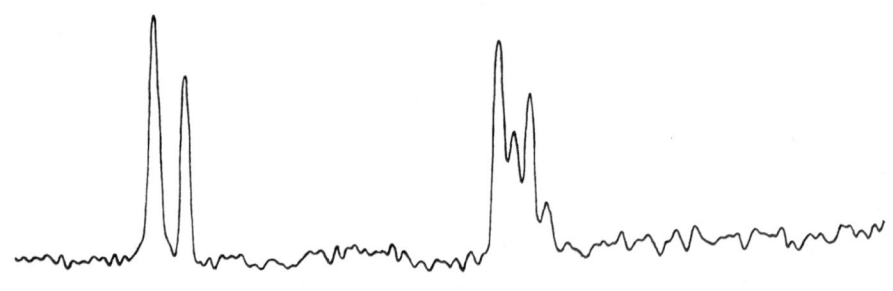

Fig. 1.   Smoothed density tracing of the Ca II H and K region; the blue shifted component of H, K, and Hε is noticeably enhanced.

Fig. 2.   Smoothed density tracing of the region containing Hγ. Note enhanced blue-shifted component, central emission, and quiescent red-shifted component.

components is a weaker, narrow third component. Radial velocity measurements show this central component to have a velocity coincident with the γ-velocity of YY Gem. Another coudé plate, taken two hours later, shows the blue-ward stellar component still slightly enhanced, but the central emission is no longer visible. The central component is definitely not visible when the stars are completely quiescent.

We conclude that the central emission feature is caused by material at the central Lagrangian point of the system, and is excited into emission by flare activity on one of the components. This configuration would not, of course, be stable; the material, if a permanent feature of the system, would have to be replenished. This could occur either via explosive ejection of material during flares or by a gradual outflow of chromospheric material. Regarding the latter possibility, Struve and Zebergs noted a systematic difference of absorption-line and emission-line radial velocities in the spectrum of YY Gem, and ascribed it to a possible outflow of material. Possible connections between this material and the occurrence of flare activity are, at the moment, highly speculative. Many hypotheses concerning flare stars and circumstellar material have been discussed, however (Greenstein, 1950; M. Johnson, 1953; Lortet-Zuckerman, 1965; Evans, 1971). Till now there has been no evidence for any association of UV Cet-type flare stars with circumstellar matter.

Leung: Are all your observations photometric?

Bopp: We have many simultaneous spectroscopic and photometric observations of YY Gem. Unfortunately, the interesting plate was obtained in 1971, February, before we started our simultaneous programme, so we do not know how bright this particular flare was.

Leung: I have some observations of YY Gem, too, and I wondered if our observing runs overlapped.

Fracastoro: Could you remind us what the light curve of YY Gem looks like?

Moffett: The light curve of YY Gem, obtained by Kron, shows nothing strange except the secondary light variations which he attributed to spots on one of the components. The two stars are practically identical in terms of mass, luminosity, etc. and the orbit is nearly circular. I would like to make a few additional comments: The word 'flare' has been used many times during this symposium to describe a wide variety of features and, as Dr. Batten has pointed out, the term 'flare' does not have a unique meaning. In discussing YY Gem, we use the word 'flare' to describe the eruptive events as observed on the UV Cet class of stars. The flares observed in YY Gem appear to be almost identical to flares occurring on other UV Cet stars. The flare shapes and colours seem to be characteristic of the UV Cet class, but the frequency of occurrence exhibits a somewhat different behaviour. In 1971, when flares were first detected on YY Gem, we observed one flare event every four hours, but in 1972 the flare activity decreased to about one every ten to twelve hours of observing. This does not occur in the other UV Cet stars; rather, their mean frequency of flare activity remains very constant year to year. The change in flare frequency on YY Gem most probably is related to the secondary light variations which Kron and Struve found some years and which were absent in others.

Smak: Have you made any estimates as to whether the expansion velocity of 7 km s^{-1} is sufficient to lift the matter from the surface of the star?

Bopp: The velocity of escape from an M0 dwarf is about the same as that from the Sun, so for material to be ejected from the system would require a velocity of 650 km s^{-1}, if we disregard the effects of the companion star. If there is a slow leakage of material some source of acceleration would be required – I do not yet know what it is. I might mention that in the ultra-violet spectrum of YY Gem, around 3200 Å, there are emission lines of FeII, which are also seen in the spectra of red giants. In these latter, the lines have been interpreted as arising from an extended atmosphere or corona.

Larsson-Leander: If there are no more comments on that topic, we will ask Dr. Robinson to speak on high-speed photometry of Z Cam.

Robinson: I have been making high-speed photometric observations of Z Cam, using Nather's photometer on the 82-inch Struve telescope at McDonald Observatory. The time resolution of the observations, which were made without filters, is between one and five seconds. The observations show that a model similar to that developed by Smak (1971) and by Warner and Nather (1971) for U Gem also explains the major features of Z Cam. The period of Z Cam is seven hours; the primary is a G1 star, on the main sequence, that fills its Roche lobe and transfers mass to a white dwarf of about solar mass. The white dwarf is surrounded by an optically thick disk of gas, and is invisible except during eruptions. A bright spot is formed where the stream hits the disk and contributes at least 20%, and probably more than 50%, of the white

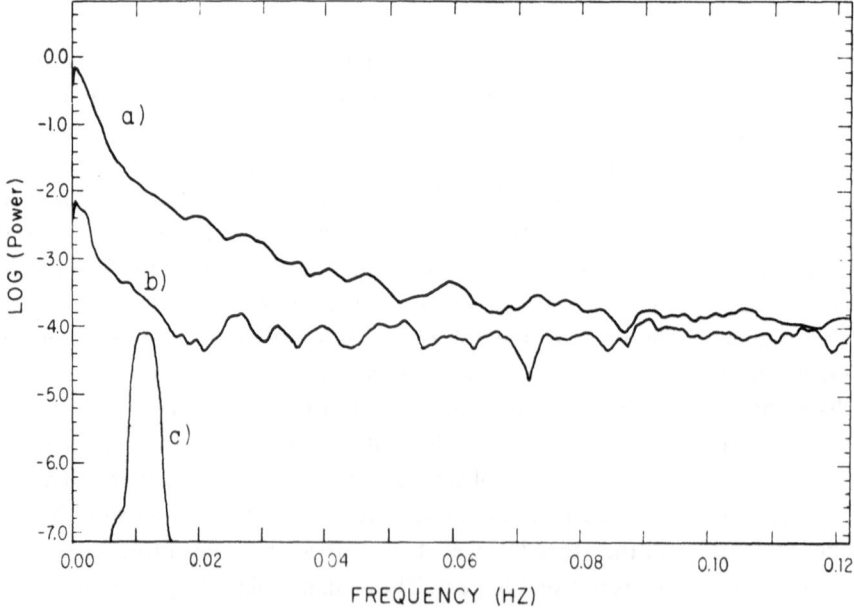

Fig. 3.   Power spectrum of light variations of Z Cam (see text for explanation).

light of the system during minimum. The brightness of the spot varies dramatically, thus giving rise to the flickering in the light curve.

I have made a time-series spectral analysis of my light curves of Z Cam. The upper curve in Figure 3 shows the power spectrum of a typical light curve near minimum light. The middle curve is the power spectrum of the 'light curve' of a comparable constant star. Both curves have been smoothed, and the lower curve is the spectral window resulting from the smoothing. For frequencies less than about 0.09 Hz (period >11s) the power spectrum is a measure of the flickering in Z Cam. It is impossible to interpret this directly, but progress can be made by looking for changes in the power spectrum, from night to night. The distribution of the power does not change, but the total power does – the curve moves up and down without changing

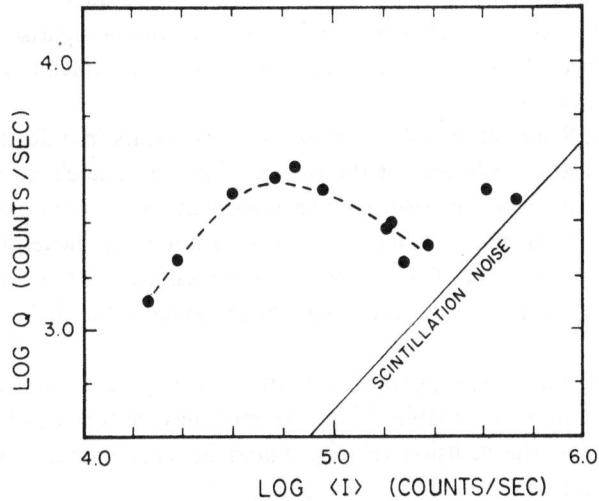

Fig. 4.   Plot of log Q against log $\langle I \rangle$ (quantities are defined in text).

shape. The frequency distribution is constant over the entire cycle of eruption. It is difficult to believe that the G-type star could erupt without affecting the process of mass transfer. For this and other reasons, I assume that the eruptions are in the white dwarf.

The light curve of Z Cam can be approximated by a cubic polynomial. Because of the preceding result, changes in the light curve can be described by a single parameter. I have chosen Q, the square root of the variance of the light curve about the polynomial: Q is proportional to the total power. Figure 4 shows a plot of log Q versus log$\langle I \rangle$ (where I is the mean brightness of Z Cam during a run). The diagonal line is the value of log Q expected from scintillation noise. Two runs are badly affected by scintillation: the remainder define a curve. This curve can be explained by the bright-spot model. When log$\langle I \rangle \lesssim 4.5$, the white dwarf is contributing little light to the system, and variations in brightness are primarily in the bright spot. If the percentage of variation in the bright spot is independent of its brightness, we expect the relation between $\log Q$ and $\log \langle I \rangle$ to be a straight line parallel to the scintillation line. If $\log \langle I \rangle \gtrsim 4.6$, the white dwarf contributes most of the light of the system – the curve becomes horizontal. The curve may even go down again for $\log \langle I \rangle \gtrsim 5.0$. This might be explained by increases in the size of the disk during eruptions (as Smak (1971) has suggested occur in U Gem). The transferred matter then has less energy when it strikes the disk. The spot is fainter and the flickering decreases in amplitude.

Leung: Is your photometer a single-channel one?

Robinson: It is two-channel, but we often use it as a single-channel one.

Leung: Do you not look at a comparison star?

Robinson: We look at a comparison star at the beginning and end of each run, and take sky readings every half hour. The external errors of our brightness measurements may be a few per cent, but they are not significant on the scale of Figure 3.

Leung: If you used both channels and looked at the comparison star and the variable simultaneously, would it be much easier to sort out which is scintillation and which are real variations?

Robinson: No. Scintillation will be different for two stars in different directions in the sky. It will also be different for the two diaphragms. There's no way you could measure the light of the comparison and variable stars through the same diaphragm.

Devinney: Do I understand that the relative amount of flickering decreases at maximum light, so the light of the white dwarf is washing out the variations?

Robinson: Yes, relative to the total light of the system, the flickering goes down dramatically!

Smak: I think that your suggestion that the disk may be larger when the system is very bright is quite reasonable. Then the spot may be formed closer to the Lagrangian point and the collision velocity would be smaller than usual. Therefore, the spot would not be so bright.

Bath: How constant is the power spectrum between maximum and minimum? You said that this constancy indicates that the red star is not the source of instability, and that there is essentially no change in the mass-transfer rate during an outburst. Can you deduce from this any quantitative values of the possible range of mass-transfer rates?

Robinson: From the flickering alone, you cannot calculate the mass-transfer rate.

Bath: But you were concluding from the power spectrum that the rate must be essentially constant.

Robinson: Possibly you could determine the exact amount of change in brightness of the spot, but the observational scatter is so large that the result would probably be misleading.

Smak: I think the point is that the luminosity of the spot, as we see it in the visual region, is a fairly complicated function of the rate of mass transfer, the velocity of collision, and the physical properties of matter....

Bath: ... and therefore I don't think you can necessarily conclude that the blue star is undergoing the outburst.

Robinson: I think you can conclude that the red star isn't erupting; because if it is, it would throw off a large amount of mass. This mass, presumably, would go towards the white dwarf, and greatly change the pattern of the flickering.

Hutchings: Would you care to comment on similar observations the Texas people have made of other stars with flickering hot spots? These spots seem to me to be a fairly important observational discovery.

Robinson: We have not yet found any inconsistency between the observations of cataclysmic variables and a hot-spot model. These systems always show flickering, and are binary systems with mass transfer. The strength of flickering varies dramatically from star to star: it is stronger in Z Cam than in most, although VV Pup flickers by almost a magnitude.

Hutchings: Does the size of the hot spot vary? Can you tell this in systems in which the hot spot is eclipsed?

Robinson: We're accumulating observations of eclipses of U Gem. We'll be able to say more about that when we have reduced them. Perhaps the size of the spot is one of the least-known parameters. The best estimate at the moment is a diameter of 3×10^9 cm.

Thackeray: I'd like to make one 'ancient-historical' remark. As some of you may know, the first observation of the flickering of U Gem was made in 1856, as I think we must believe now, with a 7-inch refractor by Norman Pogson, an assistant at the Radcliffe Observatory, Oxford. The observation was quoted by Van der Bilt (1908). I think I may claim to have seen the fluctuations of VV Pup (Thackeray *et al.*, 1950) visually with the 74-inch reflector. I could hardly believe my eyes, but these fluctuations in intervals of a few seconds were recorded on five occasions.

Smak: I would like to say a few words about systematic differences between novae and U Gem systems at minimum. The first difference was noticed long ago by Kraft and refers to the appearance of the emission lines: those in U Gem systems seem stronger but of lower excitation and ionization, though there are exceptions. The second difference refers to the relative intensity of continuous radiation coming from the hot spot. As I tried to demonstrate last year at the Bamberg Colloquium, there appears to be a systematic preference among the U Gem type systems to have relatively brighter spots, though – again – there are some exceptions to this rule. Finally, I would like to show what seems to be another systematic difference between the two types of objects and it is shown in the slide (Figure 5). This diagram is basically a reproduction from an earlier paper by Kraft (1964), except that I used different symbols for novae (filled circles) and U Gem systems (open circles). It is a plot of rotational velocities ($V_d \sin i$) against semi-amplitude (K_1). An arbitrary line $V_d \sin i \times K_1 =$ const is drawn in to help show the separation. Arrows are placed at points which represent non-eclipsing systems to indicate corrections to be applied for the inclination effect. While the effect I am talking about is not very obvious, its reality seems to be supported by the case of VV Pup, which is neither a nova nor a U Gem object. It is plotted in the diagram with the K_1 value from Herbig's work and the $V_d \sin i$ value crudely estimated from the width of lines in his Crossley spectrograms.

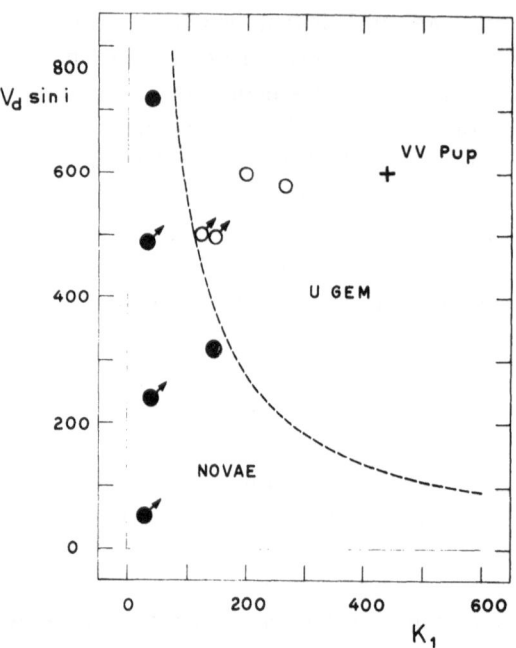

Fig. 5.   Plot of emission-line width, $V_d \sin i$, against semi-amplitude of velocity of blue component, K_1, for novae and U Gem systems.

From the location in this diagram, it would appear then that VV Pup belongs to the U Gem group. And it is really encouraging to recall that VV Pup has the longest observed shoulder in its light curve, indicating a relatively very bright hot spot and, also, that the spectrum contains very strong hydrogen lines with no helium lines being present.

To summarize, I think we can say that there are systematic differences between novae and U Gem stars, which – in all cases – refer to the properties of the circumstellar material and in the case of the two of those, which I was talking about, they appear to reflect some systematic differences in the dynamical properties of gas streams and disks in these systems. It may well be so, therefore, that the dramatic differences in their behaviour at outbursts result, at least partly, from the different properties of the circumstellar matter.

Herczeg: You mentioned that hot spots are less important in novae than in U Gem stars. Can we perhaps say that there may be novae without hot spots? In particular, what about DQ Her in which the flickering activity seems to be much weaker than the pulsation?

Robinson: The flickering in DQ Her is very strong, and frequently hides the 71 s pulsation.

Herczeg: May I ask a second question? Figure 5 indicates rather clearly that radial-velocity amplitudes are systematically smaller for novae than for U Gem stars. Does this mean, perhaps, that novae have the smaller masses?

Smak: Maybe.

Fracastoro: Does the flickering have any quasi-periodic character?

Robinson: Yes, it often looks periodic, but it is never strictly periodic. It's like any stochastic process. If you take sufficiently short runs of data, you can find very strong spikes at any frequency you choose. When you take longer runs, the spike tends to disappear.

Fracastoro: What is the sub-period of this flickering – five or ten minutes?

Robinson: The flickering varies in intervals from as short as the instrument is capable of registering up to perhaps ten minutes or more.

Kitamura: In the Elsinore meeting (Kitamura, 1970), I emphasized that the use of the Roche coordinates (with the zero-velocity curves ξ=const as one of them) is useful for hydrodynamic treatment of gas motions in close binary systems. This was already pointed out by Prendergast (1960) and Kopal (1969). With the use of Roche coordinates (ξ, η) we shall consider the case that rotating flows of inviscid gases exist around the parent star and are governed by gravitational forces only. Discussion is confined to the steady flows in the orbital plane in the case of circular orbits of the components. Taking the pressure terms into account, we start from the two scalar equations of motion on the orbital plane:

$$u_1 \frac{\partial u_1}{\partial \xi} + \frac{h_1}{h_2} u_2 \frac{\partial u_1}{\partial \eta} + \frac{u_1 u_2}{h_2} \frac{\partial h_1}{\partial \eta} - \frac{u_2^2}{h_2} \frac{\partial h_2}{\partial \xi} - 2u_2 h_1 + \frac{1}{\varrho} \frac{\partial P}{\partial \xi} = K_\xi,$$

$$\frac{h_2}{h_1} u_1 \frac{\partial u_2}{\partial \xi} + u_2 \frac{\partial u_2}{\partial \eta} - \frac{u_1^2}{h_1} \frac{\partial h_1}{\partial \eta} + \frac{u_1 u_2}{h_1} \frac{\partial h_2}{\partial \xi} + 2u_1 h_2 + \frac{1}{\varrho} \frac{\partial P}{\partial \eta} = 0,$$

with

$$u_1 = \partial \xi / \partial t \quad \text{and} \quad u_2 = \partial \eta / \partial t,$$

where h_1 and h_2 are the metric coefficients in the transformation

$$(dx)^2 + (dy)^2 = h_1^2 (d\xi)^2 + h_2^2 (d\eta)^2,$$

and q the mass-ratio $m_2/m_1 \leqslant 1.0$ between the components. The separation of the components has been taken as the unit of length, the total mass $m_1 + m_2$ as the unit of mass, and $P/2\pi$ as the unit of time. In doing so, the external force K_ξ reduces to $K_\xi = 1/(1+q)$.

As the first approximation we have neglected the velocity components along the η-constant curves orthogonal to the zero-velocity curves, because ring-like flows would not exist very far from the parent star and, if rotating flows exist around the parent star, the velocity components along the curves η=const should be small compared with the other components along the curves ξ=const.

Thus, putting $u_1 = 0$ in the above equations of motion, and solving for u_2, we can have

$$u_2 = \cfrac{1}{\cfrac{1}{h_2}\cfrac{\partial h_2}{\partial \xi}} \left\{ -h_1 + \sqrt{h_1^2 + \frac{1}{h_2}\frac{\partial h_2}{\partial \xi}\left(\frac{1}{\varrho}\frac{\partial P}{\partial \xi} - \frac{1}{(1+q)}\right)} \right\}.$$

In this equation, the pressure term occurs in the form of $\partial P/\partial \xi$.

Assuming the adiabatic relation, it follows that

$$\frac{1}{\varrho}\frac{\partial P}{\partial \xi} \varpropto \frac{\gamma}{\gamma - 1}\frac{\partial \varrho^{\gamma - 1}}{\partial \xi}, \quad (\gamma > 1),$$

where γ denotes the ratio of specific heats.

If the gases within the ring are distributed so as to have a maximum density on the η-constant curve, we may put there

$$\frac{\partial \varrho^{\gamma - 1}}{\partial \xi} = 0 \quad \text{and so} \quad \frac{1}{\varrho}\frac{\partial P}{\partial \xi} = 0.$$

Thus, for the density maximum within the ring we may neglect the pressure term. This is important.

Roughly speaking, the observed rotational velocity of the ring may be related to the u_2 velocity of gases at maximum density, and therefore, in the first approximation, the u_2 equation with the pressure term in it dropped may be directly compared with the observed rotational velocity. Inserting the observed rotational velocity of the ring in u_2, we can easily estimate the dimension of the ring, because the metric coefficients and their derivatives can be easily calculated as functions of the coordinates and the mass-ratio q.

Application to four well-known systems with gaseous rings is shown as follows:

TABLE I

Dimensions of Rings

Star	q	V_{ring}	r_{ring}	r_{ring}/r_1	$r_{\text{ring}}/r_{\text{Roche}}$
RY Gem	0.21	1.32	0.314	2.62	0.598
AW Peg	0.16	1.64	0.242	1.42	0.439
U Sge	0.38	1.11	0.352	1.68	0.754
RW Tau	0.23	1.29	0.320	1.73	0.620

In Table I, V_{ring} is the observed rotational velocity of the ring expressed in the present unit; r_{ring}, r_1 and r_{Roche} are the dimensions along the y-axis of the ring, the parent star, and the Roche limit respectively. In this computation, the data are taken from *A Catalogue of Graded Photometric Studies of Close Binaries* (Koch *et al.*, 1970) and *Sixth Catalogue of the Orbital Elements of Spectroscopic Binary Systems* (Batten, 1967). The above result indicates that the gaseous ring is kept well inside the Roche limit.

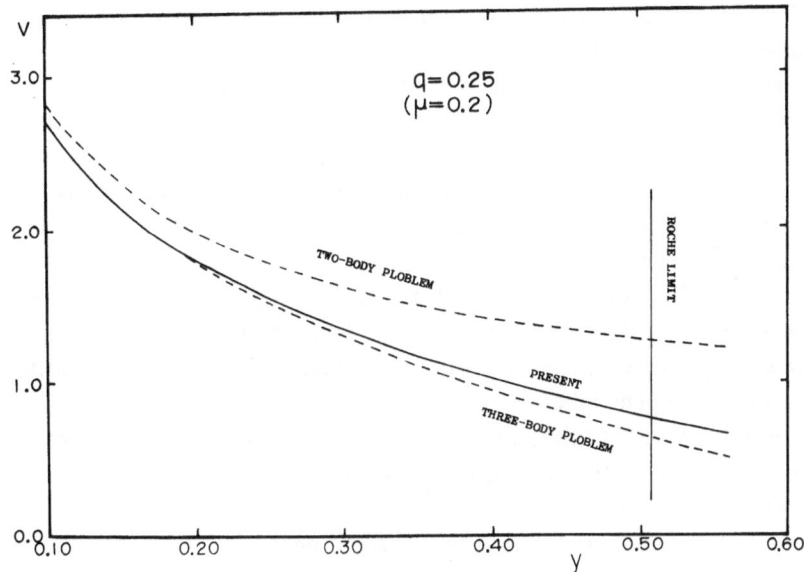

Fig. 6.   Comparison of velocity V at points on the y-axis through the primary component predicted by different approximations (see text).

Next, I would like to show you a comparison of our u_2 velocities with those from the restricted three-body problem by Huang (1967). The corresponding velocities at the same positions on the y-axis as used by Huang are calculated with our u_2 equation by dropping the pressure term. Figure 6 shows that the agreement is nearly perfect for the smaller y-values but away from the surface of the parent star the discrepancy becomes appreciable.

As the next step, we can proceed to the second approximation by taking into account the velocity component along the η-constant curves, and appropriate boundary conditions.

Biermann: I have a question about your approximations. Do you assume a disk of constant thickness, or do you make some other approximation about the thickness of flow?

Kitamura: This discussion is confined within the plane, and we need not consider the pressure term to derive a first approximation for the dimension of the ring.

Huang: The two-dimensional approximation is significant, because the gas is confined within the plane.

Biermann: But the thickness might be variable. You could treat a two-dimensional flow with variable thickness.

Kitamura: Yes, in a general treatment this should be considered.

Huang: The two-dimensional approach is allright if the z-component of velocity is initially very small, but not if it is initially large.

Biermann: I think you are quite right. This is a very good approximation as far as we can tell. However, there are really two quite different problems: a two-dimensional flow of constant thickness, with a central force, and a two-dimensional flow with varying thickness, in a binary system. You get a source term if the thickness varies – from the continuity equation. If the thickness changes a lot, the source term will be important. I think you could do the same kind of mathematics, including the source term, but it probably would not be important to this kind of flow unless the thickness varies a lot. It might be worthwhile to include the source term in other computations – for instance in treating the flow from the stream to the disk.

Huang: I agree with that.

R. E. Wilson: Did you make any trials to see how large a ring can be, and still look reasonably like a ring? I presume it can be no larger than the Roche lobe.

Kitamura: As Table I shows, the rings in RY Gem, AW Peg, U Sge, and RW Tau are all within the Roche lobes.

R. E. Wilson: Yes, but this is an observational result. What I meant was: have you run any purely computational trials to see how large a ring could be – without reference to the observations.

Kitamura: What I showed is that the u_2 equation enables us to estimate the size of the ring from the observed rotational velocity. From this equation alone we can say nothing about the 'purely theoretical' size of the ring.

Smak: Figure 6 shows a large difference between the results of your approximation and those of Dr. Huang's approximation. If I remember rightly, the numbers listed in Dr. Huang's paper refer to the rotating system of coordinates. If that is so, they should be corrected for the rotation of the system before being compared with the two-body case.

Bolton: The system HDE 226868 = Cyg X-1 is the prototype of one kind of X-ray binary characterized by X-ray variations of a factor of two or more in less than a second. There is no evidence of any periodic pulsations. The X-ray spectra of all of the known X-ray binaries are relatively flat compared to those of other X-ray sources. The identification of HDE 226868 with Cyg X-1 is now virtually certain. In the spring of 1971 a weak radio source appeared within the error box for Cyg X-1. The position of this radio source agrees with the position of the B0Ib star HDE 226868 to better than one arc second. At the same time that the radio source appeared, the X-ray source underwent a sharp decline in intensity. Thus the simultaneous X-ray and radio variability combined with the optical-radio position agreement, is strong evidence for the identity of HDE 226868 with Cyg X-1.

Soon after the identity of Cyg X-1 and HDE 226868 was first suspected, Webster and Murdin (1972) and I (Bolton, 1972) showed that HDE 226868 was a spectroscopic binary with a period of $5\overset{d}{.}6$. The velocity curve for the system is published in the references cited. The key orbital elements are as follows: $P = 5\overset{d}{.}5995$, $K_1 = 68.2$ km s^{-1}, $e = 0.09$. The velocity amplitude is large and indicates that the mass ratio cannot be more than about 3:1. The eccentricity is almost certainly real. The formal standard deviation of the eccentricity is 0.02, and great care has been taken to insure that the velocities are not affected by the contamination of absorption lines by emission.

The He II $\lambda 4686$ line is seen weakly in emission on some of the spectrograms obtained at David Dunlap Observatory. The velocity derived from this line varies in antiphase to the absorption-line velocities. If the velocities derived from the He II line represent the motion of the secondary, then a mass ratio, M_1/M_2 of 1.5 is indicated.

Emission lines of hydrogen, and sometimes of helium, are seen between apastron and periastron but not in the other half of the period. These emission lines are not strongest at periastron as one would expect if one of the stars filled its Roche lobe. Rather the emission is strongest at a point between apastron and periastron when the separation between the two stars is decreasing most rapidly. The emission velocity indicates that material is flowing from the B0Ib star toward the unseen secondary. The variation in the emission-line strengths indicates that the B0 star is losing mass to the secondary via a stellar wind.

HDE 226868 is within a degree of the center of Cyg OB3 and its radial velocity and proper motion are consistent with it being a member of the association. Its spectral type and luminosity class are also consistent with association membership, and the velocities and strengths of the interstellar Ca II features are very similar to those in the spectra of other association members. If HDE 226868 is a member of Cyg OB3, the distance of Cyg X-1 is approximately 2 kpc and this implies a mass of 20 $M_\odot$ for the B0Ib star. This places a lower limit of about 6 $M_\odot$ for the mass of the secondary, and that mass could be as high as 14 $M_\odot$ or 15 $M_\odot$. Since it appears that a collapsed stellar object is necessary in order to produce the high X-ray flux observed from this system, HDE 226868 is thus an excellent candidate for a binary system containing a black hole.

R. E. Wilson: Can you say from the rotational broadening of lines in the B-type spectrum whether the star should be rotating synchronously?

Bolton: I haven't checked that. The $V \sin i$ is fairly low – less than 100 km s^{-1}, I think.

Herczeg: Is there any indication of an X-ray eclipse? I think someone just said one comes at the wrong phase.

Bolton: Yes, I'm afraid that I'm one of those that contributed to the impression that there is an X-ray eclipse. The early optical results by Webster and Murdin and myself only covered about a three-month period so that it was impossible to derive an accurate period for the system. There were indications in literature that there was an X-ray eclipse showing up in the high energy ($E > 20$ keV) X-ray data. Therefore we took these data and tried to find a period within the range allowed by the optical observations that would satisfy all of the X-ray 'eclipse' points. This led to an eclipse in the wrong place.

A more careful analysis of the X-ray results now seems to indicate that no eclipse is shown, and in any event the improved period now available from the optical data precludes the phasing together of the X-ray 'eclipse' data.

Leung: Your estimate of 20 $M_\odot$ for the mass of a B0Ib supergiant agrees with evolutionary computations by Stothers and myself (Stothers and Leung, 1971). We believe our values of masses for supergiants are quite reliable. They are in satisfactory agreement with those inferred from main-sequence turn-off. We believe we have quite reliable values for class Ib stars, but class Ia stars show a very wide spread in mass.

Bolton: For my purposes, it won't make much difference what mass is chosen for the supergiant. There doesn't seem to be any way that the mass of the supergiant can be forced low enough to permit the secondary to be a normal white dwarf or neutron star. Now, how high you want to put the mass is up to you.

Underhill: Class Ia certainly has a wide spread. The secondary could be a Wolf-Rayet star – they are much less luminous than a B0 supergiant, say $-3^{m}5$ to $-4^{m}0$.

Bolton: A star with absolute magnitude -4 might be visible in the spectrum, particularly if it had emission lines. I think that we are about to hear something that tends to support your suggestion after a fashion. I have the following objection to invoking a Wolf-Rayet star or even a normal main-sequence star as the secondary even though such objects cannot be rejected on observational grounds. It does not seem possible to obtain the required X-ray fluxes from such objects without invoking unknown or *ad hoc* physics while the required fluxes can arise quite naturally from collapsed stars. Furthermore, if you are to say that X-rays are produced in this case by a non-collapsed star, then you must explain why other similar objects in similar situations are not X-ray sources.

Hutchings: Astronomers from the University of British Columbia and myself have been observing Cyg X-1 at Victoria, using the Image-Isocon spectrum scanner at the coudé-focus of the 48-inch telescope. The three chief advantages of this method over conventional spectrophotometry are (i) high signal-to-noise ratio, (ii) linear light response, and (iii) digital data, which enable easy and direct comparisons to be made with standard stars. We have obtained mean, rectified scans of the $\lambda 4686$ region at 8 phases through the $5^{d}6$ period. I should mention at the outset that these results are preliminary, being noisier and less carefully rectified than we hope the final results will be. The velocity changes with phase are clear, showing the $\lambda 4686$ emission feature to vary out of phase with the absorption lines. These velocities agree with Bolton's values. The $\lambda 4686$ emission is blended with $\lambda 4686$ absorption from the B star, and we do not feel there is definite evidence as yet for any variation in the strength or profile of the line.

We have further obtained the mean Cyg X-1 spectrum for the region, corrected for the primary star's orbital velocity, and the mean spectrum of the chief comparison star HD 204172. These mean spectra have a very high signal-to-noise ratio and their difference shows up as a remarkable emission-line spectrum. This must be explained either as the Cyg X-1 absorption spectrum being weakened (by a featureless continuum) or a real emission-line spectrum from the secondary. In this connection, we note that the positions of most of these lines coincide with N III, C III and some other weak emissions seen in similar observations of the Of stars HD 108 and HD 188001. We are working on the data further to test these conclusions. I would also like to discuss another X-ray source – Her X-1. By way of a brief introduction, the object shows three types of periodic variation in the X-rays and two of these have been seen in the optical region. They are (i) a 36-day cycle, for 9 of which the system is active in the X-ray region of the spectrum (the optical object has no known corresponding behaviour); (ii) a 1–7 day cycle showing a complete, sharp X-ray eclipse lasting ~0.2 of the period. Several groups of workers have derived optical light curves which indicate a continuous light variation, covering nearly 2 mag., which shows a large $U-B$ change, and have a minimum coincident with the middle of the X-ray eclipse, (iii) a pulsation in the X-rays of period ~1.24, which shows (Tananbaum *et al.*, 1972) a time-of-arrival periodicity (1.7) which leads to values for the orbital radius ($\sim 4 \times 10^6$ km) and velocity (~ 170 km s^{-1}) of the X-ray source. Very faint ($< 0.^m002$) pulses have been claimed by a group at Berkeley (Davidsen *et al.*, 1972).

David Crampton, at the Dominion Astrophysical Observatory (among others, of whom I believe Mr. Bopp is present) has obtained some dozen spectra of the object to date and these are very curious. They show a strange mixture of spectral features, some of which may be variable within hours, but whose mean behaviour suggest a change from early B-type at maximum light to late A-type at minimum. We have measured these for radial velocity and after rejecting several discrepant or poor lines, obtain a velocity curve of amplitude ~50 km s^{-1} in antiphase with the X-ray source. We have a tentative model for your consideration. The light-curve suggests a contact or ellipsoidal type of system with a small, dark companion. The former possibility raises insoluble difficulties in regard to the non-appearance of a secondary X-ray eclipse, and for this and other reasons, we explored the possibilities of a single star distorted and heated by a small, hot companion using the light curve synthesis program of Hutchings and Hill. It is possible to reproduce the range by heating a star, at its Roche limit, by a companion of size ~0.05 or smaller. The minimum is rather broad but can be filled in by (i) small flux from the hot star (there is an ultraviolet excess which disappears at minimum light), (ii) electron-scattered light in a tenuous envelope about it, and (iii) convective heating of the very highly heated hot side. We are unable however to choose between a (cool star) polar radius 0.25 and q (hot mass/cool mass) ~4.0 and a radius ~0.45 and $q < 0.5$, or anywhere between these. If, however, we assume i close (within ~10°) to 90°, the duration of X-ray eclipse suggests a radius ~0.4 and $q \sim 0.5$. Using the absolute values of orbital parameters given by Tananbaum *et al.* (1972), we find the cool star could be an early A-type star of $2R_\odot$ and $2M_\odot$, making the X-ray source $\sim 1\ M_\odot$. From this, we expect a velocity amplitude of the visible star of ~85 km s^{-1}, but calculations of line profiles accounting for the very large heating effect show that this would appear as an observed amplitude of only ~50 km s^{-1}. The present velocity estimates, which are admittedly poor, are in keeping with this. (This is described in full in *Astrophys. J. Letters*, Dec. 1, 1972.)

Smak: I don't quite understand how you got that synthetic light curve. It appears to me that for almost one-half of the period, when we are looking only at the back of the large star, we should see only as much variation as results from the temperature difference between 7000 K and 6200 K, unless you make convection so efficient as to transfer quite a lot of heat from the small star all round the big one.

Hutchings: You have a temperature distribution all the way over the star. I have just given extreme values. As the stars move, you see a constantly changing temperature distribution.

Smak: I am talking about the phases when you see only the back of the large star – about $0^{P}_{.}2$ on either side of zero phase. You see only the large star – the bright one is hidden – and yet there are variations of about one magnitude, although on your model the luminosity should change by roughly only 10%.

R. E. Wilson: Dr. Smak is precisely right, because I have made very much the same calculations as Dr. Hutchings has made and, in fact, the phase variation is essentially as Dr. Smak says it should be, and significantly different from that found by Dr. Hutchings. My theoretical light curve is much flatter around the time of the X-ray eclipse.

Bopp: I have some spectra of HZ Her (Her X-1) also. Dr. Hutchings said the type varied from B to nearly F. This is not quite correct: the behaviour is much more complicated: the strength of the K line of Ca II varies from about A3 on our plates to about A7 or later around phase $0^{P}_{.}9$. Absorption lines of He I are always visible on our spectra, however. The strength of the helium lines varies, but the variation does not seem to be correlated with phase. In addition, blended N III and C III emission at $\lambda\lambda 4640$ to 4650 and He II emission at $\lambda 4686$ are visible on some of our plates, though the strength of these features varies in thirty minutes or less. How does Dr. Hutchings' model account for the 36-day X-ray period?

Hutchings: It doesn't! The suggestion by Tananbaum *et al.* (1972) of some sort of pulsation seems to me the most reasonable yet made. We looked for possible longer periodicities in our radial velocities, but have not found any yet.

Leung: Do the short-period variations originate in the small component or in the big one?

Hutchings: They are in the X-rays, so I assume they originate in the X-ray source.

Leung: If the short-period variation is associated with the 'white dwarf', it should be of the order of seconds, like the fundamental period of white dwarf.

Herczeg: Can the X-ray pulsation be observed during eclipse, or does it disappear and reappear as it should do?

Hutchings: You can see no X-rays at all duirng the eclipse.

Bolton: As I understand it, the optical observations of pulses do not always corre-late with the observed X-ray pulses in the way that they should. Since the optical pulse amplitudes are so small, I would consider the whole business very shaky. The X-ray pulsations are very regular but it is claimed that these $1^{s}_{.}2$ optical variations appear when there are no similar X-ray variations, and vice versa. The situation is not at all clear.

Oliver: I can't speak for the Lick observers, but I have spoken to those at Rochester and they did not make a very definite statement about the reality of these pulses. We have looked for the same thing at Florida. We may have detected something, but this kind of observation is extremely difficult.

Hutchings: Various observers tried at Victoria too, and found nothing stronger than $0^{m}_{.}002$.

Robinson: Synchronous photometry, which many people are doing, is very danger-ous when you're trying to detect $0^{m}_{.}002$: almost anything can introduce 'wiggles'. You should obtain a long string of data, without summation, and then find the power spectrum, so you can have an idea what you're seeing at other frequencies too.

Scarfe: Does your model predict a detectable eclipse of the small object in ultraviolet light?

Hutchings: I have to check that. It certainly would give you a different light curve.

Lloyd Evans: I'd like to ask Dr. Bolton whether he can tell us anything about a rather different system which might have a black-hole secondary component. I believe Nolan Walborn has the observations.

Bolton: I believe you are referring to V453 Sco. I'm sure that Dr. Sahade could say more about this system than I can, since he has studied it spectroscopically. The system apparently consists of two stars of approximately equal surface brightness, according to the photographic photometry by Gaposchkin (1939). However, there are absorption lines visible for only one component. These lines indicate that the star that produces them is a supergiant belonging to the class of stars with exceptionally strong nitrogen lines that were defined by Walborn (1971). There are strong emission lines of hydrogen which shift in antiphase to the absorption-line spectrum. If these hydrogen lines are representative of the motion of the unseen star, then the unseen star is slightly more massive than the absorption-line object. Walborn has recently submitted a note on this system to the *Astrophys. J. Letters.* Shortly after that paper was submitted, I received a preprint of a paper by Shakura and Sunyaev in which they discuss the optical appearance of a black hole accreting mass through mass exchange in a binary system. Several of their predictions for a black hole accreting mass rapidly fit the observed properties of V453 Sco. These properties include a disk of high surface brightness, which produces emission lines but no absorption lines, and mass outflow from the disk. I pointed out the similarities of the observed properties of V453 Sco to the predictions of Shakura and Sunyaev, and Dr. Walborn has added a note about this to his paper.

One might expect that a black hole in V453 Sco would be a strong X-ray source. However, the disk is probably too thick in the line of sight to permit any X-rays to escape. It seems likely to me that Cyg X-1 differs in its mass-transfer rate and orbital inclination but that these two systems are fundamentally similar.

Note added in proof: Hutchings reports subsequent work on He II emission in the spectrum of Cyg X-1 in *Astrophys. J.* (to be published).

References

Batten, A. H.: 1967, *Publ. Dominion Astrophys. Obs.* **13**, 119.
Bolton, C. T.: 1972, *Nature* **235**, 271.
Davidsen, A., Henry, J. P., Middleditch, J., and Smith, H. E.: 1972, *Astrophys. J. Letters* **177**, L97.
Evans, D. S.: 1971, *Monthly Notices Roy Astron. Soc.* **154**, 329.
Gaposchkin, S.: 1939, *Astrophys. J.* **89**, 125.
Greenstein, J. L.: 1950, *Publ. Astron. Soc. Pacific* **62**, 156.
Huang, S.-S.: 1967, *Astrophys. J.* **148**, 793.
Johnson, M.: 1953, *Observatory* **73**, 109.

Joy, A. H. and Sanford, R. F.: 1926, *Astrophys. J.* **64**, 250.

Kitamura, M.: 1970, in K. Gyldenkerne and R. M. West (eds.), *Mass Loss and Evolution in Close Binaries*, Copenhagen University, p. 194.

Koch, R. H., Plavec, M., and Wood, F. B.: 1970, *Publ. Univ. Pennsylvania, Astron. Ser.* **XI**.

Kopal, Z.: 1969, *Astrophys. Space Sci.* **5**, 360.

Kraft, R. P.: 1964, *Astrophys. J.* **139**, 457.

Kron, G. E.: 1952, *Astrophys. J.* **115**, 301.

Lortet-Zuckermann, M. C.: 1965, *Kleine Veröffentl. Remeis-Sternw. Bamberg* **4**, 30.

Lucy, L. B. and Solomon, P. M.: 1970, *Astrophys. J.* **159**, 879.

Mayall, M. W.: 1937, *Harvard Ann.* **105**, 491.

Mendoza, E. E.: 1958, *Astrophys. J.* **128**, 207.

Moffett, T. J. and Bopp, B. W.: 1971, *Astrophys. J. Letters* **168**, L117.

Paczyński, B.: 1967, in J. Dommanget (ed.), *On the Evolution of Double Stars,* p. 111; *Commun. Obs. Roy. Belgique*, Ser. B, No. 17.

Prendergast, K. H.: 1960. *Astrophys. J.* **132**, 162.

Roxburgh, I. W. and Strittmatter, P. A.: 1965, *Z. Astrophys.* **63**, 15.

Schild, R. E.: 1966, *Astrophys. J.* **146**, 142.

Smak, J.: 1971, *Veröffentl. Remeis-Sternw. Bamberg* **9**, 248.

Stothers, R. and Leung, K.C.: 1971, *Astron. Astrophys.* **10**, 290.

Struve, O.: 1952, *Publ. Astron. Soc. Pacific* **64**, 117.

Struve, O., Herbig, G., and Horak, H.: 1950, *Astrophys. J.* **112**, 216.

Struve, O. and Zebergs, V.: 1959, *Astrophys. J.* **130**, 783.

Tananbaum, H., Gursky, H., Kellogg, E. M., Levinson, R., Schreier, E., and Giacconi, R.: 1972, *Astrophys. J. Letters* **174**, L143.

Thackeray, A. D., Wesselink, A. J., and Oosterhoff, P. Th.: 1950, *Bull. Astron. Inst. Neth.* **11**, 193.

Van der Bilt, J.: 1908, *Rech. Astron. Obs. Utrecht* **3**.

Von Gent, H.: 1931, *Bull. Astron. Inst. Neth.* **6**, 99.

Walborn, N. R.: 1971, *Astrophys. J. Letters* **164**, L67.

Walker, M. F.: 1972, *Astrophys. J.* **175**, 89.

Warner, B. and Nather, R. E.: 1971, *Monthly Notices Roy Astron. Soc.* **152**, 219.

Webster, B. L. and Murdin, P.: 1972, *Nature* **235**, 37.

OBSERVATIONS OF STELLAR SPECTRA
RELATED TO EXTENDED ATMOSPHERES

K. O. WRIGHT

Dominion Astrophysical Observatory, Victoria, B.C.

Abstract. Although the Struve Symposium was planned to discuss conditions in binary stellar systems, it is necessary to consider the observations of extended atmospheres in single stars in order to compare them with those of binary systems. The Sun, being the closest star, has the first chromosphere that was observed, and we can assume that the many features observed on the Sun will be much enhanced in giant stars with more extended atmospheres. Among the early-type stars, the hydrogen emission lines in the spectra of Be stars are known to vary and this suggests the atmospheres are unstable. Similar, even more pronounced effects, are seen in the spectra of shell stars such as 48 Librae. The supergiant stars have also been shown to have spectral lines that change, often quite rapidly, with time. Smolinski has observed in the spectra of some F and G-type supergiants much-broadened absorption lines that also vary with time. Wilson's systematic studies of the H and K emission lines in late-type stars have greatly extended our knowledge of stellar chromospheres, and the ionized magnesium lines at 2800 Å should give additional information as more extra-terrestrial observations are obtained. Numerous observations of the ζ Aur stars have been obtained near eclipse; ζ Aur and 31 Cyg seem to be similar systems and show similar phenomena – a rise in temperature outward from the photosphere and an extensive, variable calcium chromosphere, but a viable theory for these chromospheres has not yet been given. VV Cep and 32 Cyg probably are larger and more complex systems. The emission lines in the ultraviolet spectra of the M-type stars may be indicators of coronae in these cooler atmospheres. Most variable stars seem to have extended atmospheres and, when detailed studies are made, show changes in both light and spectrum. Shock-wave phenomena are undoubtedly important in the analysis of many of these stars, as also are fluorescent, selective-excitation processes. Flare stars, though generally dwarfs, are of great interest in studies of stellar atmospheres.

Extended atmospheres can be detected from observed gross deviations from local thermodynamic equilibrium, as in the study of emission lines, large velocity fields, and dilution effects. It is probable that the presence of a companion affects the atmosphere of the primary star in close double stars, but in stars like ζ Aur and 31 Cyg, the atmospheric effects produced near eclipse may be among the best clues for the interpretation of extended stellar atmospheres.

The study of binary stars has progressed rapidly in recent years, from the observations of light and radial-velocity variations and their analysis in terms of simple binary motion to the detailed analysis of the light curves in terms of the physics of stellar atmospheres and the radiation and gravitational effects of one star on the other, and to the detailed study of spectrum variations that are the result of these interactions. The spark for much of the latter type of investigation came from the work of Otto Struve who utilized the time he had on the McDonald telescope thirty years ago to observe with great efficiency the most interesting eclipsing binary stars, and to lay the foundations for our symposium here at Parksville.

We have heard a great deal about Struve's work, and about more recent work with other spectrographs which only now can improve on Struve's observations. However, it is desirable that we review again the observational data related to extended stellar

Batten (ed.), Extended Atmospheres and Circumstellar Matter in Spectroscopic Binary Systems, 117–133.

atmospheres in order that we may distinguish, if possible, the data obtained from observations of double stars, in which the components are interacting, from phenomena observable in normal stellar atmospheres. The analysis of the basic spectroscopic observations is difficult and the decision whether the observed lines come from the photosphere, an extended atmosphere, an expanding atmosphere or from matter moving between the stars is often most challenging. Just as Plavec (1967) has asserted that nearly all spectroscopic binaries must be considered 'close' by his definition, so Thomas (1970) and his colleagues have come to the conclusion that a stellar atmosphere has an almost limitless extension and that for our dwarf Sun, even the solar wind, which affects the planets, is part of this atmosphere.

Therefore, in this review, I feel that we can consider any observational topics that have not yet been discussed. I should like to consider stellar chromospheres as the principal subject of this review, but will mention other types of outer atmospheric phenomena in case some of the participants wish to talk about them later. These will cover the extended atmospheres of early-type stars, supergiants of all types, and something about the many kinds of variable stars, including the flare stars which seem to be dwarfs. Mass loss in general has already been discussed in previous sessions, as have expanding envelopes of all types, but especially the large-scale expansions observed in some of the hot B-type stars observed by Morton and others in the far ultraviolet region of the spectrum. It is hoped that by recording data for normal stars it will be possible to separate such data from those that are peculiar to and the result of binary interactions. I should mention a few earlier conferences related to this subject: *The Sun Among the Stars*, Riverside, Cal. 1966 (unpublished); *Spectrum Formation in Stars with Steady State Extended Atmospheres*, Munich, 1969, *NBS Special Publ.* **332** (H. R. Groth and P. Wellmann, eds.) and *Stellar Chromospheres*, Goddard, *IAU Colloq.*, No. 19, February 1972 (to be published). There have been many reviews related to extended stellar atmospheres, but I shall mention only a few in addition to the above – Feast (1970), Deutsch (1970) and Thomas and Gebbie (1971).

The chromosphere of the Sun, of course, was the first observed; it was seen as an emission-line spectrum at the eclipse of 1870. The importance of observing the solar chromosphere as a means of studying the structure of the outer atmosphere was quickly recognized, and great efforts were made to obtain flash spectra at each favourable eclipse, especially by the British Royal Society expeditions, and those by Mitchell in the United States. Pannekoek and Minnaert (1927) analyzed the 1924 eclipse data and Menzel (1931) the Lick eclipse data, but the analysis of the Khartoum 1952 spectra by Thomas and Athay (1960) and their colleagues has given us the greatest insight into conditions in the 'medium' layers of the solar atmosphere. The helium lines, especially that at $\lambda 10830$, the emission of the ionized calcium lines, H and K, and, of course, the corona are now well-documented evidence for the extended atmosphere of the Sun; the studies of Linsky (1968, 1972) are among the most important papers outlining investigations of the solar calcium lines. Yet, according to all other evidence, including its mass and luminosity, the Sun is a dwarf star and one would therefore expect that its outer atmosphere would be less extensive than those of the vast majority of stars of greater luminosity. Lower level effects, such as sunspots, granulations, plages, etc., are phenomena that may be common in stars, though they would be difficult to observe, and they may produce some of the line-broadening effects, but they are not applicable to our present discussion. However, prominences and spicules do extend above the 'normal' photosphere and as such could be introduced as topics for discussion at the appropriate time. We shall not discuss the Sun further here, but will consider it only as a basis for comparison with other stellar atmospheres. In this connection, I agree with Thomas who has always tried to develop theories applicable to the Sun and extend them to the stars.

1. Early-Type Stars

Underhill (1966) has covered the observed data for early-type stars in her book on the subject, and she has also noted the salient features of these stars that can be related to extended atmospheres in her survey paper at the Munich Colloquium (1970). Under this topic we shall consider all stars earlier than type F, since this class is where the convective zone seems to appear and where large rotational velocities almost disappear. We shall not go into detail concerning the relevant observations for each type of star, but will mention in this category Wolf-Rayet stars, Of stars, P Cygni and other B and O supergiants, Be and shell stars.

Most of the Be stars have emission lines in their spectra that are produced in a rapidly rotating envelope surrounding the main star, and these lines also have a central absorption core. The hydrogen lines are usually the strongest lines in the spectra of Be stars, with $H\alpha$ having much the strongest emission and other lines in the Balmer series becoming successively weaker with the result that by $H\gamma$ and $H\delta$ the emission is often only a double hump near the centre of the strong absorption line. The Burbidges (1953) showed that the hydrogen emission is a recombination spectrum modified by self absorption and by electron scattering. These atmospheres are unstable since the emission in many stars changes with time, the most frequent change being a variation of the relative strengths (V/R) of the two emission lobes, with a resultant change in the observed radial velocity of the central absorption core. Struve and others concluded that the angular momentum of these stars is conserved and, from measurements of the rotational velocities, found that the extended atmospheres had radii about $3 R_{\odot}$.

Probably most of the hot O-type stars have extended atmospheres. The Of stars usually show emission at $H\alpha$ and at C_{III}, $\lambda 5696$; some show He_{II} emission at $\lambda 4686$, as do lines of N_{IV}, N_{V}, Si_{IV} and He_{I}. 9 Sagittae, O7f, may be considered a typical star of this class, and observations of its spectrum indicate that the emitting shells are variable and moderately extended. Some of the lines may be excited by the Bowen fluorescence mechanism by lines at and below the Lyman continuum.

The very hot Wolf-Rayet stars, interesting though they are, will not be discussed in detail here since they have been the subject of two recent conferences, at Boulder in 1970 and at Buenos Aires in 1971. The discussion about the binary character of Wolf-Rayet stars has not yet been completely resolved, though a number of these stars are certainly components of binary systems. It has been suggested that there are two types of these objects, though they in turn are subdivided into the carbon and the nitrogen sequences. The masses seem to be between 5 $M_\odot$ and 10 $M_\odot$. Underhill has proposed, as a model for a single Wolf-Rayet star, a photosphere with a continuous spectrum like that of an O star with radius 7 $R_\odot$–10 $R_\odot$, an inner compact atmosphere where the rounded-top emission lines are formed, of thickness 1 $R_\odot$–2 $R_\odot$, and an outer, expanding, low-density atmosphere, where dilution effects and monochromatic fluorescent effects occur with a radius, if V 444 Cygni is typical, of 17 $R_\odot$.

Typical shell stars seem to have B-type central stars. They were studied extensively by Struve and he was probably the first to discuss dilution effects, which make lines arising from metastable levels stronger in these low-density envelopes than the other lines produced in the photosphere. Struve and Wurm (1938) commented on the great strength of the metastable lines of Fe II, Ni II, Cr II, etc., relative to Mg II, $\lambda 4481$. One feature of shells is that the spectra correspond to conditions of lower temperature and lower density than the underlying absorption lines, as might be expected. Thus shell lines are usually deep and moderately sharp, somewhat similar to those in α Cygni (A2p), though the underlying spectrum may be that of a B star. In the early studies of shell stars it was not realized that the shells are unstable and variable. However, Merrill and Sanford (1944) noted variations in the spectrum of 48 Librae beginning in 1938. Its spectrum has since varied in a period of about 10 yrs (Underhill, 1966). The radial velocities all follow the same general trend but the hydrogen, Fe II and Ti II lines have successively greater ranges corresponding to high, intermediate and low levels in the atmosphere. In stars like these, the line profiles are often asymmetrical, with the asymmetries on the side of the line to which the velocities are progressing. There is also a progression of velocities for the hydrogen lines, with the higher-member weak lines showing larger variations, which are interpreted as formation in the deeper layers of the atmosphere. Another shell star with a long history is γ Cassiopeiae. Hutchings (1970), during the course of his investigations of stars with rotationally extended envelopes, has found short-period changes in the emission line profiles with a period of 0^d7, but over the years its brightness has varied by more than a magnitude, and its shell has appeared and disappeared.

Supergiant stars are usually considered to have extended atmospheres. The data were surveyed and up-dated at the 1971 Trieste Colloquium on Supergiant Stars. The early-type supergiants are considered here separate from those of later types because of the Wilson-Bappu effect, among other considerations. Underhill has noted that supergiants earlier than O9 cannot exist because their effective gravity would be too low ($\sim 5 \times 10^3$) to balance the large radiative pressure gradient. P Cygni has usually been considered a prototype for early-type supergiants with its fairly sharp emission lines and strong, shortward-displaced absorption lines; it may not be typical since its luminosity has changed by several magnitudes over the centuries, but it does seem to be a fact that the supergiants do vary in light and also show small variations in radial velocity. In P Cygni itself, N II and He I lines have moderately strong emission, while N III and Si IV lines, which have higher ionization potentials, appear only in absorption. Ghobros (1962) agrees with the estimate by Struve and Roach (1939) from dilution effects that the radius of the expanding shell is about 2.5 times the stellar radius. For other B-type supergiants, Chentsov and Snezhko (1972) have studied β Orionis (B8Ia) and have found differential velocity shifts of about 10 km s^{-1} for He I and ion lines, and a Balmer progression of 10 km s^{-1}–20 km s^{-1} similar to that observed for shell stars. Hutchings (1971) has recently obtained high-resolution line-profile scanner observations of B- and A-type supergiants and found that these stars have outward-accelerating envelopes and that the profiles change with time, sometimes in a few hours. There is a velocity-excitation relation with a velocity reversal which indicates a temporary, unstable atmospheric structure. The general picture agrees with other observations that there may be pulsations that propogate and weaken outwards from the stellar surface. He suggests that the A-type stars have evolved from mass-loss B-type stars. Rosendahl (1970) studied line profiles in 64 B- and A-type Ia and Iab supergiants. He concluded that both rotation and macroturbulence contribute to the line broadening, with rotation the principal factor up to middle B types, and turbulence becoming more important thereafter. He concluded that, though the stars may be losing small amounts of mass, there is little evidence for loss of angular momentum. Since the spectrum of α Cygni is often referred to in shell spectra, reference should be made to the analysis by Groth (1960), who obtained $T_{eff}=9170$ K, and gravity, $\log g = 1.13$, assuming radiative equilibrium except for the outermost layers of the atmosphere; the line widths and intensities correspond to a macroturbulent velocity of 22 km s^{-1} and a microturbulent velocity of 7.8 km s^{-1} in the lower layers and 20 km s^{-1} in the outer regions; the radial velocity variations suggest small irregular pulsations, and the atmosphere in general seems to be similar to the lower solar chromosphere.

In his article on supergiants and Cepheids, Kraft (1960) considered that line profiles in spectra of A- and early F-type stars of luminosity classes Ib and II could be explained by a combination of microturbulence and rotation (which was confirmed later by Rosendahl, 1970) assuming the conservation of angular momentum as stars evolved across the top of the H–R diagram. He also found that Cepheids with periods of less than 10 d also fitted this hypothesis; for G- and K-type stars of class Iab, and Cepheids with periods more than 10 d, large-scale turbulence may be more important than rotation. There is an additional absorption at Hα and a few other lines in the spectra of some Cepheids; the Hα satellite is similar to that observed in 89 Herculis (F8 Ia, Böhm Vitense, 1956), where hydrogen and sodium components appear, though the displacements are different. The study of Cepheid variables as examples of extended atmospheres is certainly valid, but because of their position in an instability area of the H–R diagram which, presumably, is the reason for the pulsations, they cannot be considered normal stars. Other F-type supergiants that show evidence for mass motions by the appearance occasionally of double absorption line components are ϱ Cassiopeiae, F8 Iap (Bidelman and McKellar, 1957) and RW Cephei, G0 Ia

(Merrill and Wilson, 1956). More recent work on F0 to K5 supergiants by Smolinski (1972) has shown that a few Ia stars have Hα emission; in HD 217476, [N II] emission has also been observed. Other stars, including HD 231195, F5 Ia, and HD 12399, G5 Ia have lines that seem to be turbulence broadened and that are twice as broad as those in δ Canis Majoris; some lines also show evidence of variable structure.

2. Late-Type Stars

For the late-type stars, Deutsch (1970) at the Lunteren Symposium summarized most of the data concerning chromospheric phenomena. He reviewed the observations that the hydrogen lines vary in many giants and supergiants, and that there does not seem to be a direct correlation with changes in the Ca II H and K lines. Adams and Russell (1928) had noted the strengthening of the Balmer lines relative to the metals, which was confirmed for α Orionis, M2 Iab, by Spitzer (1939) when he found $T_{exc} = 2100$ K for iron lines and 17000 K for the Balmer lines; the hydrogen lines are probably formed in a high-temperature region similar to the solar chromosphere. Time variations in the Ca II H and K lines have been observed in several K-type stars by Griffin (1963), Liller (1968) and others. A period of 350 d for such variations in α Tauri, suggested by Kraft (1967) would fit the hypothesis that α Tauri has evolved from a main-sequence A-type star which has lost its angular momentum through a stellar wind. Further evidence for chromospheric activity in late-type stars is the presence of helium lines in some spectra. Wilson and Aly (1956) identified the He I absorption line at λ 5876 in the spectra of several stars that had strong H and K emission. Vaughan and Zirin (1968) have observed the helium line at λ 10830 in the spectra of a 'substantial' number of the 75 F- to M-type stars they observed. The absorption intensity increases in general with the intensity of the Ca II K-line emission; it appears in emission in a few stars. It is not present in F- or M-type stars. The observations suggest that the helium line indicates the presence of a hot chromosphere, and that it originates in discrete clouds or streams rather than in a homogeneous layer.

Much of our knowledge about stellar chromospheres has come from Wilson's work on the Ca II H and K lines. The basic results were given by Wilson and Bappu (1957), who found that these lines appear in emission in nearly all giants and supergiants of spectral types G, K and M and that there is an almost linear relation between K-line emission *width* and luminosity over a range in absolute magnitude from $M_v = +10$ to $M_v = -6$. Similar emissions are observed in the spectra of some main sequence stars of types as early as F5, and the frequency increases to later types. The lines are usually double-peaked, as in the spectrum of the Sun, though they cannot be seen in the solar spectrum at the dispersions used for the stars; thus it is presumed that chromospheres similar to the Sun are present also in the stars. As Deutsch (1970) notes, the intensities of the peaks vary, and fine structure in the K lines is observed in the spectra of many red giants, from which prominence activity can be inferred. An excellent summary of our knowledge of stellar chromospheres, based on these data, is given by Wilson (1966). With the advent of observations beyond the Earth's atmosphere, much more information about stellar chromospheres can be obtained, though the resolution for most of the observations to date is quite low. Doherty (1971) has shown that the Mg II doublet, $\lambda\lambda 2795$, 2802 can be seen at a resolution of 25 Å mm^{-1} for giant and supergiant stars later than K2, and Kondo *et al.* (1972) have obtained rocket observations of these lines for five stars with a resolution of 7 Å mm^{-1}. The latter find a fairly good correlation between the Mg II line widths and the Ca II data of Wilson and Bappu; the widths of the Mg II lines are definitely broader than those of Ca II, possibly as a result of the greater abundance of magnesium.

Wilson concludes from solar data that the chromosphere is not in hydrostatic equilibrium, that it and the corona are made up of material expelled upward from below, and that they are dynamic rather than static structures. The most likely source of the required energy is the convection zone where the hydrogen is partially ionized and where the energy flux is carried by the turbulent mass motions of convection. The turbulent elements generate acoustic waves which, as the energy decreases, become magnetohydrodynamic, and are thus related to magnetic fields which produce shocks to dissipate energy into the chromosphere. Wilson surmizes that all stars with surface temperatures lower than the Sun have convective zones. The observations that chromospheric K lines are observable only as early as the late F-type stars, and that stellar rotation decreases markedly at the same types, are used as the basis for the conclusion that nearly all late-type stars have chromospheres. For a number of years, Wilson has been expanding his interpretation of these results and the relation between luminosity and the width of the K line, to determine its connection with the age of the stars, the masses of evolved stars, and the rate of star formation.

3. The ζ Aurigae Stars

Probably the most detailed information concerning the outer atmospheres of late-type giant stars should be obtained from eclipses of these stars when they are observed just prior to, and just after, total eclipse. The best examples are the ζ Aurigae stars. The principal component of these systems is a supergiant K- or M-type star and the secondary component is a main-sequence, early-type B or O star. The best data available are for ζ Aurigae (Wilson and Abt, 1954; Groth, 1957), for 31 Cygni (Larsson-Leander, 1957; Wright, 1959) and for VV Cephei (Goedicke, 1938; Peery, 1966). Wilson (1960) has summarized the available publications to 1958; Wright (1970) has discussed the chromospheric K line with its observed multiple structure for these atmospheres, and Groth (1970) has considered the data from the point of view of steady-state extended atmospheres.

The basic observations for these stars are fairly well-known, though more data are needed, and, probably, new analyses should be made in order to clear up the present uncertainties in the interpretation of the results. There is an inner chromosphere, that can be defined as the region where chromospheric absorption lines of neutral and ionized metals can be observed near eclipse. These lines are superposed on the composite spectrum of the two components and are produced by absorption in the inner chromosphere as the radiation from the early-type star passes through the extended late-type atmosphere. These lines can be observed from about 3200 Å to 4500 Å, where the energy from the late-type star becomes markedly greater than that from the early-type star; the lines can be detected to about a stellar radius from the limb (defined as second and third contacts of the eclipse). The strong ionized metallic lines can be observed somewhat farther from the limb, and multiple structure in the Ti II lines at 3759Å –61 Å can be seen at times. The Ca II H and K lines have been detected nearly five stellar radii from the limb in 31 Cygni (Wright and Odgers, 1962). Multiple components of these lines have been detected on high-dispersion spectra that have been obtained at each eclipse since they were first observed for 31 Cygni at Victoria (McKellar *et al.*, 1952). These components are sometimes well separated and almost equal in intensity to the principal component, but more often are weak and barely distinguishable from the grain of the plate. The intensities of the Ca II lines do not necessarily follow the same pattern from eclipse to eclipse (Wright, 1970), which suggests that the outer atmospheres may vary in extent as well as in density. Certainly the multiple components suggest large-scale activity, such as prominences or condensations of some kind.

The increase in temperature outwards in the inner chromosphere has been found from curve-of-growth studies of the chromospheric lines in both ζ Aurigae and 31 Cygni and therefore seems to be well documented as similar to the solar chromosphere. The intensity data on which these temperatures are based are difficult to derive accurately because of the complexity of the K-type spectrum on which the chromospheric lines are superposed, and because the relative intensities of the continua of both stars are required (although the continuum of a K-type star is difficult to define) for the determination of the chromospheric intensities. The interpretation of the atmospheric densities is still uncertain since it is not known whether the source of the excitation in the chromosphere is from the B star or from some 'super-excitation' phenomenon in the outer atmosphere of the K star. Both interpretations seemed to required large clouds rather than a smooth density distribution in the outer atmosphere to explain the observed ionization phenomena. Kawabata (1957) studied the same problem, using Wilson and Abt's data, and found that an electron temperature of 12000 K–14000 K. might satisfy the data assuming that ionization by electron collisions is more important than radiative processes. More recently, Magnan (1965) has considered departures from LTE for a three-level model for the Ca II atom and found that, with an electron temperature of 15000 K and $N_e = 10^{10}$, an atmosphere with homogeneous density can be derived from the Victoria data for 31 Cygni. He suggests that the 'satellite' lines can be produced by large jets or prominences, possibly similar to those in the Sun, which could be cold elements in the hotter, outer chromosphere. The disappearance of the other metallic lines about 0.5 R_K would be the result of a rapid increase in electron temperature at this height, and thus the cold elements would not be observed.

Motions in the atmospheres of these stars can be obtained from radial-velocity measurements of the chromospheric lines. Near eclipse the velocities of the metallic lines definitely deviate from the velocity curve derived from observations obtained far from eclipse. However the velocities do not follow the pattern of the rotation effect, such as is observed for Algol when one star gradually passes behind the other. While the velocities do not seem to be entirely random, since at a given eclipse, they tend to be chiefly positive or negative before and/or after totality, the pattern for different eclipses and for different stars does not seem to remain the same. As Wilson (1960) notes, the radial velocities probably indicate a combination of random and systematic motions of gases in the chromosphere. Saito (1970) made a most detailed study of the velocities observed for ζ Aurigae and concluded that the egress velocities could be explained as a combination of ejection from the chromosphere superposed on a slow rotation and non-steady motion of the gas clouds.

The general motions in the atmospheres of these stars can also be inferred from the microturbulent velocities derived from curves of growth and macroturbulent velocities obtained from the line profiles. The microturbulent velocities found by Wilson and Abt (1954) for ζ Aurigae range from 6 km s^{-1} to 13 km s^{-1}; they seem to be slightly greater for ionized than for neutral atoms and the velocities may increase slightly with height in the atmosphere; there is some evidence for similar results from Wright's (1959) analysis of the inner chromosphere of 31 Cygni. Macroturbulent velocities derived from line profiles were calculated by Kitamura (1967) from an ingress plate of ζ Aurigae and found to be between 15 and 19 km s^{-1}. McKellar *et al.* (1959) studied the K-line profiles of 31 Cygni for the 1951 egress and obtained a good fit with a velocity of 20 km s^{-1} for the lower chromosphere and 10 km s^{-1} for the upper chromosphere.

The dimensions of the ζ Aurigae stars are summarized by Wright (1970). Although the spectroscopic and photometric data do not entirely agree, the diameters for the late-type components are probably 200 to 1000 solar diameters and their masses are 8 to 20 solar masses; the mass ratios are not greatly different from unity. The separation of the components is from 2 to 10 astronomical units, and thus large-scale effects of the B star on the K star should not be expected.

Much work remains to be done on the analysis of these important systems; considerable material has been accumulated at Victoria and elsewhere during the past few years. ζ Aurigae was favourably situated for observations during its eclipse in mid-winter, 1971–1972 and a good series of observations was obtained by Simon in Hawaii. The system of 31 Cygni just came out of eclipse in July, 1972. The K line does not show as much detailed structure as in 1951 or in 1961, but sharp features superposed on the broader chromospheric feature are easily seen. The spectrum of 32 Cygni shows more supergiant characteristics than those of 31 Cygni or ζ Aurigae, but the system is not so well suited for precise analysis because the eclipse is almost grazing, and its extent seems to vary from eclipse to eclipse. A good series of plates showing the chromospheric spectra of 32 Cygni was obtained at Victoria in 1968 and are shown in Figure 1. A less extensive series was obtained in 1971; it is hoped to make an analysis of the chromosphere from these data. Similarly VV Cephei is being followed around its orbit beginning with the chromospheric phases in 1956–1959;

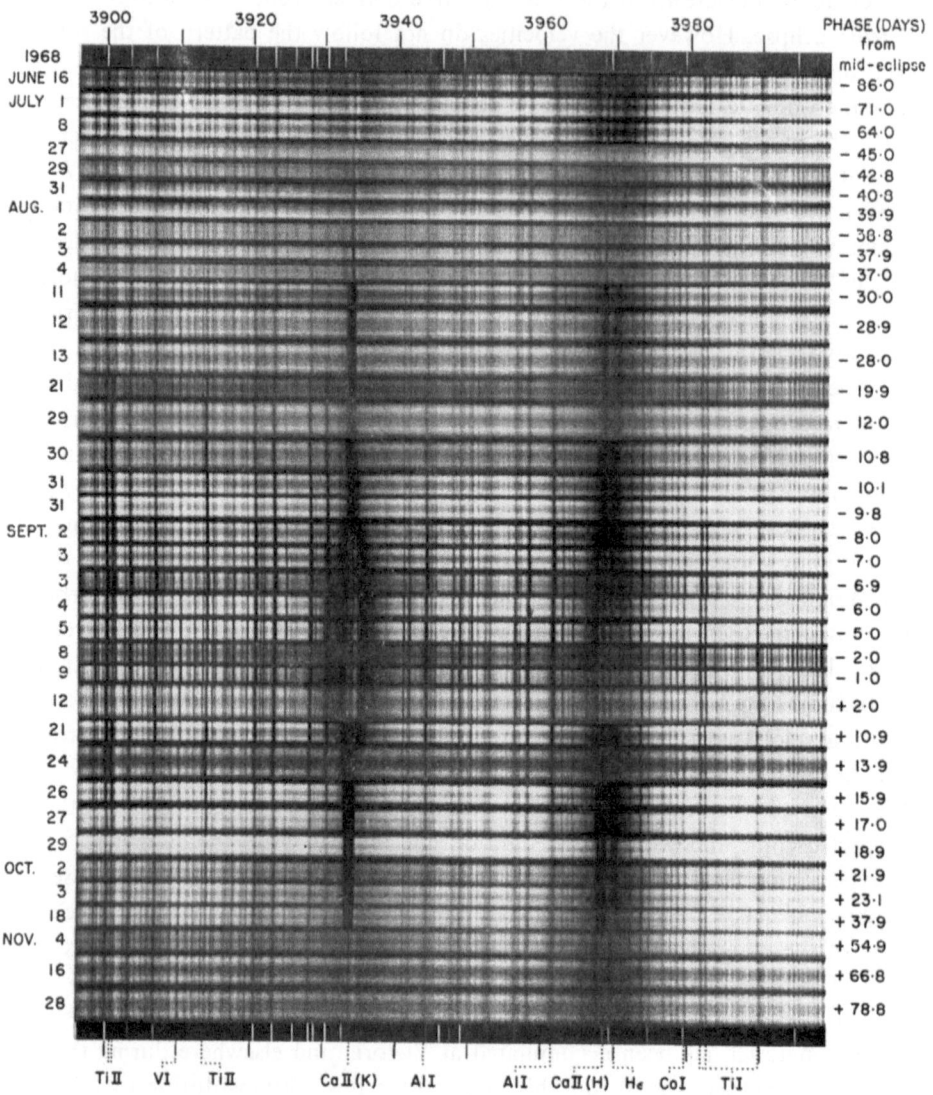

Fig. 1.   Victoria Observations of 32 Cyg at the 1968 Eclipse.
These high-dispersion (2 to 6 Å mm⁻¹) spectra cover the period when changes in the Ca II K line
were observed. Inner chromospheric effects on the lines of Fe I, Ti II, Al I, H, etc. can be observed
from phases about $\pm 25^d$ Mid-eclipse is assumed to be J. D. 2, 440, 109.8 based on photometric
observations up to 1971.

although it does not go into eclipse again until 1976, the chromosphere is so extensive
that some effects are expected to appear in a year or two. The velocity curve now
seems fairly well defined (Hutchings and Wright, 1971) from which a mass of 20 $M_\odot$
for each star has been obtained. The system ε Aurigae has frequently been considered
as one of the ζ Aurigae stars (Wright, 1970). The spectrum certainly shows chromo-

spheric effects at the time of eclipse, but the dimensions derived from the spectroscopic and photometric data are quite discrepant, and a completely satisfactory model for the system has not yet been found, though many have been suggested.

A number of other systems similar to VV Cephei have been described by Cowley (1969); for a long time, VV Cephei was considered to be 'one of a kind', and its spectrum was considered so complex that it was thought that it would be almost impossible to analyze. The most similar systems seem to be WY Geminorum and Boss 1985, which are being studied by Cowley; it is not yet certain that either system undergoes a total eclipse. The principal feature of these spectra is the 'reverse P Cygni' emission indicating a negative velocity relative to the absorption lines, which has been interpreted by Faraggiana and Hack (1969) in terms of a hydrogen envelope around the hot star which is rapidly rotating and also contracting. However other emission lines, chiefly of ionized metals, from both permitted and forbidden transitions, appear distinctly and with approximately the same velocity, in the ultraviolet region of the spectrum, and are the lines that frequently appear in the spectrum of M-type supergiants. Reverse P Cygni profiles are unusual in stellar spectra, but have been noted by Conti (1972) for the He II $\lambda 4686$ line in the spectrum of θ^1 Orionis C. They have also been observed by Walker (1972) in YY Orionis stars which are still in the gravitationally contracting phase, and therefore very young.

Many of the symbiotic stars are similar to the VV Cephei stars. For instance, CH Cygni showed remarkable changes in the past few years, having had outbursts in 1964 and 1967. The forbidden Fe II emission lines seemed to be formed under somewhat similar conditions to those in VV Cephei – but probably in a region far beyond the extended atmosphere that we are discussing (Faraggiana and Hack, 1969) Most observers seem to consider the symbiotic stars as binaries and, in particular, Boyarchuk (1969) has shown that the simplest explanation for these stars is a combination of a giant M-type star with a hot companion, both of which are embedded in a nebula with the hot star providing the excitation for the high-excitation lines. However, according to Boyarchuk's diagram, the Fe II emission lines may be found in the outer atmosphere of the M star.

The presence of metallic emission lines in the ultraviolet region of the spectra of luminous late-type stars was noted by Herzberg (1948) for α Herculis and α Scorpii. For many years they have been known to appear in the spectra of long-period variables (Merrill, 1960), chiefly as singly-ionized lines, both permitted and forbidden. These are the lines that appear in the symbiotic stars (Sahade, 1960) though in the spectra of some stars such as CI Cygni and AG Pegasi, lines with excitation as high as those in the solar corona have been observed. As Herzberg (1948) observed, in late-type stars, the corona would be expected to have a much lower temperature than that of the sun so that Fe II lines rather than those of Fe X might become prominent. As noted above, many of the investigations of the ultraviolet absorption and emission lines have been undertaken for the VV Cephei stars (Cowley, 1969). Struve (1944) listed many of the lines in the spectrum of VV Cephei, and Swings (1969) has tabulated the lines identified in the spectrum of Boss 1985 in the region 3144 Å–3626 Å.

A more detailed study of VV Cephei covering most of the cycle is being carried out by Wright (unpublished). The emission lines may, in some cases such as α Herculis (Deutsch, 1956), be produced around and between the stars, but the spectra of α Scorpii A and α Orionis seem to indicate that they arise in the outer atmosphere of the M star. Observations of these lines in the spectra of other M-type stars have been reported by Bidelman and Pyper (1963).

4. Variable Stars

Under our topic of extended envelopes we should probably include most of the other variable stars since, except for the eclipsing stars, variations in light and spectrum are related to changes in the atmospheric parameters. We usually consider that any eruptions, major or minor, must pass through the outer, extended atmospheres which themselves are involved, even though the ongoing radiation may arise from inter-action with a ring, disk, shell, envelope, circumstellar envelope or dissipation into the interstellar medium. Nearly all types of variable stars have members that are com-ponents of binary or multiple systems. In some cases, the possibility that the vari-ability may be directly related to the duplicity is becoming accepted. We shall mention a few types of variables that come under the topic of extended atmospheres; the time spent on further discussion will depend on the interests of participants. Perhaps I should note here that while most variable stars have been discovered from their ob-served light variations, detailed spectroscopic analysis frequently discloses changes in their spectra. In most cases variable stars are rather faint. High-dispersion studies will undoubtedly prove most fruitful as the necessary telescopes and instrumentation become available.

Early-type stars, including Wolf-Rayet stars, shell stars, etc. have been discussed earlier in this review. Contact binaries have been discussed in previous sessions – though I should like to mention Larsson-Leader's paper (1971) given at the Bamberg Colloquium last year in which he discusses the effects on a double-star system of gravitational fields, radiation flux and gas streams. I have barely touched on these effects which affect light curves as well as spectra, and which are now being studied very effectively, especially by Hill and Hutchings (1970, 1971). Fitch (1969) has con-sidered gravitational and tidal effects on a pulsating star and related the phenomena to β Cephei, δ Scuti and RR Lyrae stars.

The Cepheid variables certainly have extended atmospheres – and some 15 per cent of them are in binary systems (Lloyd Evans, 1968). Extensive theoretical investigations of shock-wave phenomena related to the RR Lyrae stars and the Cepheids have been made by Christy (1967) and the observational results have been summarized by Kraft (1967). Intensive investigations of southern Cepheids have been made by Bell and Rodgers (1964, 1967, 1968). It has recently been suggested (Kraft, 1967) that at minimum radius the atmosphere corresponds to an F-type supergiant and that as the shock wave moves outward, the characteristics shift to those of later type.

The long-period variable stars undoubtedly have extensive atmospheres and, among others, Odgers and Kushwaha (1960) have proposed that a weak shock wave travelling through the atmosphere could produce the radial-velocities observed for R Hydrae over its light cycle; for this star there does not seem to be any mass loss, but the ejected material falls back on the star. Maehara (1971) has recently studied the velocities of the two components of the absorption lines in the spectrum of χ Cygni, observed particularly in the infrared. The long wavelength components seem to be formed in the outer, low-temperature region in front of the shock wave; the short wavelength components are formed in a higher-temperature region immediately behind the shock front which moves outward; the emission lines in the violet region of the spectrum are also formed behind the shock. Many of these lines were explained by Thackeray (1937) in terms of fluorescent, selective-excitation processes involving lines situated very close to strong MgII, FeII and SiI lines between 2000 Å and 3000 Å; emission lines such as FeI, $\lambda\lambda4202$ and 4308 were accounted for in this way. Willson (1972) has taken these results and derived possible electron densities and temperatures for these stars. As Thackeray (1937) noted, the structure of the hydrogen emission lines shows that they arise from fairly low levels in the atmosphere, below the regions forming the TiO, FeI and CaII absorption lines. Until the advent of large telescopes it was almost impossible to observe long-period variable stars at minimum light with sufficient dispersion to identify the many lines. However Merrill (1953) measured Palomar plates of χ Cygni, taken by Bowen and by Wilson very near minimum, and found some 60 emission lines most of which could not be identified until Herbig (1956) showed that they are selected lines of the AlH molecule arising from levels near the rotational predissociation cutoff, as observed at low pressures in the laboratory. These atmospheres remain a very fruitful field for further study.

The RV Tau stars have been studied by Preston (1962, 1964) following observations by Abt (1953) that the hydrogen lines appear in emission and double lines also appear near maximum light. Preston showed that the violet absorption components are formed in hotter regions than are the red components. He thought that an emitting layer lies over the layer that produces the violet absorption lines and that the hydrogen emission layer is below that which produces the red absorption lines, but all are produced in an extended atmosphere. In the spectrum of U Monocerotis emission lines of TiI ($\lambda6554$) appear at rising light as does HeI ($\lambda5876$) and Hα, though the latter appears in emission at all phases.

Eruptive variables have been studied extensively by Smak (1971) who reviewed the field and listed previous surveys. He discussed novae, U Geminorum-type stars and nova-like objects and considers that the principal feature of all such variables is a gaseous disc rotating about the primary component. As these stars have been discussed earlier, and as they may not have extended envelopes such as we are considering here, we shall not develop this subject further.

Flare stars should be mentioned since they seem to be related to solar flares from considerations of their lifetimes as well as their optical and radio emissions. Recent surveys have been given by Haro (1968) and by Gershberg (1969). Most of these stars are UV Ceti stars, which are Me dwarfs. Of 25 known stars in this group, 19 are known to be binaries, though 14 of these are visual binaries. The UV Ceti stars seem to have large convective zones and therefore considerable magnetohydrodynamic activity, which produces the solar flares, might be expected for these stars. As Gershberg remarks, Poveda showed that convection must be strong in low-luminosity stars up to class K1, and Haro found flare activity in precisely the same classes. At the most recent Bamberg Colloquium Ambartzumian and Mirzoyan (1971) gave a further summary of recent studies of flare stars. They are considered to be very young stars, the T Tauri and UV Ceti groups being the youngest. All flare stars seem to be late-type dwarfs and a great many are observed in associations and clusters – which led Ambartzumian and Mirzoyan to speculate, from observations of the Pleiades, that nearly all dwarf stars pass through the flare stage in the course of their evolution. The youngest stars are G- or K-types and evolve into M-type stars as shown by their relative frequency in young, and older, clusters. The presence of ionized calcium and hydrogen emission lines in their spectra can be taken as indicators of extreme chromospheres in these stars.

In summary, let us remind ourselves of conditions to be expected for observations of extended atmospheres. Immediately we talk about extended atmospheres, we imply that they are different from those of 'normal' stars. For about twenty years it was thought that spectra of normal stars might be explained and computed by model atmosphere techniques assuming local thermodynamic equilibrium and the Saha and Boltzmann equations relating temperature, pressure and ionization equilibria. The theoreticians now feel that non-LTE may be important for at least some lines even in normal stellar atmospheres. However for the Sun and similar stars, such effects are relatively small. Therefore we feel justified in assuming that gross deviations from the simple first approximations are indicators of extended atmospheres.

   Observational criteria that have been used to detect extended atmospheres include: (i) geometrical effects: emission lines in stars of all types; (ii) velocity fields: broad, steep-sided absorption lines that can be recognized as having shapes different from the Stark-broadened wings of the hydrogen lines, the bell-shaped profiles of rotating stars, and the Voigt damping profiles of saturated lines; these are indicators of macro-turbulence. Microturbulence can be detected from the gradient effect, first observed by Struve and Elvey (1934), where the weak lines of a multiplet are strengthened relative to the strong lines by curve-of-growth effects; (iii) dilution effects, which result from an accumulation of atoms in metastable levels when the pressure is low.

   The evidence for extended atmospheres from observations of stellar spectra is overwhelming, though most of it has been accumulated over less than thirty years as a result of intensive investigations using the best equipment available. The evidence for differences between extended atmospheres of normal stars, and those of stars belonging to multiple systems is very small. In most of the categories that have been discussed here, there seem to be both single and multiple stars and there is very little evidence that there are major differences in the *chromospheres* resulting from the presence of another star. Even in the ζ Aurigae stars, where the effects can be analyzed best, the theory has not been developed sufficiently to decide about the effect of the hot B star on the atmosphere of its companion, which is less than 10 AU away. There is evidence from the radial velocities that the chromosphere is not moving with a uniform rotation, such as might be expected for stars in binary systems, but the velocities do not suggest that material is moving systematically towards the secondary star. The temperature does seem to increase outwards for these stars, but there is a similar increase in the outer layers of the Sun. Thus data obtained from the chromo-spheres of components of double stars may give important information about the atmospheres of normal stars – and vice versa. Giant components of the ζ Aur stars seem to be the most suitable stars for such studies, as they are less affected by the secondary components and do not seem to have additional envelopes or shells.

It would appear that further investigations to compare the chromospheres of single stars with those in multiple systems would be a worthwhile and challenging task. Perhaps some of the participants at this Symposium have some of the answers.

References

Abt, H. A. : 1953, *Astror. J.* **58**, 210.
Adams, W. S. and Russell, H. N.: 1928, *Astrophys. J.* **68**, 9.
Ambartzumian, V. A. and Mirzoyan, L. V.: 1971, *Veröffentl. Remeis-Sternw., Bamberg* **9**, 98.
Bell, R. A. and Rodgers, A. W.: 1964, *Monthly Notices Roy. Astron. Soc.* **128**, 365.
Bell, R. A. and Rodgers, A. W.: 1967, *Monthly Notices Roy. Astron. Soc.* **138**, 23.
Bell, R. A. and Rodgers, A. W.: 1968, *Monthly Notices Roy. Astron. Soc.* **139**, 175.
Bidelman, W. P. and Pyper, D.: 1963, *Publ. Astron. Soc. Pacific* **75**, 389.
Böhm-Vitense, E.: 1956, *Publ. Astron. Soc. Pacific* **68**, 57.
Boyarchuk, A. A.: 1969, in L. Detre (ed.), *Non-Periodic Phenomena in Variable Stars*, D. Reidel, Dordrecht, p. 395.
Burbidge, G. R. and Burbidge, E. M.: 1953, *Astrophys. J.* **118**, 252.
Chentsov, E. L. and Snezhko, L. I.: 1972, in M. Hack (ed), *Trieste Colloquium on Supergiant Stars*, p. 51.
Christy, R. F.: 1967, in R. N. Thomas (ed.), 'Aerodynamic Phenomena in Stellar Atmospheres', *IAU Symp.* **28**, Academic Press, London and New York, p. 105.
Conti, P.: 1972, *Astrophys. J. Letters* **174**, L 79.
Cowley, A. P.: 1969, *Publ. Astron. Soc. Pacific* **81**, 297.
Deutsch, A. J.: 1956, *Astrophys. J.* **123**, 210.
Deutsch, A. J.: 1970, in D. C. Morton (ed.), 'Ultraviolet Spectra and Ground-based Observations', *IAU Symp.* **36**, D. Reidel, Dordrecht, p. 317.
Doherty, L. R.: 1971, *Phil. Trans. Roy. Soc. London A* **270**, 189.
Faraggiana, R. and Hack, M.: 1969, 'Les Transitions Interdites dans les Spectres des Astres', *Mem. Soc. Roy. Sci. Liège, 5ème Série* **17**, 317.
Feast, M. W.: 1970, in D. C. Morton (ed.), 'Ultraviolet Spectra and Ground-based Observations', *IAU Symp.* **36**, D. Reidel, Dordrecht, p. 187.
Fitch, W. S.: 1969, in L. Detre (ed.), *Non-Periodic Phenomena in Variable Stars*, D. Reidel, Dordrecht, p. 287.
Gershberg, R. E.: 1969, in L. Detre (ed.), *Non-Periodic Phenomena in Variable Stars*, D. Reidel, Dordrecht, p. 111.
Ghobros, R. A.: 1962, *Z. Astrophys.* **56**, 113.
Goedicke, V. A.: 1938, *Publ. Univ. Michigan Obs.* **8**, 1.
Griffin, R. F.: 1963, *Observatory* **83**, 255.
Groth, H. R.: 1957, *Z. Astrophys.* **43**, 185.
Groth, H. R.: 1955, *Z. Astrophys.* **37**, 261.
Groth, H. R.: 1961, *Z. Astrophys.* **51**, 206, 231.

Groth, H. R.: 1970, in H. G. Groth, and P. Wellmann (eds.), *Spectrum Formation in Stars with Extended Atmospheres*, NBS Spec. Publ. **332**, p. 259.

Haro, G.: 1968, in *Stars and Stellar Systems* **5**, 141.

Herbig, G. H.: 1956, *Publ. Astron. Soc. Pacific* **68**, 204.

Herzberg, G.: 1948, *Astrophys. J.* **107**, 94.

Hill, G. and Hutchings, J. B.: 1970, *Astrophys. J.* **162**, 265.

Hutchings, J. B.: 1970, *Monthly Notices Roy. Astron. Soc.* **147**, 161.

Hutchings, J. B.: 1970, in A. Slettebak (ed.), *Stellar Rotation*, D. Reidel, Dordrecht, p. 283.

Hutchings, J. B. and Hill, G.: 1971, *Astrophys. J.* **166**, 373.

Hutchings, J. B. and Wright, K. O.: 1971, *Monthly Notices Roy. Astron. Soc.* **155**, 203.

Kawabata, S.: 1957, *Publ. Astron. Soc. Japan* **9**, 72.

Kitamura, M.: 1967, *Publ. Astron. Soc. Japan* **19**, 194.

Kondo, Y., Giuli, R. T., Modisette, J. L., and Rydgren, A. E.: 1972, *Astrophys. J.* **176**, 153.

Kraft, R. P.: 1960, in *Stars and Stellar Systems* **6**, 370.

Kraft, R. P.: 1967, *Astrophys. J.* **150**, 551.

Kraft, R. P.: 1967, in R. N. Thomas (ed.), 'Aerodynamic Phenomena in Stellar Atmospheres', *IAU Symp.* **28**, Academic Press, London and New York, p. 207.

Larsson-Leander, G.: 1957, *Stockholm Obs. Ann.* **19**, No. 8.

Larsson-Leander, G.: 1971, *Veröffentl. Remeis-Sternw. Bamberg* **9**, 185.

Liller, W.: 1968, *Astrophys. J.* **151**, 589.

Linsky, J. L.: 1968, *Smithsonian Inst. Spec. Report*, No. 274.

Linsky, J. L.: 1973, in S. D. Jordan and E. H. Avrett (eds.), *Stellar Chromospheres*, NASA SP-317, p. 48.

Lloyd Evans, T.: 1968, *Monthly Notices Roy. Astron. Soc.* **141**, 109.

McKellar, A., Odgers, G. J., Aller, L. H., and McLaughlin, D. B.: 1952, *Nature* **169**, 990.

McKellar, A., Aller, L. H., Odgers, G. J., and Richardson, E. H.: 1959, *Publ. Dominion Astrophys. Obs.* **11**, 35.

Maehara, H.: 1971, *Publ. Astron. Soc. Japan* **23**, 503.

Magnan, C.: 1965, *Ann. Astrophys.* **28**, 512.

Menzel, D. H.: 1931, *Publ. Lick Obs.* **17**, 1.

Merrill, P. W.: 1953, *Astrophys. J.* **118**, 453.

Merrill, P. W.: 1960, in *Stars and Stellar Systems* **6**, 509.

Merrill, P. W. and Sanford, R. F.: 1944, *Astrophys. J.* **100**, 14.

Merrill, P. W. and Wilson, O. C.: 1956, *Astrophys. J.* **123**, 392.

Odgers, G. J. and Kushwaha, R. S.: 1960, *Publ. Dominion Astrophys. Obs.* **11**, 253.

Pannekoek, A. and Minnaert, M. J. G.: 1927, *Proc. kungl. Akad. Weten. Amsterdam* **30**, 921.

Peery, B. F.: 1966, *Astrophys. J.* **144**, 672.

Plavec, M.: 1968, *Adv. Astron. Astrophys.* **6**, 201.

Praderie, F.: 1970, in H. G. Groth and P. Wellmann (eds.), *Spectrum Formation in Stars with Steady-State Extended Atmospheres*, NBS Spec. Publ. **332**, p. 241.

Preston, G. W.: 1962, *Astrophys. J.* **136**, 866.

Preston, G. W.: 1964, *Astrophys. J.* **140**, 173.

Rosendahl, J. D.: 1970, *Astrophys. J.* **159**, 107.

Sahade, J.: 1960, in *Stars and Stellar Systems* **6**, 466.

Saito, M.: 1970, *Publ. Astron. Soc. Japan* **22**, 455.

Smolinski, J.: 1972, in M. Hack (ed.), *Trieste Colloquium on Supergiant Stars*, p. 68.

Smak, J.: 1971, *Veröffentl. Remeis-Sternw. Bamberg* **9**, 248.

Spitzer, L.: 1939, *Astrophys. J.* **90**, 494.

Swings, J. P.: 1969, *Astrophys. J.* **155**, 515.

Struve, O.: 1944, *Astrophys. J.* **99**, 170.

Struve, O. and Elvey, C. T.: 1934, *Astrophys. J.* **79**, 409.

Struve, O. and Roach, F.: 1939, *Astrophys. J.* **90**, 727.

Thackeray, A. D.: 1937, *Astrophys. J.* **86**, 499.

Thomas, R. N.: 1970, in H. G. Groth and P. Wellmann (eds.), *Spectrum Formation in Stars with Extended Atmospheres*, NBS Spec. Publ. **332**, p. 259.

Thomas, R. N. and Athay, R. G.: 1961, *Physics of the Solar Chromosphere*, Interscience Publ. Co., New York.

Thomas, R. N. and Gebbie, K. B.: 1971, *Menzel Symposium*, NBS Spec. Publ. **353**, p. 84.
Walker, M. F.: 1972, *Astrophys. J.* **175**, 89.
Willson, L. A.: 1972, *Astron. Astrophys.* **17**, 355.
Wilson, O. C.: 1960, in *Stars and Stellar Systems* **6**, 436.
Wilson, O. C.: 1966, *Science* **151**, 1487.
Wilson, O. C. and Abt, H. A.: 1954, *Astrophys. J. Suppl. Ser.* **1**, 1.
Wilson, O. C. and Aly, M. K.: 1956, *Publ. Astron. Soc. Pacific* **68**, 149.
Wilson, O. C. and Bappu, M. K. V.: 1957, *Astrophys. J.* **125**, 661.
Wright, K. O.: 1959, *Publ. Dominion Astrophys. Obs.* **11**, 77.
Wright, K. O.: 1970, in A. Beer (ed.), *Vistas in Astronomy* **12**, Pergamon Press, Oxford, p. 147.
Wright, K. O. and Odgers, G. J.: 1962 *J. Roy. Astron. Soc. Can.* **56**, 149.

FOURTH DISCUSSION SESSION

(Friday morning; 8 September, 1972)

(following the review paper by K. O. Wright)

Chairman: O. C. WILSON

O. C. Wilson: We thank Dr. Wright for his well-presented review, and invite ques·
tions and discussion of his paper.

Bolton: I have some comments and a question. First, in regard to these reverse
P Cyg profiles in the spectrum of θ Ori C, I think the situation there is not quite so
simple as Conti seemed to believe in his publication. Walborn at David Dunlap
Observatory has some lower-dispersion spectra that show a 'regular' P Cyg profile.
Therefore, we cannot be dealing with simple infall of matter and the relationship to
YY Ori stars is tenuous at best. Second, it has been my impression from reading the
literature recently, particularly the works by Mihalas and collaborators, that the
Bowen fluoresence mechanism does not give an adequate explanation of the Of-star
emission but that these phenomena find a natural explanation in the non-LTE reference frame. In connection with this, Mihalas* has shown that some of the emission
features are luminosity sensitive. Both Walborn (1971) and Conti and Alschuler
(1971) have recently given luminosity classifications for the O-type stars, and they
both define a class of supergiants. In view of Dr. Wright's comments on the O-type
supergiants in his review, I would like to hear his and Dr. Underhill's opinion of
this classification.

Underhill: Regarding reverse P Cyg profiles, they have emission to the violet of
the absorption. They are not very common, but neither are they very rare. For instance, I've obtained spectrograms showing Hα for a good many B-type supergiants
from time to time. One time the emission is a bit stronger on the red side, another
time on the violet side of whatever absorption there is. Generally, the lines are pretty
well filled in and you really don't see anything – just a little bump if you make intensity
tracings. The simple, uniform model is either an expanding or a contracting atmosphere, but, in fact, the atmosphere changes and the whole Hα line is neither absorption
nor emission, but just slight undulation up or down on one side of the nominal profile.
We have a more complex situation than a simple uniform expanding sphere, although
what Conti says is right as a very first order of approximation.

About the supergiant early O-type stars, how do you define a supergiant? It's quite
easy for late-type stars for which there is a large difference of magnitude between stars
that are very bright and have a characteristic type of spectrum, called supergiants,
and those that are on the main sequence. There are the stars called giants too: they
are all separate. But once you consider the O-type stars, those on the main sequence

---

* See Hummer's discussion on p. 277.

Batten (ed.), Extended Atmospheres and Circumstellar Matter in Spectroscopic Binary Systems, 134–147.
All Rights Reserved. Copyright © 1973 by the IAU.

have M_v about -4, maybe as bright as -5, while supergiants, no matter what type they are, are usually of M_v only about -6 or -7. The uncertainty in determining the absolute magnitude is so great that it is very hard to say, at type O5, that you have isolated different absolute magnitude groups. I think the terms 'supergiant' and 'main-sequence' are hardly relevant considering the uncertainties in the distances of these early O-type stars. Some of these stars show evidence that their atmospheres are expanding a little more than those of others, but I would hesitate to use the term 'supergiant' because it has so many other connotations for the later-type stars. In any case, the atmospheres are so expanded that you have to consider non-L.T.E. physics – real physics. You cannot argue from the equations of thermodynamic equilibrium.

O. C. Wilson: I gather that, in a sense, what you're saying is that the spectroscopic criteria are just not sharp enough?

Bolton: I don't want to let that comment go by without challenge. The spectroscopic criteria are certainly quite sharp. I have seen the spectra, and even I would have no trouble classifying the stars. I think that there can be no doubt that the spectroscopic classes of Walborn and Conti are real. Furthermore, interpretation of a luminosity effect does not require that one invoke LTE, or non-LTE, or magic. One must simply get distances or sometimes only relative distances. I don't want to minimize the difficulties involved in this, particularly the difficulties in deriving absolute magnitudes, but one frequently finds O-type stars of two or more classes in the same cluster. Then the photometry gives the luminosity differences directly and easily. It is true that spectroscopic binaries can complicate the interpretation, but they cannot account for more than $0^m.75$, and a range of at least three magnitudes is observed.

Underhill: You can do this. I don't quibble with that, but to attach a physical meaning to the class is the difficult step.

Thackeray: As to the absolute magnitudes of the O-type stars, may I suggest that a little more work on the Magellanic Clouds will give us a pretty safe answer whether the luminosity of an O5 star can reach anything like that of a later-type supergiant. Our work in Pretoria, scratching the surface, didn't show up any O-type star brighter than $m_v = 12$ or 12.5, whereas A-type stars are known with $m_v = 9.5$ or 9.2. We are doing more work in Pretoria, but I think it's very unlikely that we shall turn up an O5 star brighter than about $12^m.2$.

Popper: I'd like to ask a question about the so-called reverse P Cyg profiles, such as are found in the spectrum of θ Ori C. What are the actual velocities? Is the emission-line or the absorption-line velocity that of the star? To interpret the profiles, you must have some picture of what the velocities are, and, if there is a change, what it is that is changing.

Wright: Conti (1972) found that the absorption was redward displaced, with some variation, and the emission was violet-displaced.

Popper: It seems to me that we are not seeing infall of matter, and then outflow, but that we are seeing an expanded shell with a changing distribution of matter in it. Sometimes the part behind the star is stronger.

Underhill: This is the impression one gets from Hα in the spectra of B-type supergiants.

Wright: I was really trying to emphasize, in my review, that the reverse P Cyg profiles have not been discussed much in the literature until very recently. The hydrogen lines in the spectrum of VV Cep can be at least partly explained as the sum of red and violet components of emission; that is, when I was determining the secondary spectrum, I had to take into account both the violet and the red sides to determine the mean profile. There is also a component at about -60 km s^{-1}, which is always present and which I think is in the system itself. But I don't think we need worry very much about infall for most of these stars. I think it's a natural phenomenon that may well be at least partly explained, as Dr. Underhill suggested, by random motions in their atmospheres.

Bolton: I am not convinced that early-type supergiants appear only in binary systems of the VV Cep type. I have seen classification spectra of B-type supergiants in which the lines are clearly doubled.

Underhill: That is an extremely rare observation.

Bolton: I'm not sure I agree. I know of several examples, and I don't see any theoretical reason why binary systems containing two early-type supergiants should not exist. The surface gravities indicate that an early B-type or O-type supergiant has not expanded by a factor of more than three or four. If they start out in reasonably detached systems with periods of a few days (say five or more) they would not begin to interact until one of the stars became of spectral type B5 or thereabouts. Furthermore, the range in absolute magnitudes from the main-sequence to supergiants is small enough, for early-type stars, so that the probability of observing a two-spectra supergiant system is much higher than it is for later types.

Huang: Dr. Underhill raised a very interesting and fundamental problem just a moment ago – about the high turbulent velocities derived from curves of growth. In a recent paper in Nature (Worrall and Wilson, 1972) it was argued that these turbulent velocities do not exist. Anyway, they are very strange, because they show that within an optical thickness of unity the velocity reaches a high supersonic value. Perhaps these high values for the turbulent velocity are the result of a lot of shocks in the envelope. I used the illustration in my review paper of people smoking in a room; jet streams come up to collide with the envelope, and I identify this envelope with the extended atmosphere. Now, when each jet stream interacts with the medium, it produces a shock. So there are a lot of shocks around to give the large curve-of-growth velocity. Do you think this is possible?

Underhill: It's a very nice idea to try to account for the broadened line profiles by the velocity fields created by shocks. But the problem of microturbulence is fundamental to the theory of curve of growth. I think the real point is that it uses the wrong theory for line formation. One obtains a curve of growth by assuming a certain relationship, between the absorption coefficient and the equivalent width of a line, that is based on some theory – elementary or otherwise. If you use pure absorption so-called theory, you obtain one curve of growth, while scattering theory gives you another. For the same number of atoms, scattering theory gives a higher flat portion of the curve of growth than does absorption theory – yet you started off with the same model and just used a different approximation to radiative transfer. Because the central part of the curve of growth is higher than expected from absorption theory, you deduce microturbulence. Actually, you used the wrong transfer theory to describe the formation of absorption lines. For these extended atmospheres, very simple physical arguments show that a scattering theory is much more relevant to the understanding of transfer than is an absorption one, and I think that practically all microturbulence results are figured out with reference to the absorption-type curve of growth. All this has been pointed out by Menzel (1939).

O. C. Wilson: We are talking here, I think, about things like ζ Aur and 32 Cyg which have always seemed to me the interstellar case of a star, an apparently small object, shining through some material. In that case it seems to me that the simple absorption theory of transfer is perfectly adequate.

Underhill: You can't argue that way. You're assuming that the populations of the absorption levels have a Boltzmann distribution. Actually, if you consider the radiative field, you may find that in these stars the atoms are preferentially packed down in the lower levels. If you assume a Boltzmann distribution to derive an effective number of atoms from the equivalent width, you just get problems, and I think the net result in a straightforward curve-of-growth measurement is the deduction of a large microturbulent velocity. This is just a fictitious quantity, and there are so many ways that one can account for it being positive. If one found a negative microturbulence, one would really worry – yet there's no obvious reason why one shouldn't find negative shifts. The approach to a physical understanding is complex, and I would be the first to say that what I am saying is far too simple.

O. C. Wilson: Couldn't it be that these displaced components of the K-line that Ken Wright showed us arise from the large elements in a hierarchy of all possible sizes of cells? The little fellows, all added up together, give us what we think is microturbulence, because they affect the line-absorption coefficients. Big elements often have nothing to do with the line-absorption coefficient: you see them individually. If you have a hierarchy of all possible sizes, presumably many more small ones than big ones, then it seems to me that the problem reduces essentially to the simplest case.

Underhill: I don't think we can answer this definitely one way or the other. My studies of physics incline me to reject entirely the hypothesis of microturbulence and to consider the non-LTE situation.

O. C. Wilson: That is a noble endeavour, but it's likely to be a difficult one. Do we have any further comments?

Batten: Going back to Dr. Huang's comment, I would like to ask whether shock waves between freshly ejected matter and the surrounding medium could produce discontinuous velocities in the circumstellar matter that would be observable as the steps in velocity curves, to which I referred in my review.

Huang: There will be no discontinuity of any kind if you have a lot of shocks around. They are smoothed out. That is why I propose a large number of shocks in the medium.

R. E. Wilson: I was curious about the values given for the individual masses of ε Aur, since this is single-spectrum binary.

Wright: They were based on some calculations that Stephen Morris made a few years ago. He estimated the absolute luminosity of the primary component in several ways, and then assigned a mass to it on the basis of evolutionary tracks (which run almost at constant luminosity for so massive a star). From the mass function it was then possible to compute pairs of values for the mass ratio and orbital inclination, and to derive from these values of the radii of the two components and hence (from the spectral type) of the absolute luminosity of the primary star. The values of the masses quoted are those that give an estimate for the last-named quantity consistent with the initially assumed value. Neither Morris nor I claim that the derived masses are reliable to all the significant figures quoted.

Fracastoro: It appears that the atmospheres of 31 Cyg and ζ Aur are not uniformly distributed even during the same eclipse – I'm thinking of the larger-scale clouds. I would also like to emphasize that it appears in general that the advancing hemisphere of the expanded atmosphere is more developed than the receding one. That's one thing that probably Otto Struve was the first to point out.

Wright: I'm not quite sure whether that is true or not. By 'more developed', do you mean 'more extended'? Then you would expect that there would be more elements in the advancing side and therefore we would observe more random velocities. Is that what you mean?

Fracastoro: No. I think that the ingress phase lasts longer than the egress.

Wright: Oh, I see. I think the data that I had on ζ Aur and that I published in my Vistas article (Wright, 1970) show the ingress and egress phases to be fairly symmetrical. I'm thinking of 31 Cyg and 32 Cyg, as well as ζ Aur, the last-named certainly varies. There is no question the 1937 eclipse indicated a greater extension than did any of the others. For 32 Cyg, however, if anything, the extension was a little bit larger at egress than at ingress. You would have to look at the data; there is scatter from one eclipse to the next, and the result is not too clear-cut.

Fracastoro: You showed a diagram of the intensity of the chromospheric K-line, in the spectrum of ζ Aur, as a function of time. All the curves seem to be higher on the right-hand (egress) side than on the left.

Wright: For ζ Aur, the Cambridge observation of the ingress of the 1937 eclipse showed greater intensities than have been observed at any other eclipse. In general, the intensities of the chromospheric lines in the spectrum of ζ Aur seem to be a little greater during ingress than during egress, for equal times from mid-eclipse. The reverse seems to be true for 32 Cyg, but the difference is not great.

Herczeg: I always thought this was explicable mainly in terms of the orbital eccentricity, the velocity of the tangential motion being higher before than after the eclipse. I would also like to ask Dr. Wright about the chromospheric lines seen in the spectrum of ε Aur near its eclipse. You mentioned these in your review, can you tell us more? Because of our utter ignorance of the nature of the secondary component, every bit of information may turn out to be quite important. My question is perhaps too naive. Can we find out in this way something about the nature of the object? Is it a star, or are the chromospheric absorption lines just relatively narrow additional absorption lines originating in a gas cloud?

Wright: The real problem about the chromospheric absorption lines in the spectrum of ε Aur, if I can answer this first, is that we don't really know what we are observing. They were easily observed in the 1955–1956 eclipse. I obtained numerous observations of them, and so did others (Struve and Pillans, 1957; Hack, 1957; Wright and Kushwaha, 1957). The chromospheric lines really looked quite similar to those in the spectrum of 31 Cyg, but, according to the radial-velocity curve the observed eclipse is the 'secondary' eclipse for ε Aur, whereas it is the primary eclipse for 31 Cyg and ζ Aur. We know so little about the system that theories of black holes, etc., are perhaps a little premature. The sharp lines are certainly present in the spectrum, but it seems probable that they are produced in some disk or other feature, and are not true chromospheric phenomena. Just what mechanism produces them, what the source behind the absorbing material is, we don't know yet – at least, not in my opinion. Concerning this eccentricity effect, it is appreciable, but I think it is not great enough to account for the effects observed in the spectrum of ζ Aur in 1937. However, the time of periastron is such that the intensities at a given phase could be greater during ingress for ζ Aur, and during egress for 32 Cyg – as I have just suggested.

Bolton: Have there been any investigations of the rapid variability of the spectra during these atmospheric eclipses? If there really are chromospheric phenomena such as you find in the Sun, where the life-times of spicules, and so on, are typically about a few minutes, then one might expect to find variations in the line profiles, and in the velocities, on a time scale of hours, in these more extended atmospheres.

Wright: Dr. Odgers at Victoria has for many years attempted to obtain a number of plates of these objects – particularly of 31 Cyg – during a night. There have been slight indications of changes during a night, but they have been quite small. Over a twenty-four-hour period, however, there is no doubt that real changes occur. We are not sure what happens within a few hours, and rapid-scanning techniques will be important in helping to find out. We tried this with the Isocon scanner of the University of British Columbia last October and there were just marginal effects when we were getting a scan every fifteen seconds or so, but even when we integrated several series of 44 traces over a couple of hours, the effects were still marginal. The resolution was not quite good enough.

R. E. Wilson: Are these chromospheric features that you describe in the spectrum of ε Aur the same ones that were described by Kuiper, Struve, and Strömgren (1937)? They described satellite lines that appeared around primary eclipse.

Wright: Yes, they are the same.

R. E. Wilson: What did you mean when you said the lines were seen around secondary eclipse? There is only one eclipse, so it has to be the primary.

Wright: For ε Aur, $u=90°$ at the time of eclipse. For the other stars, $u=270°$.

R. E. Wilson: Alright. These lines show a velocity reversal between the ingress and the egress of the eclipse, so there has to be either a rotating object or an orbiting object present. An orbiting object seems more likely, so it seems unusual to call the lines 'chromospheric'. They are absorption lines, of course, not emission lines?

Wright: They are absorption lines, and they are quite sharp. They are produced in the outer chromosphere because radiation from the B-type star passes through the chromosphere. As the eclipse progresses, the lines become stronger, and are produced in the inner chromosphere, or finally the photosphere – if you want to call it that. This happens before second, and after third contacts, so there should be some rotation effect but it is not large, and the random motions are greater.

R. E. Wilson: But for ε Aur, as described in 1937, there was a progressive velocity shift from the time the lines were first seen, until they disappeared. This is exactly what you would expect if you were seeing Keplerian motion of orbiting material. At egress the shift had the opposite sense from that seen at ingress. The lines were in absorption. So it does look as if the absorption was caused by orbiting material rather than by chromospheric phenomena.

Wright: The results for the 1955 eclipse of ε Aur have not been published in complete detail. The velocities of the 'chromospheric' lines were positive during ingress and negative during egress, as shown in Figure 11 of my *Vistas* article, (Wright, 1970). I am not sure whether it really is a rotation effect, and it may not be a chromospheric effect, since we do not know the nature of the object that causes the eclipses of ε Aur.

R. E. Wilson: It not only looks like a rotation effect, but like a progressively varying rotation effect, as if the line of sight were passing through material for which the Keplerian velocity is a function of distance from the centre of the object.

O. C. Wilson: If there are no further comments on the review paper, we will ask Lloyd Evans to speak to us on "Binaries with Giant Primaries Showing Strong Ca II Emission."

Lloyd Evans: This group of about ten known members shows the following general properties:

(1) Single-spectrum binary, spectral type G-K, II–IV.

(2) Orbital period usually ~ 20 d (17–~ 80 d).

(3) The Ca II emission is very strong and in some cases is found to consist of the normal, broad, double-peaked component with a superposed sharp component. The velocity derived from the sharp component follows the orbital radial velocity fairly closely, indicating that the emission region is close to but possibly not coincident with the primary.

(4) The primary is probably close to filling its Roche lobe.

(5) HD 209813 and λ And show semiregular light variability with a period not equal to the orbital period and much greater than the period for radial pulsation.

A new example is HD 158393, G5 III: $m_{pg}=9.3$. This was found to have strong Ca II emission and proved to be a single-lined spectroscopic binary with $P=30\overset{d}{.}9$. This star shows several interesting points:

(1) The absorption lines are very broad. If interpreted as rotational broadening, $v \sin i \sim 150$ km s^{-1}.

(2) Ca II emission is also broad. This may indicate it arises from a region sharing the supposed rotation of the photosphere.

(3) Ca II emission radial velocity follows the orbital velocity but with a slightly larger amplitude. This could arise, on the present model, from a non-uniform distribution over the rotating disk.

(4) There is a suspected 10-d periodicity in the residuals of radial velocities from the orbital solution, which exceed the measuring errors.

This work was done in collaboration with B. Emerson of H. M. Nautical Almanac Office.

Plavec: I think this is a very interesting group of stars, although I don't know exactly quite which stars you include in this group. From what you mentioned, I think you mean also, for example, λ And or ζ And, but these are single-line spectroscopic binaries. Therefore I would like to ask you what is the basis for your statement that the primary component fills the Roche lobe?

Lloyd Evans: This is based on rather rough ideas, discussed in Herbst's unpublished thesis. One simply obtains the orbit and makes some guess as to the size of the star and what mass one would need for the companion.

Plavec: But we don't observe any traces of the secondary spectrum?

Lloyd Evans: Certainly not in the one I've studied.

Plavec: Otherwise the system or the group might be related to relatively shorter-period eclipsing binaries of the type of AR Lac and other stars with pronounced calcium emission in their spectra. There is no doubt that in the spectra of λ And and ζ And, the emission lines of calcium are extremely strong.

O. C. Wilson: Sanford (1951) worked on AR Lac many years ago. There are two emission components at K and their velocity shifts indicate that they arise from something moving with the two stars. They look more or less normal, I thought. I also have a number of plates of λ And at 10 Å mm^{-1}. At even higher dispersions there is a very tiny central absorption, the width of which changes during the orbital period.

Biermann: This star HD 158393 seems very interesting because its rotational velocity seems very high for a late-type star. If you accept the figure at its face value, it would indicate that when the star was on the main sequence, its rotational velocity exceeded the break-up velocity, and therefore the angular momentum must have come from somewhere else. Can the profiles be explained by anything other than rotation?

Lloyd Evans: The star should certainly be studied at high dispersion, in particular to determine the line profiles. I am not prepared to say, from my plates at 49 Å mm^{-1}, that the profile shape is what you would expect from rotational broadening.

Cowley: As most of you probably know, there is a large project going on at Michigan to classify all of the southern HD stars, and the plates have all been taken with the Michigan Schmidt. In the course of this, we found several stars of this type, which perhaps should be looked at in more detail; perhaps some of them will turn out to be eclipsing or we may even find a second spectrum. If anyone is interested, I could send them a list of the names. Nothing is known other than that they are G or K giants that have strong H and K emission; too strong for normal stars.

Bolton: We have a number of spectrograms of various stars of this type at Toronto. They were obtained at 12 Å mm^{-1} by two graduate students – Herbst and Campbell. I've looked at these, and every one of the spectra I've seen shows very fuzzy broad lines which, interpreted as rotationally broadened lines, give velocities of the order of 100 km s^{-1}. They're absolutely the worst line profiles I've ever seen in a K-type spectrum.

Wright: How were these stars chosen?

Bolton: I'm not sure. I believe that they were picked from various lists of stars for which there is evidence of photometric and radial-velocity variations with different periods.

Andersen: Are you going to study HD 158393 at high dispersion, or is it free for anyone?

Lloyd Evans: It is unlikely that I shall study it.

Hall: As Dr. Plavec said, there is some similarity between these binaries and those like AR Lac and RS CVn. Both groups of binaries show very strong H- and K-emission lines in their spectra, and at least one component is of spectral type G or K. I have already proposed that the RS CVn binaries are in a pre-main-sequence phase of evolution. These other systems can also be understood in the same way. They contain relatively luminous stars (between luminosity classes III and II), whereas the components of a typical RS CVn system are less luminous (around class IV for the larger and between IV and V for the smaller). The components in RS CVn systems do not seem to be rotating rapidly, rather they seem to be rotating approximately in synchronism with the orbital motion. Perhaps these more luminous systems are still higher on the Hayashi track, and are rotating rapidly because of their youth (as some people say the T Tau stars are doing). As they get older, the rotation of the stars in these systems will also become approximately synchronized with orbital motion. How many of these stars are rotating rapidly? Are there some which are not doing so?

Lloyd Evans: It's news to me that the others in the group are rotating rapidly. I have material only for HD 158393.

Bolton: Those of which I have seen spectra have very poor line profiles, as though the stars are rotating rapidly. The best line profiles I've seen are in the spectrum of HD 209813, and even they are none too good.

Thackeray: Can Anne Cowley obtain rotational velocities from the Michigan material? Is anyone attempting a statistical study, such as Struve himself undertook in the 1930's to see if the distribution of orientations is random, as it should be if the profiles are rotationally broadened?

Cowley: Our dispersion is 110 Å mm^{-1}. We can say if the line profiles are fuzzy, but we can't really say anything about the rotational velocities. Somebody could take our list and observe with higher dispersion.

Lloyd Evans: I had wondered, actually, whether HD 158393 might be a pre-main-sequence star. It does lie in the region of the Sco-Cen association. When we have a better idea of the radial velocity from the computed orbit, it may be possible to say whether or not it is an association member.

O. C. Wilson: If these stars are going to be interpreted as pre-main-sequence objects, it seems rather strange to me if they all turn out to be single-line spectroscopic binaries. We need more statistics.

Cowley: And their distribution should certainly tell us something, too.

Lloyd Evans: Another point is that young stars of late type usually have strong lithium lines in their spectra. At least one of the other stars in this group has been observed by Bob Fosbury of the Royal Greenwich Observatory, and he finds no strong lithium line in its spectrum.

Oliver: Perhaps all the known stars in the group are binaries because the strong H and K emission is a consequence of the binary nature.

Catalano: Fernie, Hube, and Schmidt (1968) have pointed out that HD 209813 is a high-velocity star. I think it cannot be a pre-main-sequence star. It would be interesting to determine the masses of its components: it is one of few high-velocity stars we know of to be double.

Popper: I have looked without success for the lines of the secondary component.

Plavec: Maybe we are talking about a group of stars that actually have very little in common. The lines in the spectra of λ And and ζ And are certainly not broad nor indicative of high rotation: they are very nice and sharp. As for AR Lac, and other binaries of this group, a difficulty in their interpretation is that they do not fill the critical Roche lobe. So maybe the stars we are talking about belong to several groups, with only the strong emission at H and K in common.

O. C. Wilson: Which, however, may be stimulated by the presence of a companion.

Underhill: It would be very interesting to observe the MgII resonance lines in the spectra of these stars. Those lines should be extremely strong, and if we ever get a satellite up capable of observing that region of the spectrum, I think it would be interesting.

O. C. Wilson: Well, we're counting on you, Anne!

Hall: Dr. Catalano remarked that HD 209813 is a high-velocity star. A student of mine, Robert Montel, has examined the galactic distribution of all the RS CVn binaries we could find. We looked at the distributions of the distances from the galactic plane, and of the velocities perpendicular to that plane. The average distance from the plane is about 80 parsecs, and the average velocity (perpendicular to the plane) is approximately $10 \, \mathrm{km \, s^{-1}}$. Both these figures suggest an age somewhat less than 10^8 yrs. One slight problem is that evolutionary tracks suggest the stars are even younger. Most of the individual components of RS CVn binaries fit on to pre-main sequence evolutionary tracks, calculated for single stars by Iben (1965), at points corresponding to ages of about 10^7 yrs. Perhaps we should not try to fit components of binary systems onto tracks computed for single stars. Nevertheless, the RS CVn binaries seem to be young – I think younger than 10^8 yrs. We haven't looked at the galactic distribution of the group of stars discussed by Lloyd Evans. It could be different.

O. C. Wilson: Then you make the RS CVn stars a little younger than the Hyades.

Biermann: Iben's pre-main-sequence tracks are a bit out of date. Larson (1969) showed that there is no Hayashi phase of any importance in the pre-main-sequence evolution of higher-mass stars.

Hall: These stars are around one solar mass.

Biermann: Then you can probably use Iben's tracks, but you always have to be careful when using single-star evolutionary calculations for components of binaries. You probably know that Larson's newer calculations show that you get a shell around stars, and I have heard from R. H. Koch that he looked at the possibility of finding binaries in the pre-main-sequence phase. He couldn't find any, because, as Larson showed, the shells obscure the central stars and hide the binary characteristics.

O. C. Wilson: Now we have two people who want to speak about β Lyr. Dr. Faraggiana will give a preliminary report on observations by Hack and Cester, and Dr. Herczeg will also give a brief report.

Faraggiana: Dr. Hack obtained high-dispersion spectrograms of β Lyr (12.4 and 7 Å mm^{-1}) in both the blue and red regions of the spectrum during the 1971 international campaign. Radial velocities determined from the lines of Si II agree well with those obtained by earlier observers. Measures of the shell lines do not agree so well. The radial velocity obtained from the Hα absorption core is systematically more negative by 20 km s^{-1} to 40 km s^{-1} than the velocity obtained from the Hβ core. The emission peaks appear to be part of a single emission cut by the absorption core. The half width of the emission at Hα is about 450 km s^{-1}. The radial velocity of the emission may be constant, or slightly variable. The profiles of lines in the red region also differ from those reported by Sahade (1964). The cores of the non-metastable lines of He I $\lambda\lambda$ 5876 and 6678 vary in quadrature with the orbital velocity curve; while the metastable line of He I λ 5016 shows an almost constant velocity of -120 km s^{-1}. In general, lines originating in the stellar photosphere give velocities in agreement with those found by earlier observers, while lines originating in the envelope or streams do not.

During the same period, Dr. Cester obtained U, B, V light curves, and light curves with the narrow-band interference filters recommended for the campaign.

Herczeg: I should like to present a very brief report of a spectroscopic investigation of β Lyr which is in progress at Hamburg Observatory. This is not part of the 1971 international campaign. The spectra were taken in 1969 and 1972 with the 1m telescope (blue region, dispersion 12 Å mm^{-1} at Hγ). Our intention was to measure some emission features (Hδ, Hγ, He I λ4472), their shape and, if possible, intensity in absolute units. There is surprisingly little known about the long-term behaviour of the emission lines in the spectrum of β Lyr and a remark of Otto Struve that they may show a slow secular decline seemed to be interesting enough to follow up. We hope that a comparison of our recent spectra with Miss Gill's earlier observations and with the evaluation by Svolopoulos of Hamburg spectra taken in 1962 can possibly shed some light on this question. The 1972 spectra and part of the 1969 spectra are not yet fully reduced, but a preliminary inspection did not reveal a substantial secular change of the lines, although there may be a small decline.

Batten: I'd like to say that these results that have just been presented are the first from the international campaign that we organized last year, and there are several things of interest about them. I'd like to draw attention in particular to the very much shallower eclipse in the Hα emission lines found by Cester, and also to the fact that the asymmetry of the light curve is in opposite senses for the Hα emission line and the helium emission line at $\lambda 6678$. It appears that the Hα emission was unusually strong last year during the time of the campaign. This was not our deliberate choice: it was a bonus! We seem to have organized the campaign just when the star was behaving in some unusual way. There is some evidence that this summer the Hα emission is weaker, and I think that when we get all the results from this campaign together, we may find that we have got something very interesting indeed.

Fracastoro: Does β Lyr at any time behave in the usual way?

Herczeg: I saw Professor Hack less than two weeks ago, and she showed me her spectrograms and discussed β Lyr with me. She mentioned that she had seen no clear indication of those peaks on the emission lines which are sometimes reported as showing velocity shifts opposite in phase from those of the absorption lines. The amplitude of the emission-line velocities is supposed to be about half that of the absorption-line velocities and this would indicate that the mass ratio of the B-type star to the secondary component is about one-half. How sure can we be that these peaks are real and indicate this motion? Perhaps it would be interesting to look at some of the early spectrograms obtained by Belopolsky and Campbell. I don't doubt that the secondary component is indeed the more massive since we have a good indication of the distance and absolute brightness and, consequently, of the size of the B8 star; combining this with the photometrically determined value of R_1/a and the observed value of $a \sin i$, we obtain a mass ratio below unity.

Batten: A few years ago, Dr. Sahade and I studied Hα plates that have been obtained at Victoria from about 1966 onward. We couldn't construct a reasonable line profile, except on the assumption that there were two components to the emission – a broad one, and a sharp one which we tentatively identified with emission from around the secondary star. Perhaps Dr. Sahade will say something more.

Sahade: I had not planned to talk during the sessions, but to listen and to think of what I will say on Tuesday, but I feel I should say something about this. Belopolsky and Curtiss measured the emission line at Hβ, and apparently they found that this emission shifted in anti-phase with the stellar absorption lines. Examination of blue-region plates obtained at Mount Wilson in 1955 indicated the presence of an 'emission peak' at H8, which also appears to shift in anti-phase with respect to the stellar absorption lines. More recent work on red-region plates by Batten and myself, however, showed that the emission profiles are badly affected by superposed absorption from the stellar and shell lines. The whole emission feature at Hα, for instance, can be described, as Dr. Batten has just mentioned, by the superposition of two profiles. The broader one appears to change very little, or not at all; the narrower and stronger one appears to change greatly. The so-called 'emission peak' has no physical meaning – and I am to be blamed for having introduced such a description – it results from blending of the narrower emission with absorption features.

O. C. Wilson: We have time for one more topic. Dr. Hutchings, would you like to present some comments on emission lines?

Hutchings: Hot gas streams and envelopes exist in several binary systems and give rise to strong Balmer emission lines. Profiles from such phenomena are essentially determined by the geometry and mass-motions of the streams and envelopes and in systems where these are eclipsed, we can derive information leading to an empirical picture of what the set-up looks like. From there on we can perhaps explain in more detail the theoretical implications.

The example I have in mind is the series of profiles of Hα in the spectrum of VV Cep, through eclipse, which Dr. Wright and I worked on recently. We assumed as a first approximation (which has subsequently been strengthened) an undisturbed out-of-eclipse emission profile. As the emission-line region is eclipsed, we observed the velocity distortion effect, well known in absorption lines in Algol systems, and other effects, as follows. The emission region is a disk or shell surrounding the hot star, in which rotational velocity is a function of radial distance. Thus, as the outermost, slowly rotating regions are initially eclipsed, we observe a dimunition in the innermost part of the emission profile. Towards central eclipse, the rapidly rotating (line-wing) section of the region became occulted. This sort of information gives us a good picture of the gross physical parameters describing the disk, and possibly the stream in systems of this sort. As another example I can quote the system AR Pav which Dr. Thackeray described briefly yesterday. Here we were able to derive similar information, unfortunately less detailed because of the necessarily low dispersion of the spectrographic material. Other systems of this type might well be observed in the same way. (The further example of β Lyr comes to mind, although here we are probably hampered by many other horrendous complications.)

Cowley: I would like to mention two stars of the VV Cep type which particularly need photometry. They are AZ Cas and HR 2902 (=Boss 1985=KQ Pup). Both stars show absorption lines from an early B-type star in the ultra-violet, so that it should be possible to derive the mass ratio of both systems spectroscopically. AZ Cas has the shortest period ($9^{d}73$) of any of the systems like VV Cep (which contain an M supergiant primary, a B star, and an extended shell which emits forbidden emission lines). There is almost no photometric information on its eclipses. The next one will occur sometime in 1975. HR 2902 has a very long period (277 days) but recent changes in the ultraviolet spectrum suggest a deep atmospheric eclipse is underway. If this is not a grazing eclipse, one might expect the possible photometric minimum within the next year. Because of the excessively long period it would be unfortunate not to observe the system now.

Hall: Why do they have variable-star names? Are they variable outside eclipses?
Cowley: Yes.

Andersen: Would *uvby* photometry be useful?

Cowley: Sure.

O. C. Wilson: I think we should call the meeting to a close.

References

Conti, P.: 1972, *Astrophys. J. Letters* **174**, L79

Conti, P. and Alschuler, W. R.: 1971, *Astrophys. J.* **170**, 325.

Fernie, J. D., Hube, J. O., and Schmidt, J. L.: 1968, *Comm. 27 IAU Inf. Bull. Var. Stars*, No. 263.

Hack, M.: 1957, *Publ. Astron. Soc. Pacific* **69**, 389.

Iben, I.: 1965, *Astrophys. J.* **141**, 993.

Kuiper, G. P., Struve, O., and Strömgren, B.: 1937, *Astrophys. J.* **86**, 570.

Larson, R. B.: 1969, *Monthly Notices Roy. Astron. Soc.* **145**, 271.

Menzel, D. H.: 1939, *Pop. Astron.* **47**, 6.

Sahade, J.: 1964, *Trans. IAU* **XIIB**, 494.

Sanford, R. F.: 1951, *Astrophys. J.* **113**, 299.

Struve, O. and Pillans, H.: 1957, *Publ. Astron. Soc. Pacific* **69**, 169.

Walborn, N. R.: 1971, *Astrophys. J. Suppl. Ser.* **23**, 257.

Worrall, G. and Wilson, A. M.: 1972, *Nature* **236**, 15.

Wright, K. O.: 1970, in A. Beer (ed.), *Vistas in Astronomy* **12**, Pergamon Press, Oxford, p. 147.

Wright, K. O. and Kushwaha, R. S.: 1957, *Publ. Astron. Soc. Pacific* **69**, 402.

THE THEORY OF EXTENDED AND EXPANDING ATMOSPHERES

KARL-HEINZ BÖHM

University of Washington, Seattle, Washington, U.S.A.

Abstract. (A) The possibilities and the difficulties of a theoretical study of extended atmospheres in binaries are briefly discussed.

(B) We try to summarize and discuss critically the present status of the theory of three types of *extended* atmospheres (i.e. atmospheres in which the average photon mean-free-path is the same order of magnitude or larger than the stellar radius):

(1) Extended atmospheres in hydrostatic and in grey or non-grey radiative equilibrium.

(2) Dynamic (expanding) atmospheres which occur if the radiative acceleration is slightly smaller than the acceleration of gravity.

(3) Stellar coronae which are formed in the presence of a mechanical energy flux.

In (1) we study the importance of the 'forward peaking' of the radiation field in the outer layers of the atmosphere. The possibilities for the solution of the non-grey transfer problem in an extended atmosphere are discussed.

In (2) we pay special attention to Marlborough's and Roy's (1970) result that the atmospheric gas cannot be accelerated directly to supersonic velocities by the action of the radiation force.

In (3) the large differences in the coronal properties of stars of different chemical composition are emphasized. We draw attention to the partially unexplored but probably very interesting properties of coronae of helium-rich stars.

1. Introduction

There seems to be no general agreement among theoreticians what they call an 'extended atmosphere'. The definition of this topic becomes even more arbitrary if we exclude problems of 'Outflow of Matter' and 'Expanding Envelopes' which have been treated already in other papers of this symposium. So, I have to beg your pardon for starting out by giving a somewhat subjective definition of my topic.

As I understand it, we want to have a survey of the theory of stellar atmospheres for cases in which the atmosphere can no longer be considered as thin in comparison to the stellar radius. In such a case obviously we have to take into account the change of gravity with radius (which is rather trivial) and we have to solve the problem of radiative transfer for (at least) the case of spherical symmetry and possibly (in the binary case) for more complicated geometries. This increases the difficulty of the problem very considerably. Moreover, during the last few years (Bisnovatyi-Kogan and Zel'dovich, 1968; Kutter *et al.*, 1969; Schmid-Burgk, 1969; Finzi and Wolf, 1971; Cassinelli and Castor, 1972) it has become increasingly clear that once an atmosphere is extended in the above sense we can rather easily have a situation in which hydrostatic equilibrium no longer holds and we get a stationary expansion of the atmosphere. This statement seems to agree rather well with many observational facts. Consequently, we are forced to study not only static spherically symmetric atmospheres but also dynamic ones.

Let us look (from a naive theoretical point of view) at the possible causes for the formation of extended atmosphere. Let us first consider the case in which radiative

Batten (ed.), Extended Atmospheres and Circumstellar Matter in Spectroscopic Binary Systems, 148–170.

acceleration is *not* important. Then the local pressure scale height H must be of the same order of magnitude or larger than the radius:

$$\frac{RT}{\mu g} \gtrsim r_0; \; 8.3 \times 10^7 \frac{T}{g} \gtrsim r_0, \tag{1a}$$

or

$$\frac{1}{g} \left\{ \frac{RT}{\mu} + \frac{\xi^2}{2} \right\} \gtrsim r_0, \tag{1b}$$

at least somewhere in the atmosphere (R=gas constant, μ=mean molecular weight, g=local acceleration of gravity, ξ=turbulent velocity). As is well known there are very few normal atmospheres of nonrotating single stars in which condition (1) is fulfilled. Only if the effective value of g is strongly reduced by the presence of a strong centrifugal force in a fast rotating star or by the presence of a neighboring star in a close binary can the condition (1) be fulfilled easily for an atmosphere with radiative energy transport. However, (1) can be relatively easily fulfilled for chromospheres and coronae which have a high temperature due to the heating by a mechanical energy flux. (Cf. Kuperus, 1965, 1966; Ulmschneider, 1967; Nariai, 1969; de Loore, 1970; Böhm and Cassinelli, 1971; for a summary of the semiempirical aspects see Praderie, 1970.)

As pointed out by Underhill (1949, 1966) and Mihalas (1969) for normal early type stars and discussed by Böhm (1969) and Cassinelli (1970) for central stars of planetary nebulae, the theory predicts extended atmospheres for hot stars in which the radiative acceleration

$$|g_{\text{rad}}| = \frac{\pi}{c} \int_0^\infty (\kappa_\nu + \sigma_{\text{el}}) F_\nu d\nu \tag{2}$$

becomes comparable to the local gravity. (κ_ν=monochromatic absorption coefficient, σ_{el}=Thomson scattering coefficient, F_ν=monochromatic radiative flux).

It turns out that typically the atmospheres become extended if

$$|g_{\text{rad}}| \approx \frac{\sigma T_{\text{eff}}^4}{c} \bar{\kappa} \gtrsim 0.8 \, g. \tag{3}$$

in the case of central stars of planetaries.

Moreover, if g_{rad} approaches g too closely hydrostatic equilibrium is not even approximately possible (Cassinelli and Castor, 1972). It is believed that a considerable number of stars exist which fall into the range defined by (3).

According to this discussion it seems reasonable to review the theory of the following atmospheres:

(1) Hydrostatic atmospheres in radiative equilibrium which are extended because of the validity of condition (1) or (3);

(2) Dynamic atmospheres (i.e. atmospheres with continuous mass loss in which g_{rad} is important);

(3) Hydrostatic and hydrodynamic chromospheres and coronae which are due to the dissipation of a mechanical energy flux.

As far as I can see this covers the types of extended atmospheres which are at least partially understood from a theoretical point of view. Unfortunately it probably does not include the interpretation of those extended atmospheres in binaries for which we have the most detailed observational material like ζ Aur, 31 Cyg and 32 Cyg (Cf. Groth, 1957, 1970). This is, of course, sad. On the other hand, I think we will not be able to develop convincing theories of these rather complicated cases unless we understand first the theory of certain basic effects which occur in extended atmospheres.

Before we discuss points (1) to (3) in detail we have to ask the following question: To which extent do we have to change our discussion if we talk about extended atmospheres in reasonable close binaries.

In this case obviously the following effects are important (cf. the interesting and detailed discussion by Kopal, 1959):

(1) Gravity darkening as a consequence of stellar rotation as well as the deformation of the star by the gravitational field of the other component.

(2) The generalized 'reflection effect', i.e. the effect of the radiation of the other component on the atmosphere.

It is well known that these two effects do not introduce new basic difficulties if the atmosphere can be considered as 'thin', i.e. plane-parallel. (However, the required amount of computational work is increased very considerably.)

The influence of gravity darkening on the predicted spectra of *single* rotating stars has been studied successfully in recent years by many authors, e.g. Collins (1963), Roxburgh and Strittmatter (1965), Collins and Harrington (1966), Hardop and Strittmatter (1968a, b), Collins (1968a, b), Collins (1970), and others. These calculations are based on the (well-justified) assumption that the atmospheric structure can be calculated everywhere using the local value of g and the local value of T_{eff} as it follows from the gravity darkening law. Since data on gravity darkening in close binaries (at least for some cases of the Roche model) and on the reflection effect are available (Cf. Kopal, 1959; Minin, 1965; Peraiah, 1969; Rucinski, 1969, 1971) analogous calculations have been carried out recently for binaries by Buerger (1969). He has used grey, plane-parallel atmospheres defined by local values of g and T_{eff}. Gravity darkening as well as the reflection effect have been taken into account.

However, in this symposium we are concerned with extended atmospheres. A quantitative treatment of gravity darkening and of the reflection effect in these atmospheres would be much more difficult for these atmospheres and has – to the best of my knowledge – not yet been tried. Obviously the lateral radiative exchange must become very important in this case and a description using local values of g and T_{eff} becomes impossible. Since even the simple problems of extended atmospheres in single stars or in wide binaries (as formulated above) are far from being solved completely, we may doubt whether it will be possible to solve the problem of an extended atmosphere in close binaries in the near future.

2. Extended Atmospheres in Hydrostatic and Radiative Equilibrium

In this chapter we shall assume that (at least some) extended stellar atmospheres can exist in hydrostatic and radiative equilibrium. The interesting question concerning the validity of this assumption will be discussed briefly in the next chapter.

The hydrostatic equation can be written (neglecting turbulence)

$$\frac{dp}{d\tau_0} = \frac{GM}{r^2 \kappa_0} - \frac{\pi}{c} \frac{\int_0^\infty (\kappa_\nu + \sigma_{el}) F_\nu d\nu}{\kappa_0} \tag{1a}$$

or

$$\frac{dp}{dr} = -\frac{GM}{r^2} \varrho + \frac{\kappa_F \varrho L}{4\pi r^2 c} = -\frac{GM\varrho}{r^2} \left\{ 1 - \kappa_F \frac{L}{4\pi cGM} \right\}, \tag{1b}$$

with p=gas pressure, ϱ=density, M=total mass of the star, κ_ν=monochromatic absorption coefficient, σ_{el}=electron scattering coefficient, F_ν=monochromatic radiative flux, L=total luminosity, r=radial coordinate (distance from the center of the star, κ_0 and τ_0 are the absorption coefficient and the optical depth for some standard frequency. The flux mean κ_F of the absorption coefficient is defined as

$$\kappa_F = \int_0^\infty (\kappa_\nu + \sigma_{el}) F_\nu d\nu / \int_0^\infty F_\nu d\nu. \tag{2}$$

Obviously the first term in (1b) is the usual gravitational force (per cm^3 of matter), whereas the second term is the radiative force on the same amount of material. The ratio of these two forces is independent of r if κ_F=const. Equation (1b) also shows that a hydrostatic solution is impossible if

$$L > \frac{4\pi cGM}{\kappa_F}. \tag{3}$$

Radiative equibilbrium is described by equations of transfer for every frequency of the form

$$\left\{ \mu \frac{\partial}{\partial r} + \frac{(1 - \mu^2)}{r} \frac{\partial}{\partial \mu} \right\} I_\zeta (r, \mu) = -\kappa_\nu \varrho \{ I_\nu (r, \mu) - S_\nu (r) \}, \tag{4}$$

and the radiative equilibrium condition which may be written

$$L_r = \left(\int_0^\infty L_\nu d\nu \right)_r = 4\pi r^2 \int_0^\infty F_\nu d\nu = \text{const}. \tag{5}$$

The atmosphere will be really extended if

$$\frac{1}{\bar{\kappa}\varrho} \gtrsim r, \tag{6}$$

whereas the problem reduces to the plane-parallel case if

$$\frac{1}{\bar{\kappa}\varrho} \ll r,$$ (7)

(see Chapman, 1964; Cassinelli, 1970). It is important to note that in contradistinction
to the plane case neither the hydrostatic Equation (1b) nor the transfer Equation (6) can be transformed into an equation for the independent variable τ_ν only.
In other words, though we can e.g. determine the temperature stratification (apart
from a scaling factor) of a grey atmosphere once and for all in the plane-parallel
case by computing $T(\bar{\tau})$ (Hopf, 1934; Mark, 1947) this is not true for an extended
atmosphere. In this case the temperature stratification also depends on the relation
between $\bar{\tau}$ and r:

$$\bar{\tau} = \bar{\tau}(r).$$ (8)

(Another formulation of the same statement says that the T-stratification depends on
the ratio between the radius of curvature and the photon mean path.)

Additional insight into the problem may be gained by writing down the moment
equations of the equation of transfer (4). They are found in the usual way by applying
the operators:

$$\frac{1}{2} \int_{-1}^{+1} \dots d\mu,$$ (9a)

and

$$\frac{1}{2} \int_{-1}^{+1} \dots \mu d\mu$$ (9b)

to Equation (4). We find

$$\frac{1}{r^2} \frac{d}{dr} (r^2 H_\nu) = - \kappa_\nu \varrho (J_\nu - S_\nu),$$ (10)

by applying (9a) and

$$\frac{dK_\nu}{dr} + \frac{1}{r} (3K_\nu - J_\nu) = - \kappa_\nu \varrho H_\nu.$$ (11)

Obviously (10) and (11) can be specialized to the grey case by leaving out the sub-
script ν and by setting $J = S$. It follows immediately that

$$r^2 H = 4r^2 F = \text{const},$$ (12)

(12) is of course intuitively obvious. Many astronomers tend to consider (12) as *the*
important condition which distinguishes an extended from a plane-parallel atmo-

sphere. This approach has led to the use of the Milne-Eddington approximation

$$K_v = \tfrac{1}{3}J_v, \tag{13}$$

in many of the earlier papers in this field (Cf. Kosirev, 1934; Chandresekhar, 1934, 1945, 1960; see also Pearce, 1967). When (13) is used, (11) reduces to the same equation as in the plane-parallel case. However, in the more modern literature (cf. Chapman, 1964, 1966; Schmid-Burgk, 1970; Cassinelli, 1970, 1971; Hummer and Rybicki, 1971) it has been shown that (at least in very extended atmospheres) (13) is not an acceptable approximation. As is well known the validity of (13) implies that the radiation field is isotropic. On the other hand the 'real star' (defined e.g. by the surface $\bar{\tau} = 1$) covers only a very small fraction of the total solid angle as viewed from the outer parts of a very extended atmosphere. In fact in the limiting case that the solid angle covered by the 'star' becomes very small we have to replace (13) by the relation

$$K_v = H_v = J_v. \tag{14}$$

So, from our present point of view spherical symmetry in an extended atmosphere leads to two important effects:

(1) the decrease of F (or H) like $1/r^2$ and 2) the increasing 'outward peaking' (Hummer and Rybicki, 1971) of the radiation field in the outer layers of the atmosphere.

Most authors, with the exception of Cassinelli (1970, 1971), have restricted their actual calculations to strongly simplified models in which grey absorption (or pure coherent scattering) has been assumed and in which also the radiative transfer problem (4), (5) is decoupled from the hydrostatic Equation (1b). Most of these calculations are based on the assumption

$$\kappa\varrho = r^{-n} \tag{15}$$

(Cf. Chapman, 1964, 1966; Schmid-Burgk, 1970; Hummer and Rybicki, 1971; Cassinelli and Hummer, 1972). The most accurate numerical calculation of the radiative transfer in extended atmospheres is due to Schmid-Burgk (1970) and to Hummer and Rybicki (1971), Schmid-Burgk (1970) solves the problem in the following way: From (4) he derives integral equations for the mean intensity $J(r)$. He solves one of these equations by expanding it into a series of known coefficients. The functions are selected in such a way that the integration over r can be carried out analytically. After numerical integration over μ a system of linear equations for the unknown coefficients can be derived from the integral equation.

Hummer and Rybicki (1971) rewrite the momentum Equation (11) by defining the 'Eddington factor' f

$$f = K/J, \tag{15}$$

so that K in (11) can be replaced by fJ. Starting out with a guess of $f(r)$ they integrate (11) numerically and so get a first approximation of $J(r)$. Since in the grey case $J = S$, this approximation of $J(r)$ can be used to find a first approximation of $I(r, \mu)$

from Equation (4). From $I(r, \mu)$ we can recalculate the moments and find an improved value of the Eddington factor. The procedure is iterated. Schmid-Burgk's (1970) and Hummer's and Rybicki's (1971) results for assumption (15) are useful from two different points of view:

(1) The newly developed methods may be applied to more realistic cases of extended atmospheres (though the generalization may not always be trivial).

(2) The numerical results give us considerable insight into the mechanism of the radiative transfer in an extended atmosphere.

As an illustration of point 2 we reproduce in Figure 1 Schmid-Burgk's (1970) results for the angular dependence $I^+(\tau, \mu)/I^+(\tau, 1)$ of the 'outward' intensity in the case $n=3$. The diagram shows very clearly the very strong 'forward peaking' of the radiation field in high layers of the extended atmosphere. It also shows to which extent

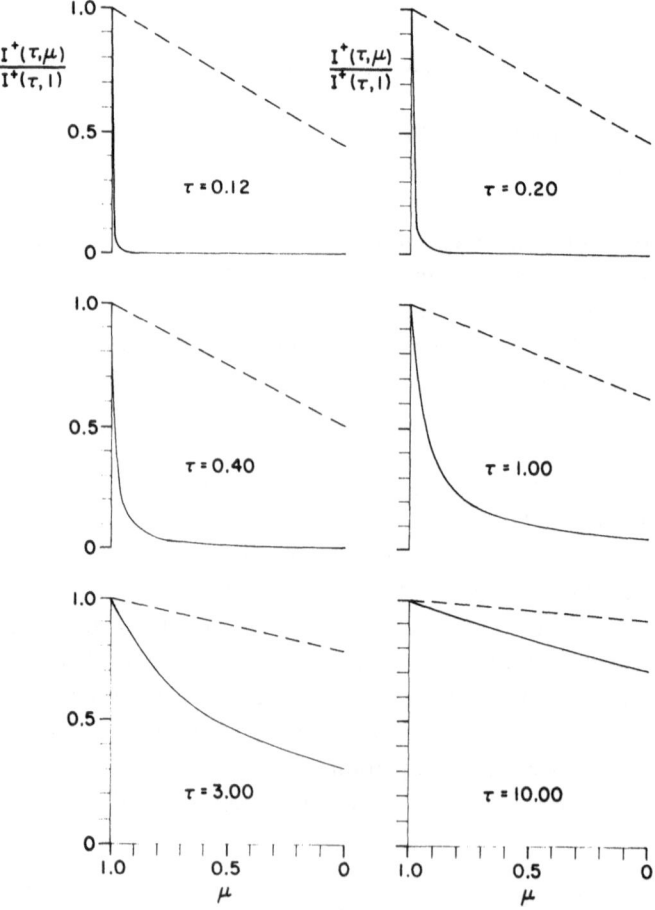

Fig. 1.   The outward intensity I^+ as a function of μ for different optical depths τ. The solid curve refers to the spherically symmetric grey atmosphere for the case $n=2$. (See formula 15.) The broken line corresponds to the plane parallel case. The diagram has been taken from Schmid-Burgk's (1969) work (with minor modifications).

Eddington's approximation leads to incorrect results. It is also interesting to note (Schmid-Burgk, 1970) that the spectral energy distribution of the emergent flux from an extended grey atmosphere is much 'flatter' (not so much peaked around one frequency) than the energy distribution from the corresponding plane parallel atmosphere.

Let us now look at the determination of more realistic models of extended atmospheres. Obviously the following points will be important:

(1) Assumption (15) has to be replaced by a realistic opacity law $\kappa(\varrho, T)$. This leads immediately to a coupling between the hydrostatic and the transfer equation. This in turn requires a critical consideration of the boundary conditions at the 'surface' of the atmosphere.

(2) We would like to include the effects of 'nongreyness' in the radiative equilibrium calculation.

(3) As soon as possible deviations from local thermodynamic equilibrium (which are of course usually more important in extended than in plane-parallel atmospheres) should be included (Thomas, 1970).

Cassinelli (1970, 1971) had tried to solve the difficult problem in which points (1) and (2) have been taken into account.

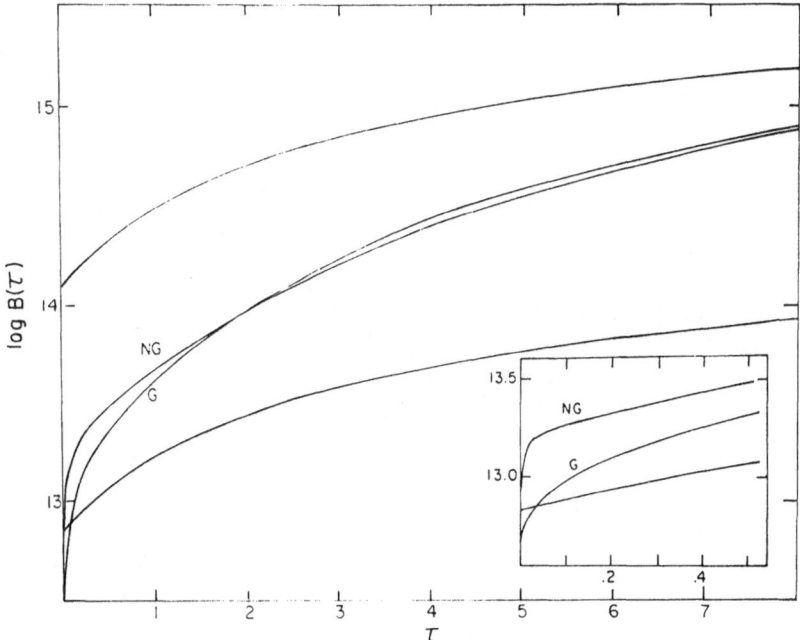

Fig. 2. $B(\tau)$ in an extended atmosphere of a star of $M = 0.6\,M_\odot$ and $L \simeq 2.05 \times 10^4\,L_\odot$, corresponding to $T(\bar\tau = \frac{2}{3}) \simeq 37\,600$ K. The geometrical radius at $\bar\tau = 10^{-3}$ is about 4.3 times as large as at $\bar\tau = 10$. The two inner curves give the grey and the non-grey models taking into account effects of spherical symmetry. The outer curves are plane-parallel stratifications drawn for comparison purposes. After Cassinelli (1970).

In the formulation of the outer boundary conditions he makes the useful assumption that the hydrostatic atmosphere has to be cut off at the point where the thermal velocity becomes equal to the velocity of escape. Such a precaution is necessary because a spherically symmetric atmosphere with a sufficiently small temperature gradient in its outer layers has a finite density at $r = \infty$. For the solution of the transfer problem Cassinelli uses the so-called S_N-method, a discrete ordinate method developed by Carlson and Lathrop (1968) for the treatment of neutron transport problems. In order to fulfill the condition of radiative equilibrium

$$r^2 \int_0^\prime F_\nu d\nu = \text{const}.$$ (16)

Cassinelli developed a temperature correction procedure which is a generalization of the well-known Unsöld-Lucy method (Unsöld, 1951; Lucy, 1964) to the spherically symmetric case. A temperature correction procedure permits us to calculate a correction to a given approximate temperature stratification if we know the derivation of the

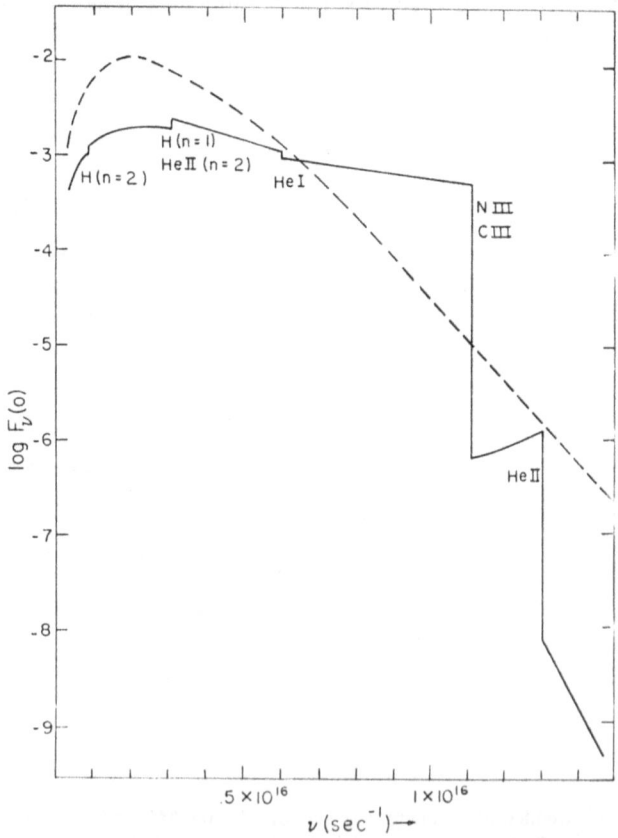

Fig. 3.   The energy distribution of the emergent flux $F_\nu(0)$ for the non-grey extended atmosphere shown in Figure 2. After Cassinelli (1970, 1971). (By permission of the *Astrophysical Journal*; copyright 1971; The University of Chicago.)

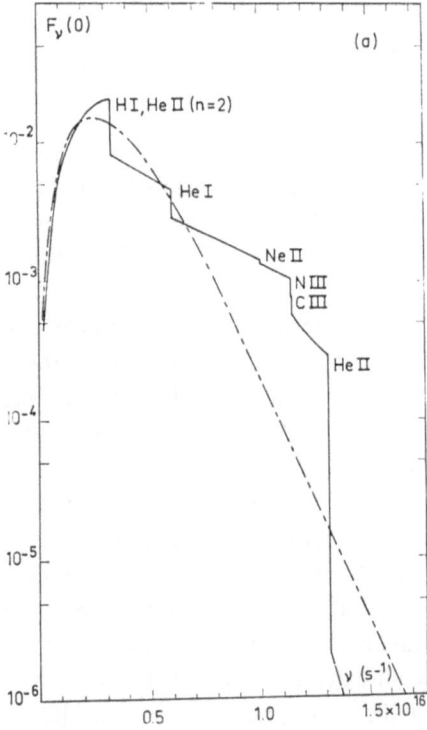

Fig. 4. The emergent flux $F_\nu(0)$ in a plane-parallel non-grey atmosphere of similar temperature ($T_{eff} = 4.3 \times 10^4$ K, $g = 6 \cdot 10^4$ cm s^{-2}) for comparison with Figure 3. (After Böhm (1969); by permission of the Springer-Verlag.)

total flux for this stratification from the correct value as required by condition (16). Unfortunately Cassinelli (1970) succeeds only in deriving temperature correction formulae in a simple form for the two limiting cases given by the conditions (13) and (14). Suprisingly, the two expressions turn out to be very similar. Moreover, as is well known temperature correction formulae do not have to be very accurate. After every iteration step one checks how well condition (16) is fulfilled and so one can judge the quality of the solution independently. Cassinelli was able to achieve a flux constancy of 1.5% or less in his models.

His calculation shows that an extended atmosphere covers a much larger temperature interval in a given optical depth range than a plane-parellel atmosphere as one would expect. This is illustrated in Figure 2 which shows the grey and the nongrey temperature stratifications for an extended atmosphere with a luminosity $L \approx 2.05 \times \times 10^4\, L_\odot$ and a mass of $M \approx 0.6\, M_\odot$. (Note that the concepts of the effective temperature and surface gravity are of course no longer useful in an extended atmosphere.) The grey and nongrey stratifications have temperatures of 34710 K and 37500 K at $\bar\tau = \frac{2}{3}$. Their 'surface temperatures' are 21890 K and 27140 K (with the grey surface temperature lower). The atmosphere is about four times as thick as the 'radius of the star' provided we set the boundary between atmosphere and star at $\bar\tau \approx 10$.

The critical luminosity

$$L_{\text{crit}} = \frac{4\pi c G M}{\langle \kappa + \sigma_{\text{el}} \rangle},$$

(17)

is about $2.3 \times 10^4 \, L_\odot$. The energy distribution of the emergent flux is given in Figure 3. In comparison to the situation for a plane-parallel model (cf. Figure 4) for a similar $T\left(\frac{2}{3}\right)$ the extended atmosphere shows a somewhat 'flatter' energy distribution and leads to an emission edge at $\lambda = 912$ Å in contradistinction to the absorption edge found in the plane-parallel case.

Finally we might ask how these more realistic models compare to the schematic ones described by condition (15). We find that usually the inner parts can be described by condition (15). We find that typically the inner parts can be described by $n \approx 2.5$ whereas the outer parts $(\bar{\tau} < \frac{2}{3})$ require a much larger value of n $(n \approx 14$ according to Cassinelli, 1971).

So far we have discussed the determination of the model atmosphere and of the continuous spectrum only. The calculation of the line spectrum could in principle be based on the same transfer Equation (4) or the corresponding momentum equations. (This was done rather early, cf. McCrea, 1928.) However, today many people feel that such a procedure would not be applicable in many cases. From an observational point of view practically all extended atmospheres do show some motion and even if a considerable part of the atmosphere can be considered to be approximately in hydrostatic equilibrium many lines will probably be influenced by the differential Doppler effect of different layers. After a study of line profiles from moving plane-parallel atmospheres (cf. Abhyankar, 1965), of thin spherical shells (Beals, 1931; Rottenberg, 1952) and certain simplified models (Sobolev, 1960) the complete transfer problem in an expanding and extended atmosphere has recently been considered by several authors (Cf. Rybicki, 1970; Lucy, 1971). Lucy has pointed out that the transfer equation for an expanding extended atmosphere can be brought into a relatively simple form provided

(1) only terms of the first order in (v/c) are retained,

(2) one uses the 'narrow line limit', i.e. one assumes that the thermal velocities are much smaller than the hydrodynamic velocities.

In this case one gets

$$\left\{ \mu^2 \left[\left(\frac{dv}{dr} \right) - \frac{v}{r} \right] + \frac{v}{r} \right\} \frac{dI'_{v'}}{dv'} = \kappa'_{v} \varrho \, \frac{c}{v_0} \, (I'_{v'} - J'_{v'}),$$

(18)

where I', J' and v' refer to the comoving frame. Lucy discusses a very effective method for the solution of transfer problems described by Equation (18) or by (4). It is analogous to the classical Schwarzschild method of describing the angular dependence of the intensity by setting the outward intensity $I^+(\mu)$ equal to a constant and inward intensity $I^-(-\mu)$ equal to another constant. However, the switchover from I^+ to I^- does not occur at $\mu = 0$ as in the plane-parallel case but at $\mu = \mu^*$ where μ^* is defined by the angle under which the 'stellar limb' is seen from a point in the extended atmosphere. This method seems to be very promising since it takes into account the 'forward concentration' of the radiation in the high layers in the simplest possible way. The result of such calculations depends of course on the assumed velocity law which can be determined (at least in principal) from hydrodynamical calculations. (See next chapter.)

The numerical calculations lead to P Cygni-type profiles in most cases.

When talking about extended atmospheres we have so far mostly emphasized effects which can be attributed directly or indirectly to the changing geometry from plane-parallel to spherically symmetric. However, there are also effects of simply having large regions with relatively low density instead of small regions with relatively high density. The importance of these effects close to the instability limit is illustrated in Figure 5 which shows the change of the continuous energy distribution of a very hot star as we approach the limit $g_{rad} = g$.

The interesting problem of the curve of growth in an expanding atmosphere (with constant velocity of expansion) has been investigated by Arakelian (1969).

I should also like to draw attention to the interesting studies of extended atmospheres by the Tartu astronomers (cf. Sapar and Viik, 1968) who have studied the generalization of the Avrett-Krook procedure to the spherically symmetric case.

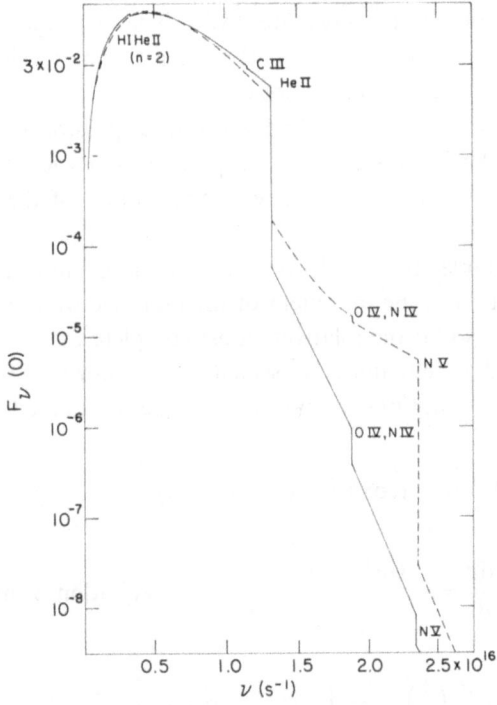

Fig. 5.   Comparison of the emergent fluxes $F_\nu(0)$ for two atmospheric models with $T_{eff} = 63000$ K. The solid curve corresponds to a value g corresponding to $\Gamma = (g_r/g)$ of 0.725. The broken curve refers to a model with $\Gamma = 0.95$.

3. Dynamic Atmospheres

Our understanding of extended atmospheres not in hydrostatic equilibrium is still very limited. Only two cases have been studied to some extent:

(1) stellar coronae which (in analogy to the solar case) may lead to a stellar wind and

(2) expanding atmospheres which occur in stars with sufficiently low g (or sufficiently high temperature) so that the instability limit $g = |g_{rad}|$ is approached but not reached.

In a certain sense case (2) is simpler than (1) because it occurs in simple radiative transfer atmospheres and does not require a mechanical energy flux to build up a stellar corona. In the present chapter we shall restrict ourselves to case (2).

Practically all theoretical studies in this field have been done with relatively hot stars, like early B-stars, O and Of stars, WR stars and central stars of planetary nebulae in mind. As is well known the interest in this field has increased very considerably because of the interesting observations of very large outward velocities in the uv lines of early-type stars by Morton and his collaborators. (Cf. Morton *et al.*, 1968.) However, as we shall see below, attempts to explain these high velocities in one simple step can be misleading.

Attempts to solve the theoretical problem have (so far) been based on the following assumptions and requirements:

(1) Look for a stationary (steady state) hydrodynamic solution.

(2) Take into account the (r-dependent) acceleration of gravity as well as the radiative acceleration in the equation of motion.

(3) Take into account radiative exchange as fully as possible in the energy equation.

(4) Treat the radiative transfer as a spherically symmetric problem and take into account terms proportional to (v/c) (due to the motion of the gas) in the transfer equation.

Because of the complexity of the problem all authors had to restrict themselves to the grey approximation in the treatment of the radiative transfer problem.

It seems to us that so far the relatively most complete treatment of the problem is due to Schmid-Burgk (1969) and to Cassinelli and Castor (1972).

Using the above assumptions the problem can be formulated as follows:

$$\frac{d}{dr}(\varrho v r^2) = 0; \quad \text{(equation of continuity)}, \tag{19}$$

$$v\frac{dv}{dr} + \frac{1}{\varrho}\frac{dp}{dr} = -\frac{GM}{r^2} + \frac{\langle\kappa + \sigma_{el}\rangle L}{4\pi c r^2}; \quad \text{(equation of motion)} \tag{20}$$

$$v\frac{dE}{dr} + pv\frac{d}{dr}\left(\frac{1}{\varrho}\right) = 4\pi \int_0^{\infty} \kappa_v (J_v - S_v)\, dv\,;$$

(energy equation, including radiative exchange)   (21)

(cf. Cassinelli and Castor, 1972), where v is the gas velocity, E is the internal energy per gram (of the gas only). It is important to note that the radiative quantities, κ_v, S_v, J_v, refer to the comoving frame of reference. The momentum equations for radiative transfer [corresponding to Equation (10) and (11)] become somewhat complex if (v/c) terms are included. They have been given by Cassinelli and Castor (1972).

The radiative acceleration term in (20) can of course also be written as

$$\frac{\pi}{c} \int_0^\infty \kappa_v F_v \, dv . \tag{22}$$

Before saying something about the solutions of the system (19), (20), (21) we should try to be aware of one possible misunderstanding to which one is easily lead if the problem is approached in a naive way. Since these expanding atmospheres occur as we get close to the situation where the absolute value of (22) becomes comparable to the local gravity one might think that the gas is directly accelerated by the 'radiation force' (which is proportional to (22)). It was recognized by Schmid-Burgk (1969) and clearly analyzed and discussed by Marlborough and Roy (1970) and Cassinelli and Castor (1972) that this is impossible (for some early remarks on this problem see Paczyński, 1968).

The essential point is the following: as is well known from the theory of the solar wind (cf. Parker, 1963; Holzer and Axford, 1970; Brandt, 1970) the *interior* boundary conditions for the flow can be fulfilled only by solutions which are subsonic throughout or by the so-called critical solution which makes a smooth transition from subsonic to supersonic flow at the sonic point. Moreover the *outer* boundary condition ($p \rightarrow 0$ at sufficiently large r) can be fulfilled only if the solution is supersonic for large r (see Figure 6). Consequently the critical solution is the only one which fulfills both boundary conditions. In other words the correct solution always has to pass through

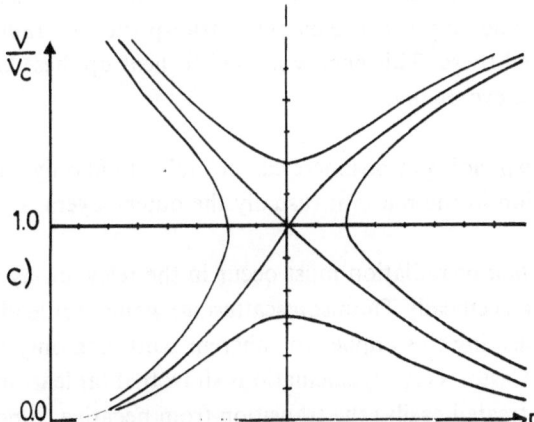

Fig. 6. $v(r)$ for the typical stellar wind solutions, neglecting radiative acceleration. After Schmid-Burgk (1969).

a point $v = c_s$ (c_s = local velocity of sound) continuously. Let us now look at a case in which the radiative acceleration term is important but in which the radiative exchange term in (21) can be neglected (adiabatic case). Then Equations (19), (20), (21) can be reduced to the following relation, provided the perfect gas law holds. (Marlborough and Roy, 1970; Cassinelli and Castor, 1972.)

$$\frac{r}{v}\frac{dv}{dr} = \frac{2c_s^2 - \{GM(1-\varGamma)/r\}}{v^2 - c_s^2},\tag{23}$$

when the $\varGamma$ is the ratio of the radiative acceleration to the local gravity:

$$\varGamma = \frac{\langle\kappa + \sigma_{el}\rangle L}{4\pi cGM}.\tag{24}$$

Obviously $\varGamma = 1$ corresponds to the instability limit $|g_{rad}| = g$. In order to get a smooth transition from subsonic to supersonic velocities the numerator in Equation (23) must go from negative to positive values exactly at the point $v^2 = c_s^2$. (dv/dr is positive everywhere). This condition can be fulfilled only if $\varGamma < 1$ in the subsonic region (See Equation 23). In other words, the radiative acceleration cannot be used to accelerate the gas in the subsonic region. However, as pointed out by Marlborough and Roy (1970) the flow can be occeterated in the supersonic region by radiative acceleration (as described e.g. by Lucy and Solomon, 1970). Cassinelli and Castor (1972) call this the 'afterburner' mechanism.

How can we then accelerate the gas up to the sonic point?

As we saw in the preceding chapter a *hydrostatic* atmosphere becomes more and more extended (it is less and less bound) as we approach the limit $\varGamma = 1$. Consequently less and less energy is needed to drive an outflow of the gas. It is implicit in part of the earlier work (Bisnovatyi-Kogan and Zel'dovich, 1968; Schmid-Burgk, 1969) and it was very clearly discussed and emphasized by Cassinelli and Castor (1972) that the acceleration is possible only because energy is transported by radiation to the outer layers and deposited there. This energy is used to heat up these layers sufficiently so that they can escape eventually.

This shows that:

(1) this type of expanding atmosphere can be understood only if the absorption and emission of radiation in the relevant (usually the outer) layers is treated in sufficient detail,

(2) *'true' absorption* of radiation must occur in the relevant layers, because a pure scattering process (specifically Thomson scattering) would not lead to a deposition of energy. (Strictly speaking this applies to coherent scattering only.)

Cassinelli's and Castor's (1972) calculations show that (at least in the optically thin case which can be treated easily) the transition from negative to positive total energy of the gas (i.e. kinetic energy plus gravitational energy plus enthalpy) occurs just somewhat below the critical (sonic) point in all interesting cases.

Cassinelli and Castor (1972) find that strictly speaking all spherically symmetric stellar atmospheres show an outflow of matter. However, if Γ is not sufficiently close to 1 the sonic point will occur at such a large distance from the star that the total outflow is completely negligible. Only as we approach $\Gamma = 1$ does the outflow become considerable and the atmospheric structure becomes very different from the hydrostatic case. Atmospheres which are not in hydrostatic equilibrium show in general a flatter temperature stratification especially in their upper layers, than hydrostatic atmospheres. (Cf. Bisnovatyi-Kogan and Zel'dovich, 1968; Schmidt-Burgk, 1969; Cassinelli and Castor, 1972; see also Böhm, 1968.)

As especially Schmid-Burgk (1969) has pointed out the topology of the stellar wind solutions $v(r)$ can be much more complicated than in the simple cases which are usually discussed provided Γ is not too much smaller than 1 and $\bar{\kappa}(r)$ is not a monotonic function of r. The topology of the flow solutions for the case in which $\bar{\kappa}(r)$ has a maximum somewhat above the critical point is illustrated in Figure 7 (taken from Schmid-Burgk's thesis).

Finally I should like to emphasize again that all the models discussed here refer to situations in which $\Gamma < 1$ everywhere. In other words, we get dynamic atmospheres (with outflow) though the effective gravity

$$g_{\text{eff}} = g - g_{\text{rad}}, \tag{25}$$

is directed inward everywhere.

4. Stellar Coronae

Instead of the above chapter title the observer would rather see one indicating a chapter on *chromospheres and* coronae. However, it seems that we really understand too little about the formation of chromospheres to include it in our discussion unless

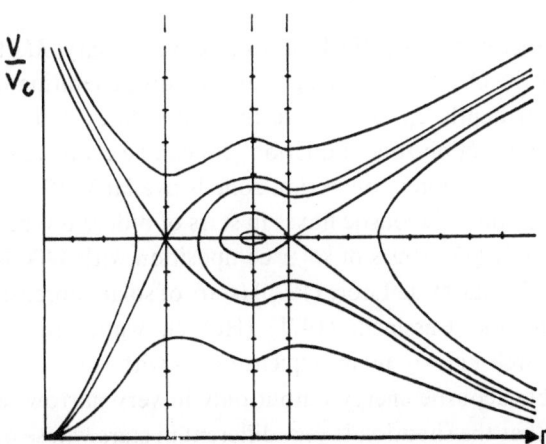

Fig. 7. Typical stellar wind solution $v(r)$ for an atmosphere in which the radiative acceleration has its maximum somewhat above the sonic point. After Schmid-Burgk (1969).

you would just call the transition region from the photosphere to the corona a chromosphere. On the other hand, it seems rather obvious that the extended chromospheres observed in stars like ζ Aurigae are something very different from such a transition region.

The prediction and computation of stellar coronae is based on ideas developed for the calculation of the transition region and the corona in the solar case which seems to work surprisingly well there (at least according to the opinion of many astronomers). However, recently some astronomers (cf. Ulrich, 1972) have expressed some doubts concerning the validity of the 'standard' theory of coronal heating. It is not yet quite clear how serious these objections are (see below). Moreover, the standard (shock-wave heating) approach is the only one which has been worked out at least in some detail and which consequently can be applied to stars other than the sun. Therefore, we shall restrict ourselves to the discussion of coronae (and transition regions) which are due to the heating by shock waves. The shock waves are thought to come from the continuous steepening of acoustic waves which are generated in the stars outer convection zone. Obviously, a necessary condition for the existence of a corona is the presence of an outer convection zone and the generation of a sufficiently strong acoustic noise flux in this zone. The calculation of the acoustic noise flux has usually been based on the Lighthill-Proudman theory (cf. Lighthill, 1954, 1955; Proudman, 1952). Though Lighthill (1967) himself has raised some objections against the astrophysical application of his theory and has strongly urged us to consider the generation of gravity waves by the convection zone, it is true that only acoustic waves have been seen in the solar atmosphere. As pointed out by Souffrin (1966) this must be due to the short radiative relaxation time of the gravity waves. With these facts in mind we can easily calculate the acoustic ('mechanical') flux

$$F_m \simeq 19 \int_{z_0}^{z_1} \varrho M_*^5 \frac{v^3}{l} \, dz. \tag{26}$$

F_m is the acoustic energy flux, v is the local convective velocity, M_* the corresponding Mach number, l is the characteristic length of the flow generating the acoustic noise. z is the geometrical depth, z_0 and z_1 are the coordinates of the upper and lower boundaries of the convection zone. Equation (26) can be evaluated easily for any star for which a model of the outer convection zone is available. (See e.g. Biermann and Lüst, 1960). Calculations of acoustic fluxes in stars of different type have been carried out e.g. by Kuperus (1965), (stars of solar composition with 4400 K $\leqslant T_{\text{eff}} \leqslant$ 7000 K), Nariai (1969) (He-rich stars), de Loore (1970) (stars of solar composition, 2500 K $\leqslant T_{\text{eff}}$ $\leqslant$ 16630 K), Böhm and Cassinelli (1971) (He-rich white dwarfs, 5790 K $\leqslant T_{\text{eff}} \leqslant$ $\leqslant$ 30000 K). As is well known, main sequence stars with solar chemical composition show a considerable acoustic energy output only in very narrow temperature range. It is worth noting that the situation is very different in stars whose outer layers consist mostly of helium (Nariai, 1969). Probably the most extreme objects in this respect are the white dwarfs with helium-rich outer layers (Böhm and Cassinelli, 1971) in which acoustic fluxes can be reached which are considerably larger than the radiative flux of the Sun. The situation is illustrated in Figure 8, which shows the completely different behavior of the mechanical flux as a function of the effective temperature for main sequence stars on the one hand side and for helium-rich white dwarfs on the other side. It is to be expected that these helium stars should have considerably denser and hotter coronae than main sequence stars.

   Let us now look at the problem of the computation of coronal densities and tem-peratures for stars of different effective temperature and surface gravity.

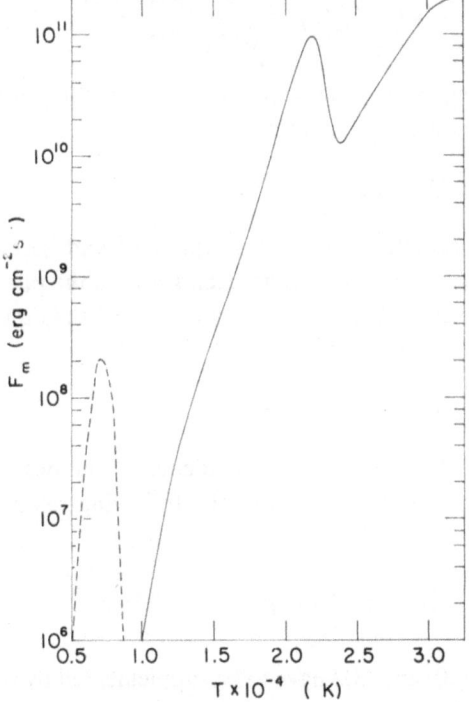

Fig. 8.   Comparison of the acoustic flux F_m in main sequence stars (broken line) and in helium-rich white dwarfs (solid curve).

   The basic theory has been outlined very clearly in the Ph.D. thesis of Kuperus (1965; see also de Jager and Kuperus, 1961). Very considerable improvements in the details of the physical theory have been made later (cf. Ulmschneider, 1967, 1971a, 1971b; Stein, 1968). However, we feel that the basic ideas can be most easily under-stood if we restrict ourselves mostly to the simple approach described in Kuperus (1965, 1966) work.

   We have to start from the assumption that the mechanical flux, F_m, above the convection zone has already been calculated [Equation (26)]. One now makes the very plausible assumption that the (spatial) decrease of F_m due to shockwave dissipation can be described by a local 'absorption coefficient' $\tilde{\kappa}(z)$. One finds

$$F_m(2) = F_m(0) \times \phi(z) \times \exp\left(-\int_0^z \tilde{\kappa}(z)\right) dz, \tag{27}$$

z is the geometrical height counted from the point where the shock wave dissipation starts, $\phi(z)$ is a correction factor which takes into account the losses in the radial mechanical energy flux due to the refraction and reflection of shock waves. After logarithmic differentiation (27) can be written as a differential equation for F_m:

$$\frac{dF_m(z)}{dz} = -\tilde{\kappa}(z) F_m(z) - \frac{F_m(z)}{\phi(z)} \frac{d\phi(z)}{dz}. \tag{28}$$

The local energy dissipation E_d, (available for the heating of the gas) is related to $\tilde{\kappa}$ and F_m according to the simple relation:

$$E_d(z) = \tilde{\kappa}(z) \times F_m(z). \tag{29}$$

Let us now call the radiative energy loss (per cm^2 and second) E_r. The increment dF_c in the conductive energy flux (going back towards the photosphere) will be equal to the difference between the radiative energy loss and the input of heat due to shock wave dissipation:

$$dF_c(z) = \{E_r(z) - E_d(z)\} dz. \tag{30}$$

Taking into account the well-known dependence of the heat conductivity of a fully ionized gas on the temperature we have the following relation between conductive flux and temperature

$$\frac{dT}{dz} \simeq 6 \times 10^5 \times T^{-5/2} \times F_c. \tag{31}$$

The Equations (28), (30) and (31) have to be supplemented by the hydrostatic equation which Kuperus (1965) writes as

$$\frac{dn}{dz} = -n\left\{\frac{1}{H} + \frac{1}{T}\frac{dT}{dz}\right\}, \tag{32}$$

with n=total particle density.

In the simplest type of problem one assumes that $\phi(z) \approx 1$ (i.e. refraction effects are unimportant). It turns out that in many cases F_m and $\tilde{\kappa}$ can be expressed in terms of the local Mach number, the local velocity of sound, the 'period' P of the shock waves and the density. For instance, if the Mach number does not get too large Kuperus finds (this is based in part on the work of Landau and Lifschitz, 1959):

$$F_m = \tfrac{4}{3}\varrho c_s^3 \frac{(M_*^2 - 1)^2}{(\gamma + 1)^2 M_*}, \tag{33}$$

with $M_* =$ Mach number and $\gamma =$ ratio of specific heats. He also finds

$$\tilde{\kappa} = \frac{4\,(M_*^2 - 1)}{M_*^3 c_s P}.\qquad(34)$$

The whole problem is defined by the four coupled ordinary differential Equations (28), (30), (31), and (32) and three algebraic Equations: (29), (33), and (34). Correspondingly, we have seven unknown functions, namely F_m, $\tilde{\kappa}$, E_d, F_c, M, T, and n. Note that E_r and c_s should not be counted because they can directly be calculated from T and n.

Obviously the system (28), (30), (31), (32) can be integrated (at least in principle) as an initial value problem if n, T, F_m and e.g. M_* at the bottom of the chromosphere were known (Ulmschneider, 1967). One can also start the trial integrations from the corona (de Jager and Kuperus, 1961; Kuperus, 1965) assuming that $(\mathrm{d}F_c/\mathrm{d}z)$ in the corona is known.

In the actual calculations one encounters a number of difficulties of which we shall mention only two:

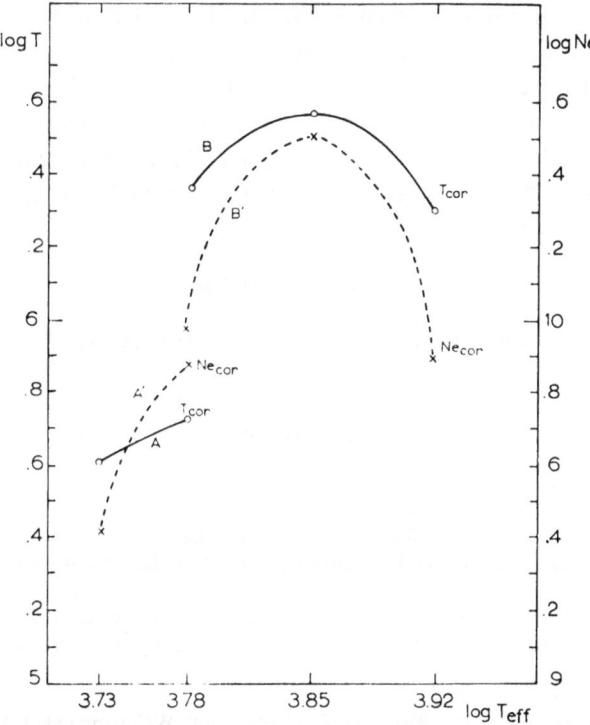

Fig. 9. Coronal temperatures (solid curves) and coronal electron densities (broken line) as a function of T_{eff}. The curves on the left side refer to $g = 10^5$ cm s^{-2} the lines in the middle of the diagram to $g = 10^4$ cm s^{-2}. After de Loore (1970). (By permission of Reidel Publ. Co.)

(1) It is difficult to describe the radiative losses E_r in the low temperature region (near the photosphere) correctly.

(2) It is important but very difficult to know what the energy distribution in the acoustic noise spectrum emerging from the convection zone will be like. [In the simple approach above this enters in the form of the factor P in the denominator of Equation (34)] Our knowledge of stellar convection as well as our understanding of noise generation by convection is still far from permitting reliable predictions in this respect. However, some very interesting studies of this problem have been made recently for the solar case. (Stein, 1968; Ulmschneider, 1971.)

The complexity of the problem is of course increased if the stellar wind influences the energy balance. (Kuperus, 1965; Shklovskii, 1965).

As an example of the type of information which one gets we reproduce some of the results by de Loore (1970) in Figure 9. The drawing shows the dependence of the coronal temperature and density on the effective temperature of the star in a range of surface gravities which are not too different from those for main sequence stars. One essentially finds that both the coronal temperature and density seem to be monotonic functions of the acoustic flux (at least in the range covered by these calculations).

It seems to me that in the field of stellar coronae a number of interesting developments are to be expected in the near future. Even a basically simple theory like that of Kuperus (1965) has not yet been applied to the objects with very high acoustic fluxes like helium stars and especially helium-rich white dwarfs. How high a coronal temperature would we expect in these objects? What kind of X-ray spectrum would we predict for these coronae? (Predictions of the X-ray emission of coronae of normal stars have been made by de Loore and de Jager 1970.) Even more drastic effects may be expected if one includes more exotic ways of coronal heating in such objects. (Strittmatter *et al.*, 1972).

Acknowledgement

This study has been supported by NSF Grant GP 28882 A, No. 1.

References

Abhyankar, K. D.: 1965, *Astrophys. J.* **141**, 1056.
Arakelian, M. A.: 1969, *Astrofizika* **5**, 75.
Beals, C. S.: 1931, *Monthly Notices Roy. Astron. Soc.* **91**, 966.
Biermann, L. and Lüst, R.: 1960, in J. L. Greenstein (ed.), *Stellar Atmospheres*, Univ. of Chicago Press, p. 260.
Bisnovatyi-Kogan, G. S. and Zel'dovich, Ya. B.: 1968, *Soviet Astron.* **12**, 192; *Astron. Zh.* **45**, 241.
Böhm, K. H.: 1968, in D. E. Osterbrock and C. R. O'Dell (eds.), *Planetary Nebulae*, Reidel Publ. Co., p. 297.
Böhm, K. H.: 1969, *Astron. Astrophys.* **1**, 180.
Böhm, K. H. and Cassinelli, J. P.: 1970, in H. G. Groth and P. Wellmann (eds.), *Spectrum Formation in Stars with Steady-State Extended Atmospheres*, NBS Spec. Publ. **332**, p. 54.
Böhm, K. H. and Cassinelli, J. P.: 1971, *Astron. Astrophys.* **12**, 21.
Brandt, J. C.: 1970, *Introduction to the Solar Wind*, Freeman and Co., San Francisco.
Buerger, P.: 1969, *Astrophys. J.* **158**, 1151.

Carlson, B. G. and Lathrop, K. D.: 1968, in H. Greenspan, C. N. Kelber, and D. Okrent (eds.), *Computing Methods in Reactor Physics*, Gordon and Breach, New York.

Cassinelli, J. P.: 1970, *Extended Model Atmospheres for Central Stars of Planetary Nebulae*, Ph.D. Thesis, Univ. of Washington.

Cassinelli, J. P.: 1971, *Astrophys. J.* **165**, 265.

Cassinelli, J. P. and Castor, J. I.: 1972, *Astrophys. J.* (in press).

Cassinelli, J. P. and Hummer, D. G.: 1971, *Monthly Notices Roy. Astron. Soc.* **154**, 9.

Chandrasekhar, S.: 1934, *Monthly Notices Roy. Astron. Soc.* **94**, 444.

Chandrasekhar, S.: 1945, *Astrophys. J.* **101**, 95.

Chandrasekhar, S.: 1960, *Radiative Transfer*, Dover Publ., New York.

Chapman, R. D.: 1964, *Radiative Transfer in Extended Stellar Atmospheres*, Ph.D. Thesis, Harvard.

Chapman, R. D.: 1966, *Astrophys. J.* **143**, 61.

Collins, G. W. II: 1963, *Astrophys. J.* **138**, 1134.

Collins, G. W. II: 1968a, *Astrophys. J.* **151**, 217.

Collins, G. W. II: 1968b, *Astrophys. J.* **152**, 847.

Collins, G. W. II: 1970, *Astrophys. J.* **159**, 583.

Collins, G. W. II and Harrington, J. P.: 1966, *Astrophys. J.* **146**, 152.

de Jager, C. and Kuperus, M.: 1961, *Bull. Astron. Inst. Neth.* **16**, 71.

De Loore, C.: 1970, *Astrophys. Space Sci.* **6**, 60.

De Loore, C. and de Jager, C.: 1970, in L. Gratton (ed.), 'Non-Solar X- and Gamma-Ray Astronomy', *IAU Symp.* **37**, 238, Reidel Publ. Co.

Finzi, A. and Wolf, R.: 1971, *Astron. Astrophys.* **11**, 418.

Groth, H. G.: 1957, *Z. Astrophys.* **43**, 185.

Groth, H. G.: 1970, in H. G. Groth and P. Wellmann (eds.), *Spectrum Formation in Stars with Steady-State Extended Atmospheres*, NBS Spec. Publ. **332**, p. 283.

Hardorp, J. and Strittmatter, P. A.: 1968a, *Astrophys. J.* **151**, 1057.

Hardorp, J. and Strittmatter, P. A.: 1968b, *Astrophys. J.* **153**, 465.

Holzer, T. E. and Axford, W. I.: 1970, *Ann. Rev. Astron. Astrophys.* **8**, 31.

Hopf, E.: 1934, *Mathematical Problems of Radiative Equilibrium*, Cambridge Univ. Press.

Hummer, D. G. and Rybicki, G. B.: 1971, *Monthly Notices Roy. Astron. Soc.* **152**, 1.

Kopal, Z.: 1959, *Close Binary Systems*, Chapman and Hall, London.

Kosirev, N. A.: 1934, *Monthly Notices Roy. Astron. Soc.* **94**, 430.

Kuperus, M.: 1965, *The Transfer of Mechanical Energy in the Sun and the Heating of the Corona*, Reidel Publ. Co., Dordrecht.

Kuperus, M.: 1966, *Trans. IAU* **XIIB**, 564.

Kutter, G. S., Savedoff, M. P., and Schuermann, D. W.: 1969, *Astrophys. Space Sci.* **3**, 182.

Landau, L. D. and Lifshitz, E. M.: 1959, *Fluid Mechanics*, Pergamon Press, London, p. 372.

Lighthill, M. J.: 1954, *Proc. Roy. Soc. London* **222**, 1.

Lighthill, M. J.: 1955, in J. M. Burgers and H. C. van de Hulst (eds.), *Gas Dynamics of Cosmic Clouds*, Interscience Publ., New York, p. 121.

Lighthill, M. J.: 1967, in R. N. Thomas (ed.), 'Cosmical Gas Dynamics', *IAU Symp.* **28**, Academic Press, New York, p. 429.

Lucy, L. B.: 1964, *Smithsonian Astrophys. Obs. Spec. Report.*, No. 167, p. 93.

Lucy, L. B.: 1971, *Astrophys. J.* **163**, 95.

Lucy, L. B. and Solomon, P. M.: 1970, *Astrophys. J.* **159**, 879.

Mark, C.: 1947, *Phys. Rev.* **72**, 558.

Marlborough, J. M. and Roy. J. R.: 1970, *Astrophys. J.* **160**, 221.

McCrea, W. H.: 1928, *Monthly Notices Roy. Astron. Soc.* **88**, 729.

Mihalas, D.: 1969, *Astrophys. J. Letters* **156**, L155.

Minin, I. N.: 1965, *Astrofizika* **1**, 275.

Morton, D. C., Jenkins, E. B., and Bohlin, R. C.: 1968, *Astrophys. J.* **154**, 661.

Nariai, K.: 1969, *Astrophys. Space Sci.* **3**, 160.

Paczyński, B.: 1968, *Acta Astron.* **18**, 511.

Parker, E. N.: 1963, *Interplanetary Dynamical Processes*, Interscience Publ., New York.

Pearce, W. P.: 1967, *Radiative Transfer in Finite Extended Stellar Atmospheres*, Ph.D. Thesis, Northwestern Univ., Evanston, Ill.

Periah, A.: 1969, *Astron. Astrophys.* **3**, 163.

Praderie, F.: 1970, in H. G. Groth and P. Wellmann (eds.), *Spectrum Formation in Stars with Steady-State Extended Atmospheres*, NBS Spec. Publ. **332**, p. 241.

Proudman, I.: 1952, *Proc. Roy. Soc. London* **214**, 119.

Rottenberg, J. A.: 1952, *Monthly Notices Roy. Astron. Soc.* **112**, 125.

Roxburgh, I. W. and Strittmatter, P. A.: 1965, *Z. Astrophys.* **63**, 15.

Ruciński, S. M.: 1969, *Acta Astron.* **19**, 245.

Ruciński, S. M.: 1971, *Acta Astron.* **21**, 455.

Rybicki, G.: 1970, in *Spectrum Formation in Stars with Steady-State Extended Atmospheres*, NBS Spec. Publ. **332**, p. 87.

Sapar, A. and Viik, T.: 1968, *Publ. Tartu Astrophys. Obs.* **36**, 120.

Schmid-Burgk, J.: 1969, *Mass Loss from Central Stars of Planetary Nebulae*, Ph.D. Thesis, Univ. of Heidelberg.

Shklovskii, I. S.: 1965, *Physics of the Solar Corona*, Pergamon Press, Oxford.

Sobolev, V. V.: 1960, *Moving Envelopes of Stars* (Engl. Transl.), Harvard Univ. Press.

Souffrin, P.: 1966, *Ann. Astrophys.* **29**, 55.

Stein, R. F.: 1968, *Astrophys. J.* **154**, 297.

Strittmatter, P. A., Brecher, K., and Burbidge, G. R.: 1972, *Astrophys. J.* **174**, 91.

Thomas, R. N.: 1970, in H. G. Groth and P. Wellmann (eds.), *Spectrum Formation in Stars with Steady-State Extended Atmospheres*, NBS Spec. Publ. **332**, p. 38.

Ulmschneider, P.: 1967, *Z. Astrophys.* **67**, 193.

Ulmschneider, P.: 1971a, *Astron. Astrophys.* **12**, 297.

Ulmschneider, P.: 1971b, *Astron. Astrophys.* **14**, 275.

Ulrich, R. K.: 1972, *The Propagation of Acoustic Waves in Stellar Atmospheres and Chromospheric Heating*, preprint.

Underhill, A. B.: 1949, *Monthly Notices Roy. Astron. Soc.* **109**, 562.

Underhill, A. B.: 1966, *The Early Type Stars*, Reidel, Dordrecht.

Underhill, A. B.: 1970, in H. G. Groth and P. Wellmann (eds.), *Spectrum Formation in Stars with Steady-State Extended Atmospheres*, NBS Spec. Publ. **332**, p. 3.

Unsöld, A.: 1951, *Naturwiss.* **38**, 525.

FIFTH DISCUSSION SESSION

(Friday Evening; 8 September, 1972)

(following the review paper by K.-H. Böhm)

Chairman: A. B. UNDERHILL

Underhill: We thank Dr. Böhm for his masterly summary. Are there questions?

Plavec: Dr. Böhm, there exist two single-spectrum binaries in each of which the component we are able to see is obviously a helium star. They are υ Sgr and HD 30353 (also known as KS Per). They are certainly different from white dwarfs, but they are helium stars probably with pure helium atmospheres. The masses might be around one solar mass, but the stars have probably evolved into the red-giant region in the H–R diagram, so, as a very rough estimate, their radii may be about 15 solar radii. The effective temperature of KS Per – according to calculations by Danziger *et al.* (1967) – is approximately 10000 K, so it is more or less an A0 star, if you can call a helium star A0. Would such a star still generate a sufficiently strong acoustic outflow so that it builds a large and dense corona? Then mass outflow from such a star need not be explained as a consequence of the body of the star itself filling the critical lobe, but rather by the facts that the star has a corona dense enough, and there is sufficient drive for the material, that mass outflow would be governed by this mechanism.

Böhm: I would say that these stars should have strongly developed coronae. I forgot to mention that calculations for helium stars that are not white dwarfs have been carried out by Nariai (1969). He finds that stars along the main sequence, up to temperatures of about 20000 K, have a very strong acoustic output. Giant stars, how-ever, will not have a strong mechanical flux up to quite such high temperatures. On the other hand, it is fairly certain that the mechanical flux will still be large at a temper-ature as low as 10000 K, and we should, therefore, expect the formation of a corona.

Wright: I was reading about emission lines in the spectra of M-type stars and Herzberg's suggestion (1948) that the coronae of these stars might contain mostly singly ionized iron, rather than highly ionized iron. Do you think that would be reasonable?

Böhm: Yes, I think it is generally accepted that stars of very late spectral types do not have true coronae, because their atmospheres have higher densities and, therefore, the convection velocities in their outer regions are lower. Consequently the acoustic output becomes smaller and the temperature of the corona is lower.

Underhill: You get into somewhat of a semantic problem whether 'corona' de-scribes only plasma with a very high temperature – hundreds of thousands of degrees – around a star, or whether it also describes an extended, cool atmosphere (10000 K or less). I hesitate to use the term 'corona' for both types of extended atmosphere.

Böhm: Maybe I should qualify my answer. It is certain that these late-type stars would not have hot coronae.

Batten (ed.), Extended Atmospheres and Circumstellar Matter in Spectroscopic Binary Systems, 171–173.

Wright: Well, we do observe these emission lines. My real question was – can we say they come from a corona?

Underhill: Well, you do have extended atmospheres that are quite cool around Be stars or Ae stars. I mean the temperatures of the atmospheres would be 10000 K to 15000 K. You don't call these coronae; you call them extended atmospheres.

Wright: I was really wondering if Dr. Böhm's theory of a corona could be used to study these late-type atmospheres.

Underhill: The key to that is whether you have a source of mechanical energy that can be transposed into temperature. Then you end up with a very high temperature because the gas cannot radiate the energy away.

Böhm: Not necessarily, it depends on the mechanical flux.

Underhill: How large it is, of course, yes. But this question of helium stars and generating a lot of energy – there are other helium stars, the normal ones, such as Popper's star.

Böhm: But they possibly have higher temperatures. Just as for the main-sequence stars, the main question is whether or not the helium stars have a convection zone which ends somewhere in the atmosphere. Only in this case will velocities just below the atmosphere be sufficiently high to support a corona. In a helium star, the convection zone occurs at higher temperatures than it does in a hydrogen star, but if the temperature is too high, then even this convection zone no longer exists. A helium star with a temperature of 30000 K, having a surface gravity comparable to that of a main-sequence star, will have no surface convection and no corona.

Underhill: Well, the normal helium stars, the original ones you were talking about, probably have a temperature of the order of 15000 K to 20000 K. Those begin to come into the temperature range for convection zones, I believe.

Böhm: Yes.

Popper: Well, I'm really out of my depth here, but you did refer in passing to some question about the acoustical heating mechanism for the energy provided to the corona. Just before coming here, I read a review of a conference on convection (Heap *et al.*, 1972) – mainly with reference to the Sun – held at Greenbelt. I got the impression that this standard mechanism had very successfully been called into question. Is my impression of that review correct?

Böhm: No. As far as I understood it, certain computational procedures (for instance, basing calculations on weak shock theory) were called into question. Ulrich, in particular, thinks that the whole mechanism might not work because, in the Sun, it would require a certain part of the acoustic spectrum (corresponding to a period of 100 s or so). According to Ulrich, these waves should have been seen in the Sun, but so far they have not been detected. As far as I know, this is one of the stronger arguments against the mechanism, but I don't know how strong it really is from an observational point of view.

Underhill: That more or less agrees with what I remember of the discussion. Let's adjourn until tomorrow morning.

References

Danziger, I. J., Wallerstein, G., and Böhm-Vitense, E.: 1967, *Astrophys. J.* **150**, 239.
Heap, S. R., Leckrone, D. S., Jordan, S. D., and Underhill, A. B.: 1972, *Earth Extra-terrest. Sci.* **2**, 69.
Herzberg, G.: 1948, *Astrophys. J.* **107**, 94.
Nariai, K.: 1969, *Astrophys. Space Sci.* **3**, 160.

SIXTH DISCUSSION SESSION

(Saturday morning; 9 September, 1972)

Chairman: M. G. FRACASTORO.

Peters: HR 2142 is a B1Vnne star which periodically displays a short-term shell phase. The period for the recurrent shell phase is 81 d. During the interval 1969–1972, I have observed eleven shell phases of HR 2142. The observations made during 1969–1971 are discussed in a recent paper (Peters, 1972). Only a brief summary of these earlier observations will be presented now.

The shell lines which periodically develop near the centres of the Balmer lines and λ3889 of He I during the shell phase are red-shifted relative to the photospheric features at first and shift toward the centre of them as they strengthen. As an example, consider the Hβ shell line which initially has a radial-velocity shift of 45 km s^{-1} relative to the centre of the emission feature, shows a shift of 30 km s^{-1} at zero phase, (defined as the time of maximum core strength) but a slightly positive shift of 5 km s^{-1} at a phase of $0^P.02$. The duration of this sequence is 5–7 d. A variation is observed in the strengths of the shell phases, however, similar shell phases do occur in groups of two to four.

The observations made during the shell phases of 1972 January and April revealed that the shell lines re-appear for a short time after an initial decline in strength. The observations made during 1972, April showed that the shell lines in all observed features except Hβ had disappeared (at 46 Å mm^{-1}) two days after zero phase. Four days later, the shell lines were again present and about 80% of the strength observed at zero phase. These shell lines had completely disappeared two days later. The shell lines which re-appear after zero phase are blue-shifted relative to the photospheric lines. A plate of dispersion 11 Å mm^{-1} on 1972, January 30 revealed a radial velocity shift of -85 km s^{-1} relative to the centre of the emission feature at Hβ.

The two parts of the shell phase are so different in character that I call the first one the primary and the other the secondary shell phase. The duration of the secondary shell phase is about 1–3 d. Previously, these double shell phases were called long-duration shell phases because fragmentary data suggested that the shell lines persisted continuously for 14 d before disappearing.

Spectrograms at 11 Å mm^{-1} obtained at Lick Observatory by D. M. Popper and M. Plavec, covering nine complete cycles, reveal a periodicity in V/R at Hβ with phase. Additional plates acquired since 1971, November confirm the Figure 4 in the paper cited and remove most of the speculative portions of the curve. The strict periodicity of the shell phases and the regular variations in V/R at Hβ suggest that HR 2142 is a binary and that the shell phase appears when the system presents a certain orientation relative to our line of sight. A model currently under consideration by Ronald S. Polidan and myself suggests that material is being transferred to the primary, from a yet undetected secondary of spectral type later than A, in the form

Batten (ed.), Extended Atmospheres and Circumstellar Matter in Spectroscopic Binary Systems, 174–191.
All Rights Reserved. Copyright © 1973 by the IAU.

of a thick stream. The shell phase occurs when a portion of the stream passes in front of the primary as we see it (Figure 1). Such a stream would explain the radial-velocity and intensity sequence observed during the primary shell phase. Whereas we look more or less along the stream when the shell lines are weak, we are observing the stream more nearly tangentially just after zero phase (when we view the stream through its thickest portion). Material leaving the secondary star from L_2 could be responsible for the secondary shell phase. Alternatively, this phase could occur when we view along a counter stream. The high negative velocity observed for the secondary shell-phase lines suggests that we are observing a different part of the stream from

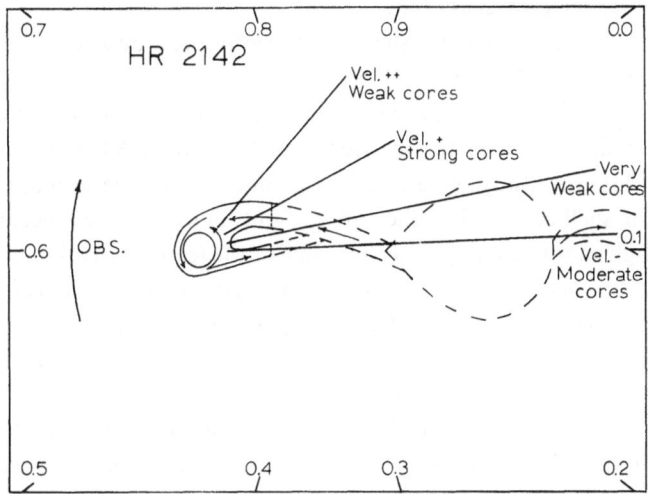

Fig. 1. Model proposed by Peters and Polidan for HR 2142. Figures around the border indicate direction of line of sight at the indicated phases.

that seen in the primary shell phase. The weak, permanent core observed in Hβ and Hγ could be formed in an expanding envelope (velocity -30 km s^{-1}) which surrounds the primary star. The rapid rotation of HR 2142 could be a result of angular momentum which has been transferred to the primary as it accreted material from the secondary.

Whereas the above qualitative model can explain many of the observed features of the shell sequence, it is difficult to explain the regular V/R variations in Hβ. The short duration of the secondary shell phase suggests that the material responsible for it is rather localized. The orbital inclination of the system must be 70°–80° (an eclipse must just be missed). It is difficult to understand how material from either L_2 or a counter stream could be so localized this far out of the orbital plane.

This system raises the wider question: are some Be stars binaries in the process of mass exchange? Would envelopes around such stars be different in type (density distribution, velocity distribution, turbulence, etc.) from those around single Be stars? What spectral features might distinguish these two possible types?

A number of Be stars do not fit the scheme of emission-line profile versus inclination described in Dr. Boyarchuk's review paper – notably HR 2142 which has a high
$V \sin i$, but whose spectrum shows only a very weak permanent core in Hβ and none
at all in Hα. Unlike some equator-on Be stars, the spectrum of HR 2142 shows only
emission lines of Fe II, and no shell lines of Fe II, Ca II, Ti II, etc. We have observed
the Ca II triplet at λ8550 in emission, although few stars show this triplet in emission.
We have also observed it in the spectra of π Agr, φ Per and υ Sgr. The first two of
these also show Fe II in emission. Could the Be stars that display Fe II and the Ca II
triplet in emission be binary systems exchanging mass?

Van 't Veer: Do you know the mass ratio of the system?

Peters: No, we can only guess at it.

Van 't Veer: Has the mass transfer only recently been observed – or has it been
observed for some years?

Peters: As far as I can tell, the system has been behaving in this manner for the
last 50 yrs. McLaughlin, reported by Merrill and Burwell (1949), found that the red
component of Hβ was always greater than the blue, and I have seen references to the
red shifts of the shell cores, when they are present, but the earlier spectrograms were
really not of high enough quality to reveal some of the features that I have found.

Chen: I find this investigation very interesting, and the binary model indeed offers a
plausible interpretation of the periodical changes of spectral lines. I still like Struve's
suggested model of a single star with circumstellar matter, however. Mr. Tom Morgan,
of the University of Florida, has investigated the hydrodynamic flow around a single
star (Morgan and Chen, 1972). He found possible periods of variations of the order
of days or months. I believe, therefore, that this is an alternative explanation of
periodic variations. Are you quite sure of the measurements of radial velocity from
the absorption lines in the spectrum of HR 2142?

Peters: I have measured some of our plates for radial velocity and looked for
variations. The lines are very diffuse: the star has a high value of $v \sin i$. Only three
helium lines have proved suitable for measurement, but they are often asymmetric
between the shell phases. The asymmetry could be caused by helium emission from
the circumstellar envelope distorting the feature, or it could possibly be absorption
lines of a secondary spectrum blending with the primary spectrum.

Smak: How serious could be the contamination of the absorption-line intensities
by the variations of the emission-line intensities? Or, to ask a more risky question,
is the secondary maximum possibly due to some strange behaviour of the emission
component?

Peters: I don't think so.

Plavec: I think that one possible difficulty in this picture of HR 2142 is that the
fan of material that is streaming away may be obscured by the fairly big body, so it is
difficult to imagine how it can be seen, projected against the body of the only star we
observe. In that case, probably what you mention as the 'counter stream' would be
more feasible, because, after all, we observe something like it in U Sge, or the ring of
β Lyr. We have no explanation for these streams that show violet-shifted lines, but

nevertheless they exist. Certainly there is a possibility that we can explain HR 2142 by a single star, but I would like to hear from Dr. Chen exactly how this periodicity arises in a single star.

Chen: Morgan applied Euler's hydrodynamical and continuity equations to the flow around a single star. The main external force is the gravitational attraction of the star. Then, as Dr. Huang indicated, the difficulty is to impose a boundary condition for the flow. He considers the first-order 'perturbation' of Limber's (1964) steady-state solution. The set of resulting time-dependent equations yields periodic solutions.

Hutchings: Has Mrs. Peters estimated $v \sin i$ from the profile?

Peters: Yes, it is 350 km s^{-1}, determined graphically from the He I lines $\lambda\lambda 4026$ and 4144.

Oliver: Is there any photoelectric photometry of the system?

Peters: No, but I wish someone would put it on their list.

Cowley: What mass ratio did you assume?

Peters: 0.25.

Cowley: Doesn't this imply a secondary component that should be visible spectroscopically?

Polidan: That value of the mass-ratio makes the secondary something around a middle A-type star. I don't think you could detect it.

Cowley: Not if it is a main-sequence star, rather than an evolved subgiant that could be seen at long wavelengths.

Polidan: If you believe the theoretical calculations, this was originally a fairly massive system, and after mass exchange you should have a B-type star and a middle A-type star.

Underhill: One B0 star with a shell would have a mass about 16 $M_\odot$, so the other one would have a mass of 4 $M_\odot$ – then its spectral type should be B6 and we ought to be able to see the hydrogen lines.

Peters: The spectral type of the proposed secondary is later than B9, so the mass ratio should be lower.

Biermann: If the cooler star is a well developed subgiant, then it has the same luminosity now that it would have had if it had evolved as a single star, because the luminosity of a subgiant is determined not by its total mass, but by its core mass. If this star now has a mass of 4 $M_\odot$, and originally had a mass of 10 $M_\odot$ or 12 $M_\odot$, it is really strange that you don't see it.

Smak: First, I think the mass ratio we are talking about was an assumed one – it could be much different. Secondly, there are many similar systems, with similar mass ratios, in which the secondary spectrum is invisible – so we should study this one in detail.

Fracastoro: Mrs. Peters also wants to read a communication from Dr. Aller, who was prevented by illness from attending this Symposium. The report is on work by S. Heap and Aller.

Peters: The star HD 45166 ($V = 9\overset{\mathrm{m}}{.}88$) may be described best as a 'pseudo-Wolf-Rayet star'. Its spectrum displays intense, very sharp emission lines corresponding to the higher-ionization stages of C, N, O, superposed on a background continuum which shows the Balmer absorption lines. It is not a true Wolf-Rayet star since the emission spetrum changed strongly between 1922 and 1933 (Anger, 1933) the emission lines are very sharp, and include lines of C, O, and N in approximately equal strength. The present investigation includes high-dispersion spectra, energy scans, and filter photometry in the satellite ultra-violet. In brief, HD 45166 appears to be a binary, consisting of a late B-type star and a hot companion responsible for the emission line spectrum; the hot companion differs from true WR stars in having a normal helium abundance and, presumably, a low mass, comparable with that of the Sun.

The width and total intensity of the Pickering lines indicate strong blending with hydrogen. The widths of the lines of C, N, O, etc. indicate velocities of only about 110 km s^{-1} a value characteristic of Of stars rather than WR stars, although the character of the emission spectrum (e.g. strong O lines in emission) differs markedly from that of classical Wolf-Rayet stars. There is no evidence for the Bowen fluorescent mechanism. The absorption spectrum corresponds to a late B, thus setting the system far apart from a normal Wolf-Rayet system; the late B component is either a slow rotator or we are looking at the system pole-on. The emission-line spectrum appears to be about the same as it was when Neubauer and Aller (1948) examined it in 1943–44, but short-term variations are likely. By combining photoelectric energy scans covering the region $\lambda\lambda$3250–7780; (kindly made in 1971 by R. Stone with the Wampler scanner at the Crossley) with ultraviolet photometry from the OAO-2 (using Code's preliminary calibration factors as modified by Code to account for degradation of the filters), the total energy distribution is obtained. Space absorption is estimated by indirect arguments, from data secured by various observers for nearby stars and clusters in this region, as well as the strength of the λ4430 band; the adopted value is $A_v = 0^m.2$.

That HD 45166 is a binary composed of a late B-type star and a hot companion is indicated by the following: (1) superposition of high-excitation emission line spectrum and low excitation absorption line spectrum, (2) relative velocity between emission and absorption lines, (3) veiling of Balmer discontinuity in a late B-type star by the hot companion, and (4) high ultra-violet flux from the hot companion. The companion appears to be responsible for the emission-line spectrum and is referred to as the quasi-Wolf-Rayet star. If it is a close binary involved in mass-exchange, the qWR star has not yet lost its hydrogen-rich envelope. Continuing mass exchange is compatible with observation of long term variations in the emission spectrum.

From the profiles of the Balmer lines it is concluded that at λ4340, the qWR and B8V stars are of about equal luminosity, hence $M_v(\text{qWR}) \sim 0^m.0$. From a distance estimate of the system of 700 parsecs, $M_v(\text{qWR}) \sim 1^m.2$. In any event the quasi-WR component is much fainter than a classical WR star. In this respect HD 45166 seems comparable to the central star of a planetary nebula, e.g. nucleus of NGC 1514 or that of NGC 6543, but HD 45166 is not the nucleus of a planetary nebula.

Paczyński (1967) has suggested that the occurrence of WR stars is a consequence of mass transfer in close binary systems, the process of mass-loss stopping here, evidently, just short of a complete loss of the hydrogen-rich envelope. The remnant has instabilities which produce a WR-type spectrum. The binary system composing HD 45166 does not belong to the same position of the HR diagram as true WR stars. The absorption line component has a later spectral type and lower luminosity than the companions to true WR stars which are usually O and early B giants. The emission-line component is also less luminous than true WR stars, and hence less massive. HD 45166 is an example showing that the atmospheric conditions producing highly excited C, N, O emission lines may arise during the evolution of low-mass stars as well as high mass stars and in stars of normal abundance as well as those of high helium abundance.

Underhill: I've talked a lot about this work with Sally Heap. Although Aller uses the term 'quasi-Wolf-Rayet', I think she feels very strongly that the emission-line object is very similar to the central stars in planetary nebulae – which have sometimes been called 'Wolf-Rayet' although the emission lines in their spectra are much narrower than those in the spectra of Walf-Rayet stars, and the composition of their spectra is quite distinctive. I would suggest that you forget the word 'Wolf-Rayet' in considering HD 45166: it's a hot subdwarf of absolute magnitude around $1^{m}.0$ or $0^{m}.0$, combined with a late B-type star.

Fracastoro: We come now to a discussion of period changes. First, Van 't Veer.

Van 't Veer: This is not a prepared paper but merely a comment on the first two days when we heard so many things on gas streams which can last for at least ten or even fifty years, as we have seen this morning. It is important to ask two questions:

(1) What is the influence of this mass transport on the orbital period of the system?

(2) Do we measure period changes which are in agreement with our knowledge on gas streams?

The first point can be roughly treated in the following qualitative way. The matter is flowing from one component to the other so the total mass of the system is conserved. Using Kepler's third law we have $\omega^2 r^3 = $ const, r being the distance between the two components and ω the angular velocity. On the other hand, from the constancy of angular momentum we can deduce that ωr^2 must decrease if the angular momentum of the outflowing matter is increasing. This is the case for a semi-detached system of mass ratio $r \geqslant 1$ with matter outflowing from L_1. The result must be a decreasing period and a decreasing orbital radius with shrinking Roche lobes which can perhaps maintain the mass transport for a reasonable time. If the star filling its Roche lobe has a much lower mass than the primary component, we will obtain the opposite result. So the period changes can give us an important information on some details of the mass transport. Unfortunately the period measured between two successive primary minima not always coincides with the real one, defined as the time elapsed between the corresponding conjunctions. Asymmetries of the light curve during minima are often the cause of erratic fluctuations which may mask the real period changes. I am afraid that we do not possess sufficient observations of semi-detached systems to permit clear conclusions; but I should be glad to hear your opinion on this point. I should like to end this short comment by asking you still two other questions:

(1) What do we know about the mass ratios of the systems with gas streams?

(2) What is the mass of the outflowing matter?

Smak: I have four comments. First, if one assumes that the orbital momentum is conserved, then the absolute dimensions of the Roche lobe around the secondary reach a minimum at a mass-ratio of about 0.8. Secondly, the main reason for the mass outflow is not the shrinking of the Roche lobe, but the expansion of the star, either on the thermal, or nuclear time-scale. My third comment is that the simple picture you have presented is modified, though only slightly, by including the rotational momenta of the components. Finally, the major trouble with the simple star-to-star mass-transfer interpretation of the period changes is that it gives rather short time-scales. The best example is provided by the W UMa systems, which show rather fast period variations, which would imply time-scales of the order of 10^6 yrs for these systems. This is definitely too short, as compared with the large number of these systems observed and their very probable connection with the main-sequence phase of evolution.

Van 't Veer: It is true that I was ignoring the rotational momentum of the individual stars. I think you will agree that this is not more than two or three per cent of the orbital angular momentum of the system. Secondly, on the changes in the Roche lobe, it would be very difficult to maintain a stream if the lobe were expanding.

Plavec: When people working in different fields meet in a conference, it must occur inevitably that discussion turns towards areas which have been discussed in the field for years. Dr. Van 't Veer has my admiration because he managed in a short time to present several problems which have been discussed for years and for example have been described by Kruszewski (1966). So let me concentrate on the problem of how the smaller star can maintain the outward stream not only for 50 y but probably for a much longer period of time. The actual reason is that the star is developing on a nuclear time scale, and because, in this case, it is a subgiant, while it is still burning hydrogen in its core. It has a natural tendency towards expansion but because there is the Roche lobe around it, it must happen that some material is being pushed across the Roche limit just because of internal forces. As a consequence, mass loss can go on for a very long period of time, defined more or less by the main-sequence evolutionary time for a star of one of three solar masses.

Batten: To answer one of Dr. Van 't Veer's questions directly, period changes of the kind you described are known. The prime example is β Lyr. I think most people would agree that the period change in that system is due to mass transfer. I think the case is almost as strong for U Cep, but of course in other systems there are many other factors that we do not understand. The picture of observed period changes is very complicated, and obviously other factors are at work. Dr. Smak recently proposed a model which might account for this although in my review paper, I did express some reservations about it.

Smak: I hope that will be recorded!

Batten: I might add that the increase in period of β Lyr has now been observed for about 200 y, and that of U Cep for nearly 100 yrs. Of course, most systems known to show period changes have fluctuating periods.

Huang: Since Dr. Batten mentioned β Lyr, I would like to add that one of the reasons why I proposed that the secondary component of β Lyr is the more massive component is the period increase, that can be explained as due to the ejection of mass as Dr. Van 't Veer has just described.

Fracastoro: Sometimes we find spurious changes of period due to third-body light-time effects or to apsidal rotation – so the problem is rather complicated. Periods increase and decrease not only in contact binaries, and in dramatic pairs like β Lyr, but in very common semi-detached systems. In order to fit the time scale, one could perhaps think of sporadic phenomena, like prominences in the Sun – not a continuous outflow of matter.

Van 't Veer: So you conclude that no binaries have constant periods?

Fracastoro: Oh, no: that is not my conclusion – although maybe sometimes there are period changes below our ability to detect them.

Oliver: You mentioned contact and semi-detached systems: it is now becoming clear that even detached systems show period changes. Dr. Van 't Veer mentioned that changing asymmetries in the primary of secondary minimum of the light curve can cause apparent period changes. In many systems, this does happen, and it causes confusion. There are many complications that completely obscure the period changes caused by mass transfer. It's a 'fun' problem.

Fracastoro: Now Dr. Herczeg has a contribution, and we will postpone discussion of that until we have also heard from Dr. Hall.

Herczeg: I am going to make two somewhat longer comments. First, a summary of the basic facts that observers have established concerning period changes of eclipsing binaries, then an attempt to interpret the period of β Lyr.

1. Period changes in general

In several cases timings of minima go back 100 yrs or more, although visual estimates (unlike visual measurements by expert observers like Wendell, Dugan, Danjon, Detre and others) are far too inaccurate to reveal more than the crudest features of the $(O-C)$ diagrams. It is only since the mid-1940's, the advent of extended photoelectric studies by the aid of the multiplier phototube, that we can detect rather minute details of the period changes. For the benefit of the theoreticians working on this field I should like to give a set of short statements, not to say axioms, summarizing the experience of decades of observations.

(a) The first of these statements sounds rather trivial. Most eclipsing binaries do show period changes. This applies specifically to so-called typical Algol configurations and W UMa-variables where constant periods are exceptional. We find more constant periods among detached systems although even here we know some surprising cases showing large period changes (e.g. TX Her).

(b) Part of the observed period changes may well be spurious caused by small, random or regular, deformations of the light-curve near minimum. While such effects certainly exist, their influence seems to be relatively small, of the order of, say, $0\overset{d}{.}01$ as judged from studies like that of Van Woerden (1957) on SV Cam or Kwee (1958) on

VW Cep. Major shifts of the minima up to $0\overset{d}{.}1 - 0\overset{d}{.}2$ may in most cases correspond to real changes of the orbital elements.

(c) However, one possibly important reservation should be mentioned here. Small, random shifts of the epoch of minimum, occurring from cycle to cycle and remaining necessarily undetectable, may sum up (this is a problem of one-dimensional random walk) to produce the observed time residuals. I hardly doubt that in a few cases this type of spurious period changes determines the character of the $(O - C)$ diagram.

(d) We know perhaps a dozen well-defined cases of apsidal motion, possibly 2 or 3 cases of light-time effect and a few binaries with linearly increasing or decreasing period, but the overwhelming majority of the $(O - C)$ diagrams remains quite irregular. With the higher accuracy of the timings, a very frequent, almost typical pattern emerges. Roughly half of the better studied cases show virtually constant or nearly constant periods for years, even decades, then a more or less sudden change occurs resulting in a new period that again remains constant for years or decades. This type of variation makes it improbable that the above mentioned random-walk mechanism would be at work.

(e) These sudden period changes mean in many cases a smooth bending over of the $(O - C)$ curve, extending the transition to the new period to several years. Often, however, the transition is rather fast and we can localize it to a few months or even weeks. In these cases we may speak of veritable discontinuities or jumps of the binary period, accompanied sometimes by observable spectral or light-curve changes, like changes of the spectroscopic behaviour of RCMa about 1914 (Wood, 1957) or the flaring of W UMa observed by Kuhi (1964). This system of VW Cep showed such a discontinuous period change just during the 1959 international campaign.

The $\Delta P/P$ ratio is practically always of the order of 10^{-5}. The sudden changes seem to occur at irregular intervals, with a mean time lapse between them of perhaps 20–25 yrs. Neither the positive nor the negative changes appear to be clearly preponderant; indeed, there is even tendency to exhibit alternating sign sequels: $+ - + -$ etc. Any long term, secular change of the period can be, of course, very effectively masked by this type of irregular, sudden period change.

2. The Period of β Lyr

Although period changes of β Lyr were indicated very early after Goodricke's discovery of the variability, and the non-linearity of the period increase was already extensively discussed by Argelander, all efforts to describe the variation of the period failed hitherto. Usually polynomial expressions were suggested, containing terms up to E^5 in the formulae for the $(O - C)$ values of the timing of minima, but predictions invariably broke down after only 10–12 yrs. The period has been found to be steadily increasing for almost 200 yrs, while the rate of increase has been steadily decreasing all the time. The situation is the more intriguing as the $(O - C)$ curve and the period changes seem to be quite smooth, without conspicuous irregularities. Since polynomials in E failed to represent the variations of the period, I tried to describe them – and this first sounds almost like a scientific joke – by an infinite series. I found a surprisingly good representation by a simple exponential expression of the form

$$P = P_0 (1 - \alpha e^{-\beta x}).$$

This also seems to make more sense from the point of view of physical interpretation than the formal polynomial representations, since a formula like this suggests a slowly but steadily decreasing stellar activity, so to speak, probably a diminishing rate of mass loss from the whole system. The $(O-C)$ curves, following from this representation of the period changes, should correspond to an $ax + c^{-bx}$ type curve which is, indeed, quite manifestly the case. On the other hand, a detailed discussion of the $(O-C)$ curve is made rather difficult, not so much by the irregular nature of the curve itself, as by large and irregular shifts of the epochs of minimum due to occasional strong distortions of the light curve. A comprehensive discussion of the period of β Lyr has been finished recently and I am going to submit the final manuscript for publication within a couple of months.

Hall: I want to propose a model to explain the large cyclical period changes which are observed in virtually all of the semi-detached Algol-like eclipsing binaries. Before beginning I must, in order to lay the proper foundation, make clear a few points.

I am restricting my attention to the semi-detached Algol-like binaries, which are now understood to be remnants of post-main-sequence mass exchange. I am excluding contact, and I am excluding the RS CVn binaries, most of which are definitely detached and are probably in some sort of pre-main-sequence phase of evolution (Hall, 1972). Both the W UMa binaries and the RS CVn binaries also experience period changes of similar size to those in the Algol-like binaries, but probably for a different reason.

Theoretically, we would expect a monotonic (if not uniform) period increase in Algol-like binaries as a consequence of mass transfer from the less massive cooler subgiant to the more massive hotter star.

In contrast, what we see observationally are very large $(\Delta P/P \sim 10^{-5})$ period increases and decreases. There is a tendency for the changes to be comparable in size and to alternate in sign, this often giving the impression of a sinusoidal variation in the $(O-C)$ diagram (a plot versus time of the observed time of primary minimum minus the time computed with some assumed constant period). However, it is quite clear now that very few, if any, of the $(O-C)$ curves are really sinusoidal or even strictly periodic. Earlier attempts to explain the cyclical period changes as a result of apsidal motion or of orbital motion around a third body must, of course, be on the wrong track if the variations are not really periodic. The characteristic time between alternate period changes is between 10 yrs and 100 yrs for different binaries, so 30 yrs is a good number to remember in the discussion which follows. One or two binaries do show the predicted slow monotonic period increase, U Cep being the best example (Batten and Plavec, 1972). But even in U Cep, large alternating variations in period are superimposed on this steady increase. In all of the other binaries, presumably the teady increase is occurring but is completely masked by the larger alternating variations. A few people, Dr. Herczeg for example, have emphasized that there is a marked tendency for the period changes to be abrupt rather than gradual, with the period constant in between these changes. This is a feature which my model should be capable of explaining.

In order to avoid confusion, we must note that, since the very small slow period increase can be neglected, there is an average orbital period P and major semi-axis a which describes each system during any given century or so. Likewise there must be an average amount of mass transfer from one star to the other and (if some of the angular momentum carried over by the mass goes into rotational angular momentum) also an average fraction describing how much goes into rotation. Therefore, we can simply look at fluctuations in these various quantities about their average values.

The fundamental basis of my model is an exchange back and forth between orbital angular momentum, J_{orb}, and the angular momentum stored in the disk around the hot star J_{rot}. Such an exchange was suggested for another reason by Paczyński (1971), and invoked later by Smak (1972), as the only reasonable mechanism to explain the alternate period changes observed in the U Gem binaries. With this exchange as a basis, I want to show that a few relatively elementary considerations make it very natural, in fact unavoidable, that the period should alternate cyclically between values larger than the average and values smaller. A complete explanation of this phenomenon must be extremely complicated, but I think now is the time to take the first step in the right direction. My hope is that this model, although incomplete and unfortunately qualitative in many places, is correct as far as it goes and is, in fact, a step in the right direction.

As Smak has shown must be the case with the U Gem binaries and as my calculations show must also be the case in the Algol-like binaries, about $10^{-5} M_\odot$ of mass in the rotating disk must be involved in the exchange. This is too much mass to reside exclusively in the optically thin outer parts of the disk and in that sense must be part of the 'interior' of the star, even though its rapid rotation makes it part of the 'disk'. Even more important to note is that $10^{-5} M_\odot$ is too much mass (by at least one or two orders of magnitude) to be accounted for by the amount of mass which comes over in the stream in the course of 30 yrs. Therefore, the stream must act only as the 'trigger' which controls the exchange between J_{rot} and J_{orb}.

In Figure 2 is a schematic representation of an $(O-C)$ diagram, not meant to be strictly sinusoidal or periodic. At point A the period P is maximum and, by Kepler's law, a is maximum also. Consequently J_{orb} is maximum. Since there must be conservation of total angular momentum in the system, J (total)$=J_{orb}+J_{rot}$, it follows that J_{rot} must be minimum. When a is maximum, the Roche lobe around the mass-losing subgiant must have its maximum size. And since it is generally accepted that the rate of mass loss is proportional to the degree to which the contact star overflows its Roche lobe, the rate of mass loss should be minimum. At point B the opposite of each of these statements can be made.

Let us start at point A, where the stream has its minimum strength. If ever the stream increases slightly, it will begin to encourage the disk to rotate more rapidly,

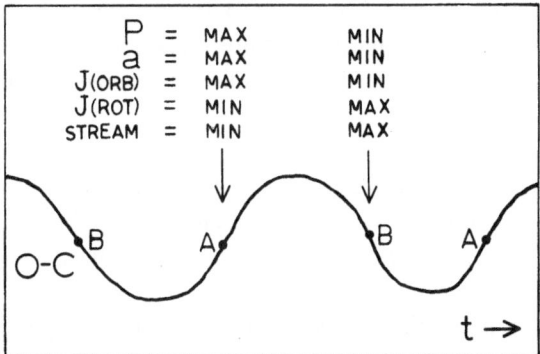

Fig. 2.   Schematic (O—C) diagram to illustrate Hall's mechanism for period changes (see text).

probably in order to have the disk velocity match the stream velocity at the point of impact. The disk thus begins to increase its J_{rot}, drawing not so much from the angular momentum in the stream itself but mostly from the J_{orb}. In order for tidal interaction to allow the exchange to take place on such a short time scale, it is clear there must be turbulent viscosity in the disk.

The crucial thing is that, once J_{rot} begins to increase and J_{orb} consequently begins to decrease, the separation a begins decreasing and the stream increases further. Thus the process must be self-accelerating and the period must continue to decrease for some time.

In order for this self-accelerating process not to continue forever, it is necessary either that (1) there exist some maximum amount of J_{rot} which the disk can contain or that (2) there exist some maximum rate of mass loss that the contact subgiant can sustain for any appreciable length of time. The existence of both of these maxima seems reasonable to me (although I cannot prove it) but one of the maxima is probably the critical factor in this situation. It is convenient for my preliminary presentation of this model that the existence of either maximum would be sufficient to prevent the self-accelerating mechanism from continuing forever.

This brings us to point B, where the stream has its maximum strength. If now the stream decreases slightly, the disk will be encouraged to begin slowing down. As the disk then transfers J_{rot} back to J_{orb}, the separation a must begin increasing again and the Roche lobe around the mass-losing star must expand. With the stream decreasing further, we now have a self-decelerating mechanism and the period must continue to increase for some time. It is easier to explain why the self-deceleration does not continue forever. The simplest explanation is that the Roche lobe expands sufficiently to shut off the stream entirely; in other words, the minimum rate of mass transfer is zero. It is, however, also possible that the critical factor in this situation is the existence of a minimum J_{rot}, maybe zero. But again it is convenient for my model that either of these minima would be sufficient to stop the self-decelerating mechanism. We are now back at point A, and the cycle can begin again.

My model could rather easily account for Herczeg's abrupt period changes. There is no obvious reason why the binary might not remain at point A for some time, especially if the stream then is turned off completely. Similarly, it is possible that the binary might remain at point B with nearly the maximum amount of mass flow in the stream for a length of time appreciably longer than the time scale involved in the actual self-acceleration or self-deceleration part of the cycle. The appearance of the $(O-C)$ diagram then would be straight line segments at point A and nearly straight line segments at point B, with the curvature between not always apparent in a typical observed $(O-C)$ diagram.

I have been examining light curves of totally eclipsing Algol-like binaries in an attempt to discover some correlation between asymmetries around totality with the $(O-C)$ diagram. Restricting my attention to light curves obtained in the visual, I tried to decide whether the bottom of totality was symmetrical (either slightly rounded, as in the case in VW Cyg and AQ Peg, or exactly flat) or asymmetrical (either slanted up as in the case of SW Cyg or slanted down as in the case of Walter's light curve of RV Oph). It is premature for me to be certain whether or not a correlation exists, but there is some indication that asymmetries always occur when the $(O-C)$ diagram is between A and B (when the disk is being speeded up by the steadily increasing stream) and that the totalities are symmetrical between B and A (when the disk is slowing down and the stream is decreasing). Note that the asymmetries do not occur when the stream itself is a maximum, i.e., on either side of B itself. Thus it is probably more relevant to think of the asymmetries in terms of an asymmetry in the distribution of mass in the disk itself, and not in terms of the stream. I did not consider old photographic light curves, whose spectral response often reached into the ultraviolet, because if my experience with SW Cyg is any indication, the U light curve can reveal the influence of the gas stream or the hot spot on one side of the hotter star, while at the same time the V light curve can be revealing an optically thick accumulation of matter on the other side of the star.

Smak: I have four comments again. First, what Dr. Hall described as slow variations of the period which are hidden in larger, faster, and more irregular ones, corresponds to the systematic effects due to mass-transfer on a nuclear time-scale. It may be worthwhile to mention, however, that there are few cases where we do observe rather fast period variations being apparently due to the rapid mass-transfer on a thermal time-scale. Three examples I wish to mention are: β Lyr, V 367 Cyg (suggested by Plavec in this context), and SV Cen (suggested by Paczyński). As a second comment, I want to suggest that it could be good to check whether the expected variations in the size of the Roche lobe around the secondary are sufficiently large to modify the rate of the mass outflow. My third comment refers to the possibility of a temporary storing of the momentum in the disk, as was first suggested by Paczyński (1967b) in his discussion of WZ Sge. It is quite obvious that in this case, we should try to learn more about the structure of such massive disks, instabilities which are likely to occur in them, and their time scales. Only good models of disks will eventually permit us to discuss this mechanism in a more meaningful way. There has been only one serious attempt in this direction, reported briefly in a paper by Prendergast and Burbidge (1968) and this brings me to my fourth comment. Numerical results by Prendergast and Burbidge imply that due to the transfer of momentum across the disk, there can be a mass outflow from the system. This might be a very powerful mechanism for period changes, although without detailed calculations we cannot make any numerical predictions.

Finally, to clarify one of my earlier remarks, I do realize that the W UMa systems also show alternating period changes. My point is only to indicate that if any such variations were interpreted in terms of secular evolution, that would result in time-scales that were definitely too short.

R. E. Wilson: I would like to ask Dr. Herczeg: what is the multiplier of t in his equation? I am interested in using his formula.

Herczeg: I have only preliminary values, and wanted only to indicate the basic character of the relation. When I have a definitive value, I will send it to you.

Milone: What about these binaries that show a migrating asymmetry in the light curve? As this travels through the minimum, surely it can distort the minimum and thus affect the $(O-C)$ values?

Hall: I think that binaries of the RS CVn type and Algol-like binaries should be considered separately. Migrating asymmetries are found in the light curves of the former type. You are correct in principle – and the RS CVn binaries do show large variations in $(O-C)$. I believe, these are the result of true changes in the orbital period, for two reasons. First, since the origin of the distortion is an uneven distribution of surface brightness on the cooler star, the distortion should only seriously affect the secondary minimum. Second, I think that the observed variations in $(O-C)$ are too large to be explained by distortion of the light curve during eclipse. Maybe this should be checked quantitatively.

Bath: When you have these alternating period changes, is there any difference between the rate at which the period increases, and that at which it decreases?

Hall: I have not looked at this very carefully, but I think there may be a tendency for increases to be larger than decreases. The $(O-C)$ plot looks a little like a Cepheid light curve – it rises more steeply than it descends.

Catalano: Sometimes we find that the $(O-C)$ residuals can be satisfied by two constant periods. If we refer the residuals to some period between these two values, we obtain a nearly sinusoidal pattern, as Dr. Hall showed us.

Hall: Yes, sometimes an abrupt period change can be made to look more like a sinusoidal change by a different choice of mean period.

Herczeg: One interesting object is VW Cep. Kwee has shown that there is a disturbance moving through the light curve in about every two years – and the $(O-C)$ values show just this period. But the amplitude of the disturbance is much smaller than that of the $(O-C)$ curve, so the disturbance certainly cannot be responsible for the whole observed variation. There are only a few cases, however, for which we have direct observations of disturbances in the light curve affecting the times of minima. I agree with Dr. Smak that there may be secular changes of the period that we could easily detect in the absence of complications in the $(O-C)$ diagram. The observed irregularities in the period, unfortunately, make it very difficult to prove secular trends within a reasonable time – say a few decades.

Van 't Veer: I basically agree with what Dr. Herczeg said about the W UMa systems. I looked at seven or eight of the best-observed systems myself. (The others are not well enough observed for us to be able to say anything about their period changes.) In these well observed systems, I have the impression that the period changes behave in two distinct ways. Either you have a continuously decreasing period, suggesting a continuous ejection, or a discontinuously increasing one, suggesting spasmodic ejection. Increasing periods are related to increasing orbital radii and Roche lobes – and conversely – so perhaps we can find a physical relation from these rather vaguely established observational effects.

Devinney: In the $(O-C)$ plot for a given system, it often happens that the upward branches are all parallel, and the downward branches are also parallel to each other. I think one of the Polish astronomers has also noted this. This suggests that some binary systems oscillate between two periods, as Dr. Catalano just said. It is very difficult, however, to interpret this physically, unless there are two modes of mass exchange.

Smak: My impression is that those $(O-C)$ diagrams show such a variety of forms that it is difficult to describe them well, to draw any obvious conclusions, or to say whether that slope is larger than the other slope. Neither do they really follow sinusoidal curves.

Walter: I think it is very difficult, or probably impossible to obtain a clear insight into the nature of period changes unless we understand the structure of the gas streams. These can work in different ways that may have quite different effects on the period. I have the impression that abrupt period changes may be correlated with changes in the structure of a system. I have spoken about the structure of TW And: the period of this system is now fairly constant. I think we should study the structure of systems, and compare it with the behaviour of their periods. I hope that we may learn about this complicated matter by such combined investigations and observations.

Huang: I agree with Dr. Walter. In order to understand the period change, we must first understand the flow pattern – which is not solved. I have discussed period changes (Huang, 1956) but as I grow older, I feel more inadequate to study this problem. In my review, I didn't even touch on this problem of the period change, because it is too involved.

Underhill: The intuitive interpretation of period changes outlined by Hall is very interesting. One of the key components is this disk of gas which is supposed to be changing in characteristics and thereby changing the orbital momentum and the period. I think spectroscopic observations of selected stars would be very useful for determining the physical properties of these disks.

Batten: This is precisely what I'm trying to do, but the observations are not easy and will take time.

Hall: I think Dr. Underhill's suggestion is very good. One fact may complicate the interpretation, however. If the disk contains 10^{-5} $M_\odot$, most of it will lie below the photosphere. It might happen that the outer photospheric layers – which are all you see – will not reflect the average angular momentum of the whole disk.

Underhill: What do you mean "the disk lies below the photosphere"? Is it inside the star?

Hall: Yes. By 'disk' I mean that part of the star and envelope which is rotating rapidly. In my model, I need 10^{-5} $M_\odot$ to be involved in the rotation. You cannot see down through that amount of matter.

Biermann: Dr. Hall's theory is very interesting. I've tried to estimate the time-scale of diffusion of angular momentum across disks. You have to assume a number of parameters, such as velocities, that you can't really check, but if you guess reasonable values of them, Dr. Hall's idea certainly doesn't seem unreasonable.

Smak: Dr. Hall, when the disk in RW Per changed, was there any strange period change?

Hall: I'm not sure that I really believe my own interpretation now of an expanding star or disk in RW Per. I have a complete *U B V* light curve and I am re-examining the older data more carefully, but it is too soon for me to tell you what the proper interpretation is. I looked at RW Per, and a dozen or more Algol-like eclipsing systems, from the point of view of my model for period changes. It seems to fit in with the others.

Plavec: In Algol-systems of short period, the disk itself may not be so terribly important because most of the material that flows from one star probably falls directly on the other one. This of course does not exclude the interpretation suggested by Hall, only you would have to consider the mass receiving star sometimes rather than the disk itself. Otherwise the picture is very attractive and one could think of the changing size of the Roche lobe and changing size of the star, although, as Dr. Smak remarked, it will be necessary to see whether these changes are large enough. Before we are willing to accept the general idea here, I would like very much to hear somebody speak about the period changes observed in Cepheids and RR Lyr stars because about ten years ago, the diagrams produced by Dr. Detre and his group for cepheids and RR Lyr stars, which are single objects, showed very much the same pattern as our picture, and at that time, the suggestion seemed very reasonable to us that this can be represented by the theory of random walk. The problem is whether we have the same situation here or not. Maybe Dr. Herczeg would know more.

Herczeg: I am not sure that the similarity of the period changes is so far-reaching A great many RR Lyr stars show periodic $(O-C)$ curves, many of which indicate very clearly the existence of several periodicities, whereas eclipsing binaries display irregular changes. I would be cautious in drawing analogies.

Goldberg: I would like to consider further this similarity between the period variation in certain single stars like the RR Lyr variables and Cepheids, and those in certain binary systems like the Algols, by discussing the period variations in the single β Cep star BW Vul and by pointing out some possible implications for binary systems.

The period of BW Vul shows the same type of oscillation about a mean period as Hall has described for the Algol-type systems. These variations have been discussed recently by Percy (1971). It is obvious that mass streaming plays no role here since the star is not a binary. There is some evidence to suggest that the oscillations are correlated with variations in the star's pulsational velocity amplitude – a significant result, as a real physical basis for the period oscillations is indicated.

If one assumes that the similarity between the period oscilations displayed by both single stars and binary systems is more than coincidence, the possibility arises that at least one of the components of certain binary stars (in particular the component undergoing mass loss) may be intrinsically variable; and the variability of this component may be the fundamental cause of the oscillation in period of the system. This being true, the role played by the actual stream or the disk surrounding the accreting component in causing period oscillations would not be particularly significant. It would seem worthwhile to investigate some of the questions that I have just discussed by making a thorough comparison of single stars and binary systems that display similar period changes and by making an effort to investigate the intrinsic variability of the components of these binary systems.

Bolton: I was not aware that BW Vul is definitely a single star. It is one of the B-type flare stars that I spoke of when we were discussing Algol. Eggen observed one flare photoelectrically: it was at least $0^m.8$.

Goldberg: I understood that observation was somewhat suspect. There is no evidence at all to suggest that BW Vul is a binary.

Bolton: This has nothing to do with mass transfer, but it does relate to period changes in close binaries. The ellipsoidal variable UU Psc has appeared in the literature several times in recent years with question marks by it. A couple of observers have suggested that it was showing period changes related to apsidal motion. I have recently been reworking the published spectroscopic data and obtaining more of my own. The orbital eccentricity is small, but I believe that the data strongly indicate apsidal motion with a period less than three years. This would make it the shortest apsidal motion period known by about a factor of ten. The stars in the system are about spectral type F0, so that this system would be the latest type system for which apsidal motion would be determined. Because of the short apsidal motion period, it is difficult to obtain the necessary spectroscopic observations. Therefore, I urge the photometrists to time the minima.

References

Anger, C. J.: 1933, *Harvard Bull.* No. 891, 8.
Batten, A. H. and Plavec, M.: 1971, *Sky Telesc.* **42**, 147, 213.
Hall, D. S.: 1972, *Publ. Astron. Soc. Pacific* **84**, 323.
Huang, S.-S.: 1956, *Astron. J.* **61**, 49.
Kruszewski, A.: 1966, *Adv. Astron. Astrophys.* **4**, 233.
Kuhi, L.: 1964, *Publ. Astron. Soc. Pacific* **76**, 430.
Kwee, K. K.: 1958, *Bull. Astron. Inst. Neth.* **14**, 131.
Limber, D. N.: 1964, *Astrophys. J.* **140**, 1391.

Neubauer, F. G. and Aller, L. H.: 1948, *Astrophys. J.* **107**, 281.

Merrill, P. W. and Burwell, C. G.: 1949, *Astrophys. J.* **110**, 387.

Morgan, T. H. and Chen, K.-Y.: 1972, *Bull. Am. Astron. Soc.* **4**, 330.

Paczyński, B.: 1967a, in J. Dommanqet (ed.), 'On the Evolution of Double Stars', *Comm. Obs. Roy. Belgique, Ser. B*, No. 17, p. 111.

Paczyński, B.: 1967b, *Acta Astron.* **17**, 287.

Paczyński, B.: 1971, *Ann. Rev. Astron. Astrophys.* **9**, 183.

Percy, J. R.: 1972, *J. Roy. Astron. Soc. Can.* **65**, 297.

Peters, G. J.: 1972, *Publ. Astron. Soc. Pacific* **84**, 334.

Prendergast, K. H. and Burbidge, G. R.: 1968, *Astrophys. J. Letters* **151**, L83.

Smak, J.: 1972, *Acta Astron.* **22**, 1.

Van Woerden, H.: 1957, *Ann. Sterrew. Leiden* **21**, 1.

Wood, F. B.: 1957, in G. H. Herbig (ed.), 'Non-Stable Stars', *IAU Symp.* **3**, Cambridge Univ. Press, p. 144.

SEVENTH DISCUSSION SESSION

(Saturday afternoon; 9 September, 1972)

Chairman: D. H. McNamara

McNamara: We'll open our session this afternoon with a few remarks by Dr. Cooper on infra-red observations of binary stars.

Cooper: I want to draw attention to the value of infra-red observations of binary systems in improving our understanding of these objects. Semi-detached systems which contain an early-type, main-sequence component and a cooler subgiant will display more nearly comparable depths of their two minima at longer wavelengths: therefore, solutions of their light curves will be more determinate. Other areas in which infra-red observations will be useful are:

(i) determination of limb-darkening coefficients and comparison of the results with those of model-atmosphere calculations (Grygar, 1972).

(ii) study of the atmospheres of components of systems that have been observed in many colours – in particular the atmospheres of cooler components of Algol-type binaries.

(iii) investigation of the possible variation of deduced values of some orbital elements (e.g. radius of eclipsed component) with the effective wavelength in which observations were made.

(iv) improvement of the determination of orbital elements so that one can also obtain a better idea of streaming effects.

(v) investigation of systems of the type of ε Aur, and of the composition and structure of the disks found in these systems; possibly detection of a secondary minimum.

Smak: I want to emphasize one point, in particular, made by Dr. Cooper: namely that one should try to observe some systems in which the complications arising from circumstellar matter are not severe, in order to study the physical properties of the secondary component.

Andersen: How faint a star will you be able to observe, and what wavelength will you use?

Cooper: I will probably observe at 0.9 μm or 2.0 μm.

McNamara: How faint can you reach with, say, a 24-inch or 50-inch telescope?

Milone: Several of us have made intermediate-band infra-red observations, including the K band, with 50-inch and 60-inch telescopes in Arizona. My recollection is that we could work on objects of at least ninth visual magnitude, and still have reasonable signal-to-noise ratios.

Chen: Before we go on to discuss ultra-violet excesses in binary stars, I would like to point out that although the Harvard and Arizona astronomers have done much work on the infra-red excesses of single stars, we know very little about infra-red excesses of close binary stars. Thus I would like to emphasize once more the need for infra-red observations.

Batten (ed.), Extended Atmospheres and Circumstellar Matter in Spectroscopic Binary Systems, 192–215.

McNamara: If there are no more questions on this topic, we will ask Dr. Hall to discuss ultra-violet excesses in close binary stars.

Hall: It is somewhat surprising that we still do not know for sure how to interpret the apparent ultra-violet excesses found in the subgiants of totally eclipsing Algol-type binaries. In our paper on SW Cyg (Hall and Garrison, 1972), we spent some time explaining why it is most reasonable to attribute at least part of the excess to contamination by light from an envelope surrounding the hot star, that is not completely eclipsed at mid-primary minimum. Briefly, the evidence in favour of this interpretation is as follows: (i) the amount of the excess is often variable; (ii) in the light curve of SW Cyg the bottom of primary eclipse appears total in V but is so rounded in U that it appears partial; (iii) some of the excesses are larger than $0^m.2$ or $0^m.3$ and hence larger than the maximum excess which would result in the limit of vanishingly small metal abundance. It is still quite possible, however, that the excesses $0^m.2$ or smaller can be the result of a true underabundance of metals, as seems to be indicated by the narrow-band photometry of Miner (1966) and McNamara (1967). Even though this narrow-band photometry was done almost ten years ago, we still cannot say for sure whether or not there is a true metal underabundance in the subgiants. The answer is of interest because there are a number of reasons to expect normal abundance or overabundance of metals, but no good reason to expect an underabundance.

The right way to settle the question would be to obtain a fairly good spectrogram during totality of some Algol-type eclipsing binary with an apparent ultra-violet excess of about $0^m.2$. Then an abundance analysis, or perhaps even visual inspection, should reveal whether or not the subgiant is truly underabundant in metals. Dr. McNamara and I together want to ask participants if anyone has such spectra and can settle the question.

Baldwin: I investigated this for the secondary component of U Cep, using a spectrogram of 40 Å mm^{-1} obtained during totality by Dr. Batten. This was the highest dispersion at which we could obtain a good spectrogram. I didn't find the difference that Miner found. I made only a qualitative comparison, with several spectrograms at the same dispersion of single stars of the same spectral type and luminosity class. Visual comparison of the spectrograms showed no significant differences between the stars, so it seems unlikely that there are significant abundance differences.

Popper: There is one observation of another system, not exactly of the same type, namely AR Lac. During total eclipse, Spinrad obtained scans in his multi-channel system and he finds a definite underabundance of metals in the subgiant secondary of AR Lac.

Hall: In my interpretation those are different types of stars, although there may be some uncertainty about that. Dr. McNamara, did you say that an underabundance as large as that implied by narrow-band photometry should 'hit you in the face' spectroscopically?

McNamara: Yes, if you have ultra-violet excesses of the order of $0^m.2$. Usually in the extremely metal-poor stars, of course, the ultra-violet excesses are of the order of

$0^m.2$. So in the eclipsing stars, regardless of whether this ultra-violet excess is caused by the filling in of lines by emission or radiation from some source, I should think that the behaviour of very strong lines would mimic what you see in the spectra of metal-poor stars. The deep absorption lines ought to be filled in, and you should be able to notice this in the spectrum of the secondary star. Now for U Cep, $\Delta m_1 \sim 0^m.10$. If this implies that the star is metal-poor, I'm almost positive that you ought to observe weak metal lines even at rather moderate dispersion.

D. B. Wood: I made some Strömgren four-colour and magnesium-index observations of some eclipsing stars. One, a W UMa system, had an m_1 index which, if interpreted as an abundance indicator, showed that the metal abundance varied, from one season to the next, by an order of magnitude. My interpretation is that the m_1 index is not an indicator of metal abundance in this case.

Baldwin: There was one difference between the secondary component of U Cep and my comparison stars. The H and K lines of calcium appeared to be filled in by emission. None of the other lines, at least on visual inspection, showed any significant difference.

McNamara: If I remember correctly, Miner's observations were based on the G-band and the CN-band, and a metal-abundance index m'. He found all the late-type subgiant components were metal-poor compared with other stars of similar temperature and luminosity class.

Andersen: You said you compared the spectrum with spectra of similar type. Could a low metal abundance disguise itself as a wrong spectral type?

Baldwin: That is something one must be careful about, but the close correspondence of all the stellar lines seemed to indicate that the assigned spectral type and luminosity class were correct.

Batten: We have the photometric solution for U Cep, and, as we now believe, a reasonable idea of the total mass of the system. This gives us the radius of the secondary component, and it is definitely a subgiant, or even a giant.

McNamara: If these stars really are underabundant in heavy elements, it does do damage to the idea of unloading mass from the subgiant components on to the brighter B-type or A-type stars in these systems. If you examine the spectra of the A-type or B-type components, you find, of course, that they have no apparent anomalies; whereas, the indices, on the other hand, if you believe that they really do measure abundances, indicate that there are some rather severe abundance anomalies in the subgiant components. If there are no more questions on this topic, I think we can turn to the contribution by R. E. Wilson and Devinney on common envelopes in close binary stars.

R. E. Wilson: Somewhat more than a year ago we published a procedure for computing light curves of synchronously rotating binaries. The scheme is quite general in that it can be used for detached, semi-detached, or fully contact systems. The model we use is very similar to that used by certain other groups and individuals (i.e. Lucy (1968), Hutchings and Hill (1973), Mochnacki and Doughty (1972) Ruciński (1969)) but the computing scheme is more accurate than the others mentioned, by at least an order of magnitude. For differential correction of such light curves, it is quite crucial to have the greatest achievable accuracy because some of the numerical derivatives (those for parameters of a geometrical nature) can be computed accurately only when the underlying basic function is extremely smooth. The computational errors in our light curves are small compared to the accidental errors of the best normal-point observations. The present work (and its continuation) has three objectives, which are:

(1) To establish whether the W UMa binaries are true contact systems, having common envelopes.

(2) To establish the gravity darkening law which holds for stars which have convective envelopes.

(3) To begin collecting fundamental data for contact systems (i.e. masses, luminosities, and radii).

The first two points are, of course, observational checks on ideas advanced by Lucy. Work on the third point has been hampered until recently by the shortcomings possessed by the classical Russell model when it is applied to close binaries. Not only does the Russell model fit poorly to observed W UMa light curves but the derived values of the parameters are systematically in error by relatively large amounts (e.g. $5°$–$7°$ in the inclination, $\sim 20\%$ in the radii). This is particularly unfortunate since many W UMa systems are double-lined spectroscopic binaries with large radial velocity amplitudes, so that the prospect for obtaining absolute dimensions is good from the spectroscopic point of view. Furthermore, the spectroscopic information can be supplemented by accurate photometric mass ratios derived from the new approach.

The adjustments presented here were produced by application of our differential-corrections program, following preliminary trial-and-error fitting. The differential-corrections program makes a least-squares solution for observations in an arbitrary number of passbands simultaneously. The solution is rigorous in the sense that only one value is found for each wavelength-independent parameter while for wavelength-dependent parameters, n values are found for the n passbands. Of course, the program computes probable errors for all adjusted parameters.

For contact binaries, constraints are imposed on the parameter values such that there are no discontinuities in surface brightness across the neck connecting the two components. The set of adjustable parameters for contact systems includes:

q = mass ratio (m_2/m_1),

i = orbital inclination,

g = gravity darkening exponent,

Ω = modified potential for photosphere,

x = limb darkening coefficient,

A = bolometric albedo,

l_3 = third light,

T_1 = polar temperature of component 1.

Only the first five parameters were adjusted for the present examples. In addition to the above list, the luminosity ratio, L_1/L_2, and polar temperature, T_2, of component 2 are computed, but these are not free parameters since they follow from the values assigned to the other quantities. Binaries for analysis were selected for having total or annular eclipses, absence of asymmetries, and accurate observations. We have solved for the gravity exponent by least squares rather than by assuming that the 'convective' value predicted by Lucy is correct, as did Mochnacki and Doughty. Short comments on our results for three systems follow:

R Z Com – Green light curves taken at two different epochs by Broglia (1960) were analyzed independently. This system seems to be only marginally contact, as the solution showed it to be slightly over contact at one epoch and slightly under at the other. The two light curves are significantly different by visual inspection. The values found for g were 1.13 ± 0.04 and 1.51 ± 0.02 for the two light curves (g is 1.00 for Von Zeipel darkening, ~ 0.32 for Lucy darkening).

R Z Tau – Contact is definitely indicated for this binary. The flat annular eclipse shown by Binnendijk's (1963) light curve could not be reproduced without destroying the fit in other phase ranges. Conceivably this apparent flat region could be due to one or two unfortunate residuals, and may not be found in future light curves. The gravity darkening, g, was 0.29 ± 0.06. The mass ratio we find is close to that found by Mochnacki and Doughty, who also studied this binary, but several other parameters are considerably different, especially the limb-darkening coefficients.

A W UMa – The components are certainly in contact for this system. g is $0.45 \pm .06$ and the mass ratio is close to that found by Mochnacki and Doughty, who did not adjust the limb-darkening for this system.

Chen: I would like to mention that models for W UMa-type stars were discussed in great detail by Lucy and Mauder at the Philadelphia meeting last year. In fact, all the items listed by Wilson and Devinney were fully discussed there. Lucy has given solutions for W UMa-type light curves. It would be interesting to compare his results with yours. Lucy pointed out, as noted by others, that the eclipse of a larger mass always gives a deeper eclipse. This is related, of course, to the problem of discrepancies between the spectroscopic mass ratio and the photometric mass ratio for some systems.

R. E. Wilson: Well, I was not at the Philadelphia meeting, and the Proceedings haven't been published yet. Perhaps you can tell me if Lucy's results were obtained by differential corrections. I presume they were based on the program that he has already published (Lucy, 1968).

Hill: Lucy wrote a special program, in differential-correcting form, for the Philadelphia conference.

R. E. Wilson: And what was the light-curve program? Basically the one that he had before?

Hill: I think so. I think it was similar to the one published recently by Mochnacki and Doughty.

R. E. Wilson: Well, that program was quite different from ours. The advantage of our program is that it has considerably higher precision than have other existing programs for light curves of close binaries, and therefore the errors are completely negligible compared to the errors of even the best normal points. This is important in a differential-corrections program, because when you're computing numerical derivatives, the basic function must be extremely smooth if you are to obtain reliable results.

Biermann: I would like to mention that this is the fourth determination of an extremely high mass ratio for AW UMa. Lucy did write a special program for the Philadelphia meeting, and for this system he found a mass ratio of about 12:1. Mauder (1972) published recently a long list of such determinations, and I believe he got a similar result. There are a number of recent papers by the Sussex group [Hazlehurst (1970), Moss and Whelan (1970) and by Dr. Thomas and myself (unpublished)] which discuss several possibilities for what really happens in the common convective envelope. In our paper, Thomas and I indicate what we should like to have observed – especially binaries in clusters, a problem which apparently John Whelan is tackling. Theoreticians, both in Europe and the United States, have provided a variety of models which occasionally disagree violently in their predictions. Observers like you could help to decide among them.

Smak: I wish to add one more comment that is extremely important for the theory, namely, you should note the degree of contact and try to interpret your result. You can get a certain degree of overflow of the inner contact surface. You don't get too close to the outer contact surface. That may indicate that in those cases any major mass outflow from the system is unlikely.

R. E. Wilson: For AW UMa, the photosphere is about half-way between the inner and outer contact surfaces.

Herczeg: I should like to put a mild objection on the record. Someone said that W UMa stars are well-behaved objects, offering no serious observational difficulties. This is certainly an over-optimistic opinion. Everybody who has studied their spectra knows that the lines are rather badly broadened and blended. For instance, in many cases we see only a single spectrum which, however, may be blended with the lines of the secondary; thus we observe a reduced amplitude of the radial velocity curve and an incorrect mass function – AK Her is perhaps an example. Also, some of the light curves have a tendency to show irregular, cycle-to-cycle changes. How does Dr. Wilson cope, for instance, with such disturbances? Using two somewhat different light curves, did you get two different sets of elements? If so, do they represent real changes of the system?

R. E. Wilson: Well, the main way in which the sets of elements differed was that the gravity-darkening parameter had a value from one of the curves which, at first sight, seemed physically unreasonable. Since that parameter is related to the brightness distribution over the stars, it seems likely that whatever happened to the system could most conveniently be described in terms of the brightness distribution.

Herczeg: Would you say that in other cases of disturbed light curves this type of variation can also be ascribed to changes in brightness distribution?

R. E. Wilson: Yes, but it's difficult to extrapolate from one example.

Herczeg: Concerning the overflow of the Roche lobe, whether or not it is possible theoretically, one should bear in mind the strange fact that there seems to be a surprisingly small amount of circumstellar matter in W UMa systems. The radial-velocity curves are symmetrical sine-like curves, as is to be expected for nearly circular orbits. There is no Barr effect observed in W UMa spectra. One observational difficulty is that you need a large telescope, in order to obtain good time resolution. Otherwise you must have an additional program to compute the line deformations due to the radial-velocity change during exposure.

R. E. Wilson: That part of the difficulty that arises from rotational broadening can be taken care of by computing the profiles from this type of model. Despite these difficulties, we have a fairly good idea of the mass functions, at least. If on top of this you can obtain a photometric determination of the mass ratios, you ought to be able to determine approximately the right masses.

Biermann: If you do the analysis in more detail, as Thomas and I did, you find that the constant in the gravity-darkening law is not really a constant but varies slightly (from about 0.06 to 0.10) for different models of stars. It might be of interest to check your different systems against our values to see whether there is any accord between these determinations. On the other hand, from the theoretical point of view, even detached, near-contact binaries must obey Lucy's type of gravity-darkening law. The law is not a property of contact binaries, but of convective zones.

Smak: I have a somewhat pessimistic remark. In spite of all these beautiful results, we should not forget that the basic models involved fail to explain several important features. One is the asymmetry of the light curve, and the other is the character of primary eclipse which must always be a transit in Lucy's model, but does not always seem to be so in practice.

R. E. Wilson: First of all, the colour effect in AW UMa is very small – probably smaller than you realize. The eclipses are very shallow and the light curve I showed was drawn on a very expanded scale. The differential effect from yellow to blue is less than $0^{m}01$. As to primary eclipse being a transit or occultation; that is associated in our program with the value of the gravity-darkening exponent. A system with classical gravity darkening has primary eclipse as an occultation: with convective gravity darkening that eclipse is a transit. From experiments with light curves, we see that this last statement is generally true. If future experience confirms this, we could tell rather quickly which gravity-darkening law to apply simply by whether primary eclipse is a transit or an occultation.

Whelan: Is the secondary hotter?

R. E. Wilson: In all systems the more massive star has the higher polar temperature.

Whelan: Really?

R. E. Wilson: It has to.

Whelan: I don't think that's necessarily right. It may be inherent in your gravity-darkening procedure, however.

R. E. Wilson: You may be taking a different approach and not be really considering the same kind of common envelope as I am. If you consider a common envelope in which there are no physical discontinuities of any kind, so that all the way from the pole of one star to that of the other all physical parameters vary smoothly, then the temperature has to be higher at the pole of the more massive star simply because gravity is higher there.

Whelan: But if you define primary minimum by the radial-velocity curve, you find that in some systems it is an occultation, and in others a transit. I don't think that you can say that, therefore, in some systems there is radiative gravity-darkening and in others convective. I think you should say: in some systems the primary is hotter, in others the secondary. I am suggesting that theory may predict something that doesn't seem to come out in your program because you represent temperature structure by gravity-darkening rather than by a physical temperature difference between the components with the secondary being the hotter.

R. E. Wilson: You would have to propose then that the physical variables are not consistent with a smooth gravity-darkening law that runs all the way from the pole of one component to that of the other.

Whelan: It's possible that there are discontinuities in the neck between the two stars. The structure of the neck is uncertain to say the least.

R. E. Wilson: You may very well be correct on physical grounds, but the structure of the neck is a problem in stellar interiors. In the first reconaissance of the problem we would like to have a smooth variation of all parameters.

Whelan: Yes, I'm not criticizing your approach to the problem. I'm extremely pleased to see people using light-curve synthesis for these systems. It has already been said here that rectification was not designed to deal with very close binary systems. The question is whether we can still get good parameters. I think the spectroscopy is the clue, but it's not easy to derive values for K_1 and K_2. The spectra are very hard to measure.

R. E. Wilson: In many systems, especially those with extreme values of the mass ratio, it's far better to take the photometric mass ratio than the spectroscopic. Nevertheless I would be pleased to have the mass function spectroscopically. That's all you need if you have a photometric mass ratio.

Whelan: That's the best way for mass ratios close to unity or to zero. I think many systems have mass ratios between 0.4 and 0.6. We don't really know because in the past analyses have been based on only a few points on the radial-velocity curve. If mass ratios do lie in that range, and the systems often show shallow eclipses because the inclinations are not very large, then I think that the spectroscopic constraints are the only thing that will enable you to obtain a reasonable solution, because there might be more than one equally good solution in the solution-space domain. The system of TX Cnc is a good example of a case for which you must have good spectroscopy before you can even undertake any sensible photometric solution. If you do have a spectroscopic mass ratio, you can get a good value for the orbital inclination and for the constant, C, (that defines the potential surface) and then you can determine parameters for these theoretical models.

R. E. Wilson: All of the systems we have studied have complete eclipses: the problem of partial eclipses does not arise.

Whelan: But some systems do have partial eclipses, and I hope that your program, and those of others, will be able to deal with them, so that we can get some very good data! The theory does seem to be in need of observational correction.

Hilditch: What are the formal probably errors of the gravity-darkening exponents you determined for these systems?

R. E. Wilson: For RZ Com we found 1.13 ± 0.04 at one epoch, at another we found 1.51 ± 0.02. For RZ Tau and AW UMa we found 0.29 ± 0.06 and 0.45 ± 0.06, respectively. These errors are probable errors, not standard deviations.

Whelan: What would happen if you fixed the exponent β at 0.08, and didn't allow it to reach 1.13? There are good physical arguments for this value of β. If you fix β, and allow discontinuities in the neck, and let the secondary temperature be defined by the minima of the light curve, would you find the temperature difference I have suggested? It's a serious problem: if the primary is in front at the deeper minimum, then it may be the cooler component in some average sense. It's important to know if the primary really is the cooler star: it's the more massive and more luminous one. If it is the cooler star, this fact must be tied in with the energy-transfer equation and the theory of these systems.

R. E. Wilson: Without a fundamental change of approach, fixing the gravity-darkening exponent would not lead to a temperature difference. The exponent controls only the amplitude of the effect, and not the sense. The sense of the effect is determined by the fact that the acceleration due to gravity is greater on the more massive component than on the secondary component.

Whelan: O.K. In Lucy's original paper (1968) in which he was trying to use his model to obtain the correct light curves, he demanded equal average effective temperatures for both components, and he found that the minima were the wrong way round. Recently Hutchings and Hill (1973) and Mochnacki and Doughty (1972) have allowed the two stars to have different temperatures. Binnendijk (1965) distinguished two classes of systems – W systems, in which the secondary is hotter, and A systems in which the primary is hotter. This is the same distinction that I made earlier by the radial-velocity curve, which shows which star is receding. Lucy's model does explain the mass-luminosity law in these systems ($L \propto M$ instead of $L \propto M^5$) but it does have this problem. If we can get good observations and good accuracy in our computations we have a very good chance of obtaining some physical insight into theoretical stellar structure.

McNamara: Are there any more questions?

Smak: With so many parameters involved, are you really sure that you obtained them all separately – or may some of them be combined?

R. E. Wilson: First, it is not necessary to use all the parameters if one doesn't want to. You can choose any subset that you would like to adjust. If you feel you have independent information that fixes certain parameters, you can specify a control integer that suppresses their adjustment. Second, we determine the probable errors, taking into account the correlation between parameters. In principle, at least, provided there are no systematic errors, these uncertainties will tell you what you want to know. Finally, we have built into our program the capability to adjust any subset of the parameters, and in each individual case we adjust only the minimum absolutely necessary.

Van 't Veer: Gravity darkening gives a relation between local temperature and local gravity. At the first Lagrangian point, the gravity is zero, but we know that the temperature is not zero. How do you solve this problem?

R. E. Wilson: Where the gravity is zero, the *effective* temperature is zero and the flux is zero. The Lagrangian point is only a point, so you get negligible flux from a differential volume.

Van 't Veer: Is this physically realistic?

R. E. Wilson: Why not? The problem only arises in a, literally, point-contact binary. Most of the systems we consider are not like that. If we ever do treat a point-contact binary, then the formal value of the effective temperature at that point is zero, but that does not mean that the local kinetic temperature is zero. We must distinguish between *effective* and *kinetic* temperatures. The former varies with local gravity, while the latter is constant (or approximately so in the convective case) over an equipotential surface.

Hutchings: Can I just make a brief comment about the spectroscopic mass ratios? These are based on line profiles which are intrinsically distorted at every phase because of the distortion of the star and because of the variation of temperature across the surface of the star. For the models which we have done for similar types of systems we have calculated this amount of distortion and we are able to reconcile the difference between the spectroscopic and photometric mass ratios. This difference is something one should be aware of and watch out for.

R. E. Wilson: And you believe that the photometric values are correct?

Hutchings: I think they are better.

McNamara: One more short remark from Dr. Whelan.

Whelan: With this sophisticated method of solving light curves, people can no longer argue that we can permit errors in our photometry because the theory is not as accurate. You need accurate photometry and good spectroscopy. If you can find systems in clusters, for which we have constraints on chemical composition and distance, then we have a good comparison for theory. So far there's only one in a cluster – TX Cnc – and that's really knocked the theory down. If we could find some more, then I think that this whole business could be placed on a firmer foundation. We have made tremendous advances in understanding the spectroscopy and in the technique of photometric solution.

R. E. Wilson: To go back to your earlier remarks: people computing models for interiors should say what the brightness distribution of the stars should be, and, if there are discontinuities at the neck, what they will be like. Otherwise we cannot know how we should modify the simple brightness distribution used in this model.

Whelan: I agree with that point.

Cooper: Dr. Wilson, when you fit all these parameters, including gravity-darkening to the light curve, was the only criterion to obtain the best possible fit – or did you, for example, choose a less well-fitting solution if, say, the gravity-darkening had a more plausible value?

R. E. Wilson: The main trouble with contact binaries is to find anything that will fit. After a few tries, the theoretical curves become fairly close to the observations, but to obtain the kind of agreement we obtained requires an enormous amount of trouble. We do not have much flexibility among these parameters. In fact there are fewer parameters than in the conventional Russell model. That has about thirteen parameters, and even when we adjust everything, including third light, we have only seven.

Biermann: In our interior calculations, we compute just the maxima and minima of the light curve. We take each component separately and say: we have this luminosity for that volume. So we have the discontinuity in our program, and this makes it possible for the secondaries to be hotter, as John Whelan has already said. It would be very important for we theoreticians if you could put another step into your program, just to see what luminosities you do get if you allow for the discontinuity. Then you would probably find all these temperature effects.

R. E. Wilson: Well, that could be done very easily, but it introduces another parameter. Recall the problem that Dr. Smak mentioned – that the yellow and blue light curves have slightly different levels at secondary eclipse. That problem will disappear when the extra parameter is added. Then you must judge whether too much weight has been lost from the determination. Of course, if there's a physical reason for believing that that parameter should be accounted for, then it should be added.

Whelan: In principle, scattering of the light at the neck might be detectable by polarization observations. Some of us tried this with Gehrels' excellent equipment at the Lunar and Planetary Laboratory. We could not find any systematic changes in polarization with phase – so we're still stuck because observations do not give the results predicted by theoreticians.

McNamara: Dr. Underhill, is *a* Cen a binary?

Underhill: Binary stars show a complex spectrum that arises from four sources: (i) spectrum of the primary star, (ii) spectrum of the secondary star, (iii) absorption lines formed during atmospheric eclipse, and (iv) lines (absorption or emission) formed in a gas stream. The problem in devising a model is to recognize which spectral details correspond to which source. One is assisted by considering the relationship of spectral changes to light changes. Klinglesmith and I would propose that *a* Cen is an eclipsing star and that the occurrence of grazing eclipses is an explanation of what is seen. The light changes are as follows: Norris (1971) has shown that *a* Cen in visible light is slightly fainter than average around phase $0^P.0$ (He I lines strong) and brighter at phase $0^P.5$ (He I lines weak). Molnar (1972) from observations with OAO-2 has shown that *a* Cen is faint in the far UV at phase $0^P.0$ and bright around phase $0^P.5$. The spectroscopic data [Norris (1971), Klinglesmith *et al.* (1971) and my own observations] show the following: (i) the Balmer lines remain practically constant in shape and strength throughout the cycle and the Balmer series breaks off at $n_m = 17$. There is no evidence of density changes; (ii) the He I lines vary in intensity in a period of $8^d.814$ from stronger than at type B2 to as weak as in type B8 or B9. The triplets maintain a sharp core; the singlets do not; (iii) a set of sharp lines (C II, O II, Si II, Si III, Mg II, Fe III) is present which varies slightly in position and strength. These lines seem to strengthen around phase $0^P.5$, (He I minimum); (iv) broad shallow lines of Fe I appear near phase $0^P.0$ and remain strong until at least phase $0^P.25$. There are no broad shallow lines of Ca II or Fe II at any time.

We suggest that the Balmer lines, the weak He I lines and the sharp lines are produced by the primary star which is a large expanded object. The strong He I lines and the Fe I lines come from the secondary star which is a compact, hydrogen-poor object bright in the far UV. The observed light changes and spectrum changes are produced as the small star passes in front of the large star near phase $0^P.0$. Around phase $0^P.5$ the small star is eclipsed, the eclipse having long partial phases. The He I lines then weaken and the sharp lines strengthen, the latter being formed as the atmospheric eclipse proceeds. We have measured a series of spectrograms obtained in 1972, May and June for radial velocity. A preliminary look at these results supports the interpretation given here. Neither star is a normal object. Possibly the small star has just finished ejecting its hydrogen rich envelope onto its companion.

Thackeray: We have just a little material on this event, which we examined after the discovery of the helium variation. We definitely found evidence that the N II lines appeared to be varying in opposite phase. This agrees physically with your remarks about C II: you seemed to be a little doubtful about that group (iii).

Cowley: What can you say about the mass ratio and the mass function?

Underhill: Nothing! The radial velocities measured on any one night show a wide scatter. A trend over the $8^d.8$ period seems to be present. The range isn't more than 15 km s^{-1} at the most. The narrow lines and the hydrogen lines vary by something like 5 km s^{-1}.

Cowley: Are your plates of high enough dispersion to determine whether the spectrum shows ^{4}He or ^{3}He?

Underhill: If there were ^{3}He, $\lambda 6678$ would have a sizeable displacement to the red. I have measured all the spectrograms of the red and yellow regions, which have a disperison of about 12 Å mm^{-1}. The He I line at $\lambda 6678$ seems to have the same sort of variation as the rest of the He I lines have. I would say there is no ^{3}He.

Hutchings: Is there any photometry on this star?

Underhill: The only photometry I know that has been published is that of Norris – which is very little. Rudolph Schild told me he had been observing the star in *U*, *B*, *V*, and that he found little, if any, variation.

Wright: Is $\lambda 4383$ the only neutral iron line that you have observed? Your plate looks as if it is fairly strong.

Underhill: No, it's not the only one. It's the easiest to put in a diagram. When $\lambda 4383$ is strong, all the major Fe I lines are present.

Wright: So you can measure quite a number of them.

McNamara: Any additional questions? Thank you for a very interesting talk. Now Dr. Andersen will speak about SX Cas.

Andersen: The eclipsing binary SX Cas was first extensively observed by Struve (1944). The period is $36^d.6$, the spectral types approximately A6 and G6 with giant characteristics. Struve's velocity curve showed abnormally large scatter and $e \cos \omega$

close to 0.5, in violent disagreement with the photometric result that $e \cos \omega = 0$. From this, Struve inferred the now well-known type of distortion of the velocity-curve by absorption from a gas stream, which also gives rise to strong emission at $H\alpha$ and $H\beta$. During the entire cycle, Struve noted the weakness of the Si II lines and Mg II $\lambda 4481$ and interpreted this as dilution effect and the spectrum as arising in a shell. Dr. Batten's collection of plates of SX Cas have been studied during a visit to the Dominion Astrophysical Observatory, both as regards selection and measurement of a number of reliable lines in order to obtain a good radial-velocity curve for the system and to detect possible differences between various ions, as well as to measure more suspect lines (H, Si II, Mg II). The profiles of the Balmer emission lines have also been obtained. A few points of interest have emerged:

(1) Strong changes in the emission profiles have been noted at the same phase in cycles separated by a year or two, and smaller but significant changes occur from one night to the next at a phase at which the geometry of the system and the velocities change the least, and where the total light is constant. Struve's observations extended over three cycles only, to minimize the effect of these changes.

(2) The secondary spectrum has been detected, and the few available points on the velocity curve indicate amplitudes of about $K_1 \simeq 45$ km s^{-1} and $K_2 \simeq 90$ km s^{-1}. These values are rather uncertain, especially that for K_1, as the velocities of the primary show a much larger scatter than the observational uncertainty justifies. The velocity curve leads to two inferences:

(a) with a mass ratio $m_2/m_1 \simeq 0.3$–0.5, the range indicated by the velocities, the radii of the Roche lobe are as given in Table I. Comparison with the fractional radi

TABLE I

Possible Parameters for SX Cas

m_1/m_2	0.3	0.5	r_{ph}
r_p	0.49	0.44	0.08
r_s	0.27	0.31	0.21

obtained from the light curve shows the primary well within its lobe, leaving space for the gas ring proposed by Günther (1959). The present data seem to indicate that the secondary does not fill its Roche lobe, despite the substantial amount of circumstellar matter in the system, but the scatter in the velocities, especially of the primary, makes this conclusion uncertain.

(b) The total mass of the system appears to be around 8 $M_\odot$, with 5 $M_\odot$ to 6 $M_\odot$ for the primary, which is too high for a main-sequence late A-type star. However, the observed spectrum of the primary is clearly not a normal stellar spectrum, but a shell spectrum with weak Si II and Mg II $\lambda 4481$, and the Balmer lines frequently visible to about H30, the higher Balmer lines being extremely sharp. The spectrum is strongly variable in character, sometimes having very sharp and deep absorption lines, at other times showing only wide, shallow lines of complicated structure. The changes can be seen on plates with only little separation in phase, but cycle-to-cycle variations may play a role. The primary thus has an extended envelope dominating its spectrum, apparently to the extent of concealing the spectrum of the primary star itself.

Underhill: Have you observed the primary spectrum in the red?

Andersen: Apart from the emission at Hα, no. The plates we have were obtained mainly in order to study this emission. Even at 30 Å mm^{-1}, a IIa-F plate needs about four hours exposure. The continuum is very weak, especially near the D lines, but there are some measurable iron lines from the secondary spectrum around λ6400.

Hall: You mentioned that the cool star apparently does not fill its Roche lobe. Whenever the radial velocity curve of the hotter star is distorted in such a way as to give a spurious eccentricity, Struve thought that in general the value of K_1 was exaggerated. Hardie (1950) tried to remove the distortion in the velocity curve of U Cep, found a circular orbit, and in the process he reduced K_1. I think that was the only quantitative investigation of this problem. If K_1 could be decreased for SX Cas, you would get a mass ratio farther away from one, and this would make the fractional radius of the Roche lobe smaller.

Andersen: As I understand it, Hardie tried to separate the spectrum of the stream from that of the star. We got high-dispersion plates of SX Cas for the same purpose. Now our two plates of highest dispersion are at 0$\overset{P}{.}$83 and 0$\overset{P}{.}$9 – just the phases where Struve found the maximum distortion. Now the plate of 0$\overset{P}{.}$83 does not show any asymmetric profiles – all the lines are wide. There is a suggestion of asymmetry on the plate at 0$\overset{P}{.}$9, but I measured only the wide lines – both plates were measured on an oscilloscope machine. It so happens that we have as yet few plates covering the distorted part of the velocity curve, and the only advantage of this is that our value of K_1 should be relatively unaffected by the distortion.

Batten: As far as U Cep is concerned, I think the value of K_1 should be increased again. I believe that the particular way in which Hardie attempted to separate the spectra of stream and star was based on a wrong assumption – at least for U Cep. I'm not at all sure that it is a safe generalization to say that removing the distortion by the stream will reduce the value of K_1.

McNamara: In the spectrum of U Sge the lines other than hydrogen give quite different velocities from those given by the hydrogen lines. Is this also true for U Cep?

Batten: There are few lines other than those of hydrogen, in the spectrum of U Cep, that are measurable at the dispersion I have used.

Cowley: I think the fact that Dr. Andersen finds the same γ-velocity for both components and the zero eccentricity suggests that most of the distortion of the velocity curve has been removed.

Andersen: I have measured 75 lines on the plates, in order to obtain decent mean velocities for each ion. There appears to be little or no difference between ions. At some phases we see emission in lines other than those of hydrogen.

Hall: This is a bad problem and unfortunately I don't have the final answer yet. But there is circumstantial evidence to suggest that the source of the problem lies somehow in the radial velocity curve of the primary star. Look at the list of Algol-type binaries in which this problem exists, i.e., in which the subgiant appears not to fill its Roche lobe. Exclude from this list the binaries of the RS CVn type, which are fundamentally different. Excluding these is important because in these binaries the subgiants very definitely, and by a large margin, do not fill their lobes. Then if you look at the remaining candidates, you find that the binaries giving you the most trouble are those in which the overall amplitude of the radial velocity variation of the primary star is smallest relative to its scatter; at least Struve's radial-velocity curves show this, and they are still the only ones available. An example is RW Per in which the subgiant seems to be smaller than its Roche lobe by quite a significant margin, but in this case there is almost no amplitude to the radial velocity curve of the primary – the variation is almost all scatter.

Thus it is my feeling that the Algol-type binaries with 'undersize' subgiants can be explained as follows: those in which the subgiants clearly do not fill their Roche lobes are not really Algol-type (because they are RS CVn-type instead) and those which clearly are Algol-type have distorted radial-velocity curves and hence erroneous mass functions which make it appear that the subgiants are smaller than their Roche lobes whereas in fact they fill their lobes.

Andersen: Although the scatter is very large, the internal mean error of the individual plate velocities is only about 2 km s^{-1}. I have tried to estimate the minimum mass ratio, that I thought I could get from the velocity curve, and even with that, the Roche lobe is at least 33% larger than the secondary component.

Thackeray: I believe that Struve's curve is based on a massive series of observations in one cycle.

Andersen: Three consecutive cycles in the summer of 1943.

Thackeray: I suppose it is possible that gaseous-stream activity was unusually strong then.

Sahade: From your description of the spectrum you seem to have a shell spectrum, and you have somehow to make sure that you are measuring the stellar line.

Andersen: I don't really believe that I am at any point measuring stellar lines – but from the kind of variation I think I am measuring the spectrum of something that moves with the primary star.

Sahade: Shell lines in a binary spectrum may move in phase with one of the components, and yet have an amplitude different from that of the orbital motion of the star.

Andersen: How can we explain the spectrum of the primary star? Can we have a massive star surrounded by a shell that makes it look like an A-type star? The shell is dense enough to hide the primary completely, but still dilute enough to show the high Balmer lines.

Plavec: Didn't Struve classify the stars as giants? He added the prefix c to the spectral types, indicating high luminosity. If the primary were a bright giant, then it could have five or six solar masses at its place in the H – R diagram.

Andersen: The prefix c indicated very narrow lines – but these are the shell lines. We don't know how luminous the star is.

Plavec: The problem is similar to that of V367 Cyg.

McNamara: Although we have deliberately let the discussion go on, three more people want to speak this afternoon. We have a series of contributions on VV Cep and the ζ Aur stars.

Wright: I should like to present a few preliminary results on measures of the Hα line in the spectrum of VV Cep as observed for the past fifteen years. An orbit for the system and analysis of the emission profile as observed near eclipse has been given by Hutchings and Wright (1971). In order to derive the emission wings in the spectrum of the early-type star, it was necessary to subtract the profile of the spectrum of α Ori from that of VV Cep over about twenty Ångstroms, but the results are remarkably consistent. In order to disentangle this complex spectrum further, the additional lines at Hα were plotted with the Hα emission line as a continuum and velocities were measured relative to the undisplaced centre of the Hα line after orbital motion was removed. The principal, fairly definite feature, is a sharp absorption line that is observed over the whole cycle with velocity close to the velocity of the system, about -20 km s^{-1}. After secondary eclipse, near J.D. 2438700, a fairly sharp emission feature can be seen on nearly every plate, with velocities about -60 km s^{-1}. Similar emission lines can be seen in the ultra-violet region of the spectrum. Fairly strong absorption lines have also been measured on all plates. The velocities are about 0 km s^{-1} from primary to secondary eclipse and increase to about 30 km s^{-1} from secondary eclipse to the present. These lines become double near mid-eclipse and are probably related to mass motions between the stars and around the secondary star, though a detailed model is not yet available. Numerous other weak lines, with velocities up to ± 250 km s^{-1} also have been measured.

Faraggiana: Several spectra have been taken since 1967 at the coudé foci of the 193 cm and 152 cm telescopes of the Haute Provence Observatory with dispersions of 9.7 Å mm^{-1} and 12.4 Å mm^{-1}, covering the region $\lambda\lambda 3500$–7000. The radial velocities have been measured with an impersonal device that gives the positions of the lines with an error of one or two microns. The radial-velocity curve of the M-type star has been derived, both in the red and blue regions of the spectrum, from the lines of the neutral metals Cr, V, Fe and Ni. The results have been plotted with the previous ones of Wright and Larson (1969) on the curve computed by Hutchings and Wright (1971). The agreement between ours and the Victoria measures appears very good. In Figure 1 we have compared our measurements with those of Peery (1966). All his radial velocities are less negative than ours at the corresponding phases. It is possible that the difference Δv at the minimum of radial velocity (with 5 km s$^{-1} < \Delta v < 10$ km s^{-1}) is real. The only ionized element belonging to the M-type star is Ba I as indicated by the radial velocity of the three lines $\lambda\lambda 6496$, 4934, and 4554.

Emission lines of Fe II and [Fe II] are present and according to McLaughlin (1951) show radial velocities close to that of the centre of mass of the system indicating that they are formed in an envelope surrounding the system. Our radial velocities from lines of [Fe II] are effectively nearly constant at -20 km s^{-1}; the radial velocities from lines of Fe II are variable. We have measured the sharp core of the emission which

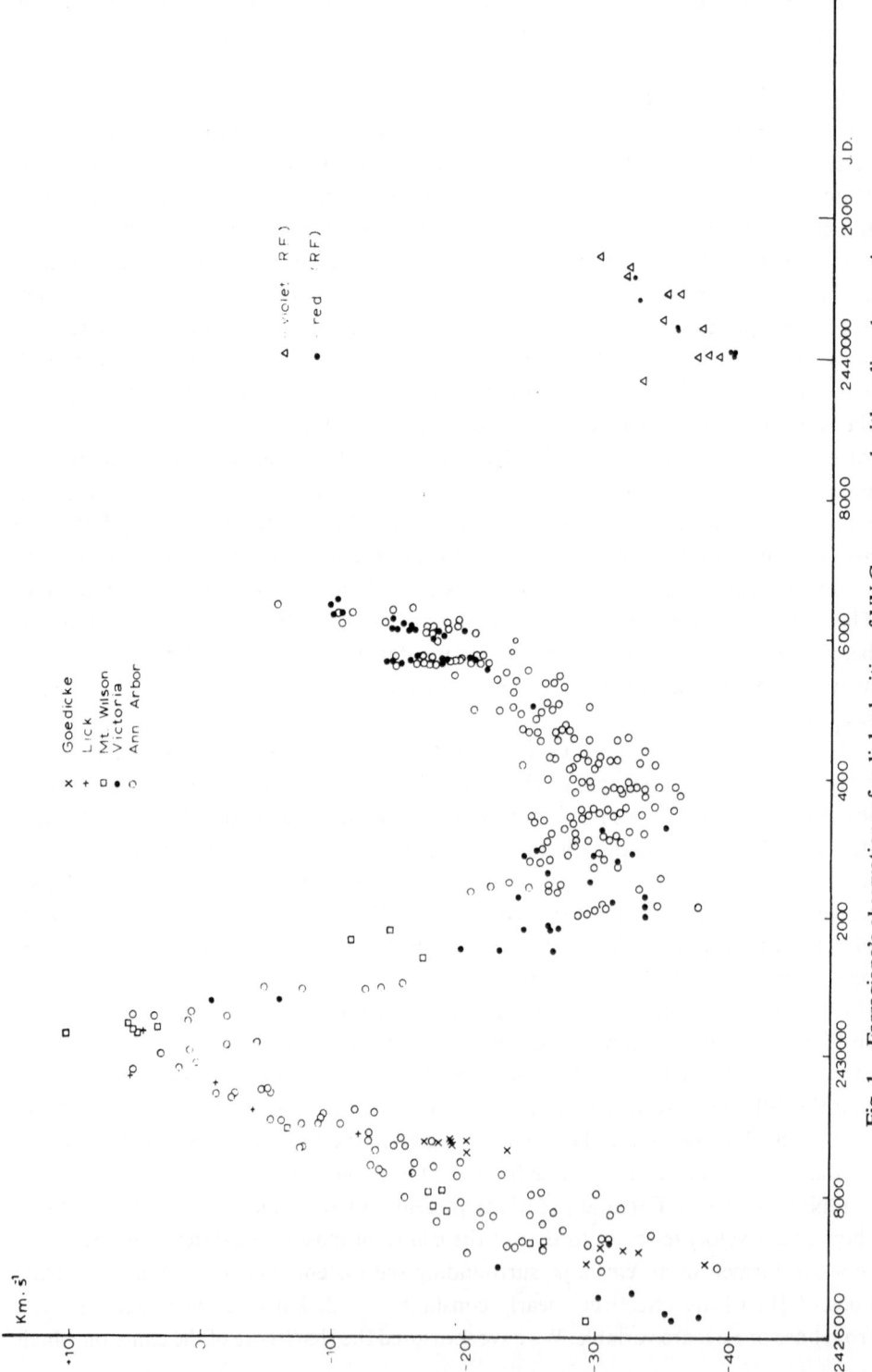

Fig. 1.   Farragiana's observations of radial velocities of VV Cep compared with earlier observations.

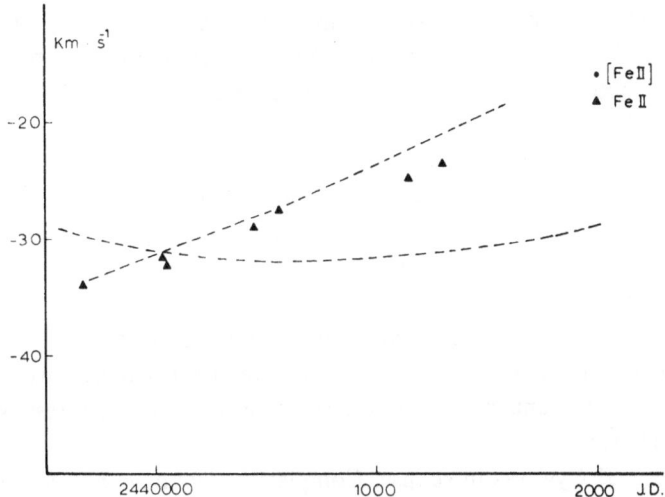

Fig. 2.   Radial velocities derived from lines of Fe II and [Fe II] in the spectrum of VV Cep, compared
with the stellar velocities.

is well visible and measurable on all our spectrograms. The Fe II curve cuts the curve
derived from the M-type spectrum at J.D. 2 440 000, and becomes increasingly more
positive with respect to that curve (Figure 2). Hence it does not seem to indicate the
motion of the B-type companion because in this case the maximum radial-velocity
should have been expected around J.D. 2 440 000. From a preliminary visual inspection
the intensity ratio $\lambda 4233$ Fe II/$\lambda 4243$ [Fe II] is variable reaching a maximum value in
1969–1970. The absorption cores of the first Balmer lines show a radial-velocity
progression, Hα giving a more negative velocity. The values of radial-velocity of the
higher members of the series (H9 to H17) are almost constant and range from slightly
more than zero (1967–1968) to about $+20$ km s^{-1} in 1972. The amplitude of the
radial-velocity progression Hα–Hn increases from about -30 km s^{-1} in 1967–1968 to
about -40 km s^{-1} in 1970–1972. Two spectrograms show a different behaviour:
GB 1246 (1971, July 25) and GB 1736 (1971, December 17): a sharp discontinuity
of about 40 km s^{-1} between Hγ and Hδ is observed and the radial-velocities of the
higher members of the series are slowly decreasing toward the mean value given by
the other spectrograms.

The radial-velocity of the absorption core of Hα has been plotted on the graph
published by Wright and Larson. The zero velocity is that of the M-type star. Our
data are very well connected with the previous Victoria observations. The emission
radial-velocity of Hα has been measured by taking the centre of the interpolated
emission contour and by measuring the shift with respect to the absorption core. The
results are plotted on the previous graph: also in this case a good connection with the
Victoria data is found.

Wright: A coordinated program for observing eclipses of the ζ Aur stars, spon-
sored by IAU Commission 42, has just been concluded. I should like to summarize

the observations and then additional comments will be made by Dr. Wood and Dr. Kitamura. The program began with 32 Cyg in 1971, October-November. The most extensive observations were obtained in Japan at Dodaira, Okayama on Akita. The Japanese observers showed that, this time at least, the eclipse was not total and, indeed, Saito and Sato showed that at mid-eclipse about half the disk of the B-type component is eclipsed by the K-type star. Using data from previous eclipses, a period of $1147^d.20 \pm 0^d.25$ has been derived. Spectroscopic observations were obtained at Okayama, Haute Provence and Victoria, and at least seven other groups, obtained photometric observations.

The eclipse of ζ Aur was also well-observed, both photometrically and spectro-scopically by the Japanese observers. The most interesting phases were from 1971, November 26 to 1972, January 7. A good series of spectra during egress was obtained by Simon in Hawaii.

In the spring, 31 Cyg was in eclipse. Complete results have not yet been received but the reported observations show that the period of 3784^d, derived from the 1951 and 1961 eclipses, is confirmed. Good series of spectra were obtained in Japan, especially during ingress and at Victoria, especially during egress.

I should like to extend special thanks on behalf of Commission 42 to all observers in this successful program.

Wood: I want to speak primarily about observations of 31 Cyg and 32 Cyg. While I'm on my feet, however, I'd like to mention the preprints of a paper by D. Schuerman

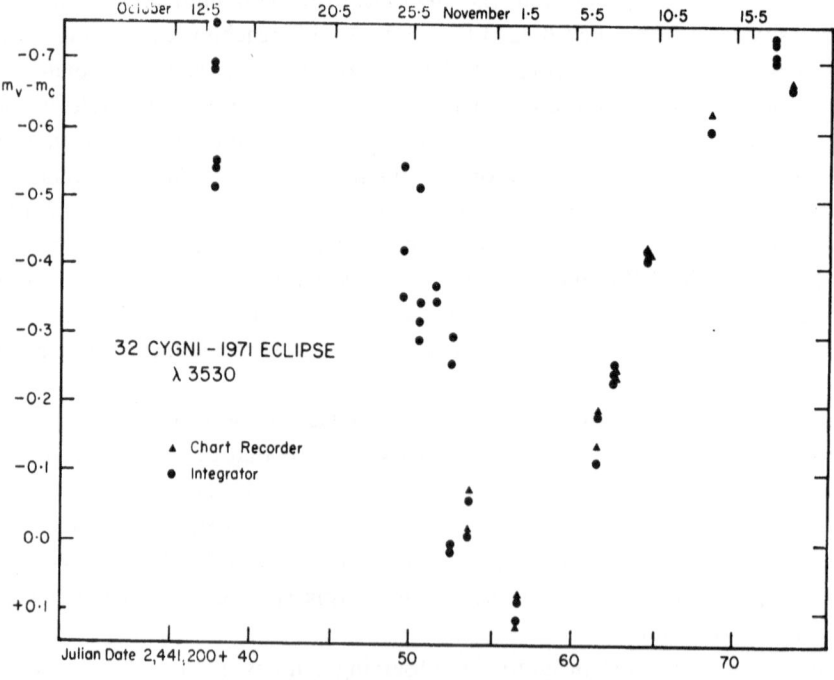

Fig. 3. Florida observations of the 1971 eclipse of 32 Cyg.

TABLE II

Eclipses of 31 Cyg and 32 Cyg

Wavelength	Depth of Eclipse	
Å	31 Cyg	32 Cyg
	m	m
3530	1.83	0.76
4240	0.67	0.28
5000	0.16	0.11

For 31 Cyg, egress from totality began between July 1.32 and 2.31 (UT). Eclipse ended
in yellow by July 6.14, but still down by
$\sim 0^m08$. Totality began between April 28.4
and May 2.4 (all dates 1972).

that several of us have received here. He considers the effect of radiation pressure on the Roche potentials and finds that some systems in which one component is very luminous may have no contact surface. This is perhaps relevant to some of our discussions. At Rosemary Hill Observatory R. H. Bloomer and others have observed 32 Cyg with three narrow-band filters prepared by Gyldenkerne for the collaborative campaign. The observations are shown in Figure 3 and Table II. These observations seem, at first sight, to agree well with Japanese observations reported by Dr. Wright. All observations together should establish well the parameters during eclipse. Apparent intrinsic fluctuations during eclipse are probably observational scatter. The duration of eclipse and the period were so poorly known that observations were made on nights of dubious photometric quality since even a crude indication of the phase of the eclipse would be better than nothing. We have already reported to Dr. Wright on possible rapid variations (in an hour or two) in the ultra-violet (λ3536) region. We have nothing to add and are greatly interested in learning of other observations.

Bloomer, D. H. Martin, and myself have been observing 31 Cyg. Our results are summarized in Table II. The observations immediately preceding eclipse are of particular interest. In Filter No. 6 (λ5000 narrow-band) on four nights in April the observations agree well. On April 26 UT (JD 2441433), the first two observations averaged 0^m02 brighter, but four others agreed with the earlier ones. On April 28 UT (JD 2441435) the first four observations average 0^m02 fainter as though eclipse had begun but the last four (looking quite good) made about 45 min. later showed the system more than 0^m1 brighter.

In Filter No. 5 (λ4240 narrow-band) again the first four nights in April were extremely congruent with the variable 1^m15 brighter. On April 26, the average of seven observations were slightly more than 0^m06 fainter, suggesting that eclipse was beginning in this wavelength. On April 28, the observation began with $C-V=1^m05$ (variable brighter), dropped in 40 min. to 0^m90 and this increased in the next 30 min. to 1^m13.

Observations in Filter No. 4 (λ3530 narrow-band) gave quite constant results on four nights from April 13–25. On April 26, there was a steady increase from a Δm of 1^m60 at 07^h00^m UT to 1^m81 at 09^h20^m; then on April 28, in a pattern similar to the other colours, the measured magnitude difference ranged from 1^m75 at about 06^h30^m UT (3 observations) to 1^m67 about 07^h20^m (5 observations) to 1^m82 at 09^h00^m (6 observations). There may have been an eclipse in the red. The readings in April were rather constant at $C - V = 1^m49$; on May 5 they were 1^m46 and on May 22 between 1^m44 and 1^m45. If this is an eclipse effect, the depth is 0^m04. The observations during totality were extremely constant on any given night in all colours. The observations in the 3530 Å region, however, suggest a slow brightening during eclipse of about 0^m15. We will discuss the evidence in more detail when the complete description of the observations is published.

Hall: I have some photometric observations of 31 Cyg and ζ Aur. They were obtained in *U, B, V* bands by Larry P. Lovell, an amateur astronomer. I know he got at least one observation on the downward branch of the light curve, three on the upward branch, and several in totality. People might like to know that these observations were made.

Kitamura: Between September of 1971 and January of 1972, more than three hundred differential *UBV* photoelectric observations of ζ Aur were made in Japan. The photoelectric observations were carried out with the 91-cm reflector at the Dodaira Station of Tokyo Astronomical Observatory, the 30-cm reflector at the Okayama Astrophysical Station of the same Observatory, the 25-cm reflector at the

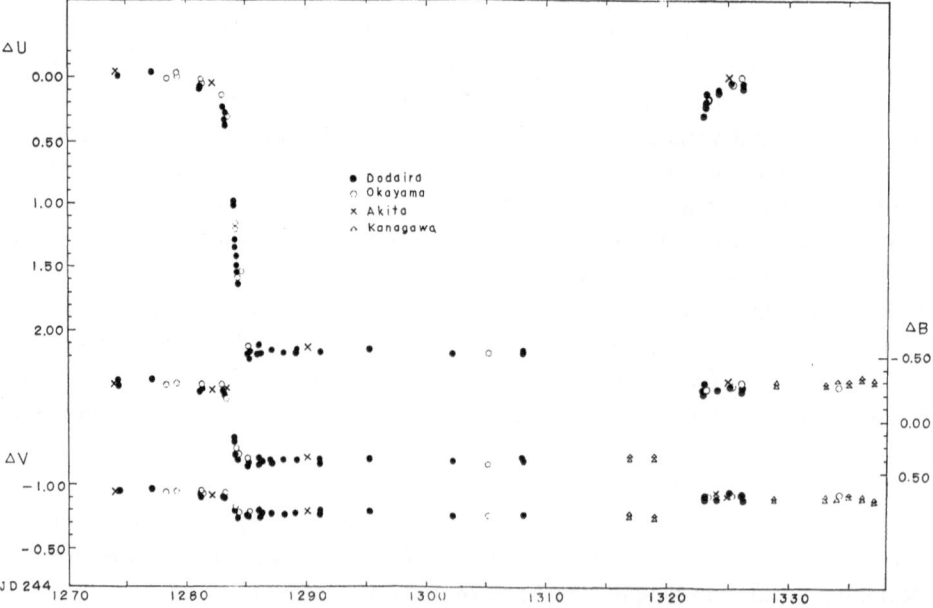

Fig. 4.   Light variations of ζ Aur compared with λ Aur between JD 2441274 (1971, Nov. 18) and JD 2441337 (1972, Jan. 20) observed from Japan.

Akita University and the 20-cm refractor at the Education Centre of Kanagawa Prefecture. These cooperative observations were undertaken with the purpose of covering as many phases as possible during the eclipse. Observations were made on forty-eight different nights, differentially with respect to λ Aur. Figure 4 shows light variations of ζ Aur $-\lambda$ Aur, in the Johnson system. General features of the light curve are similar to those in previous eclipses, but in the present eclipse the depths are systematically greater by 0ͫ04 to 0ͫ06 irrespective of the *UBV* colours.

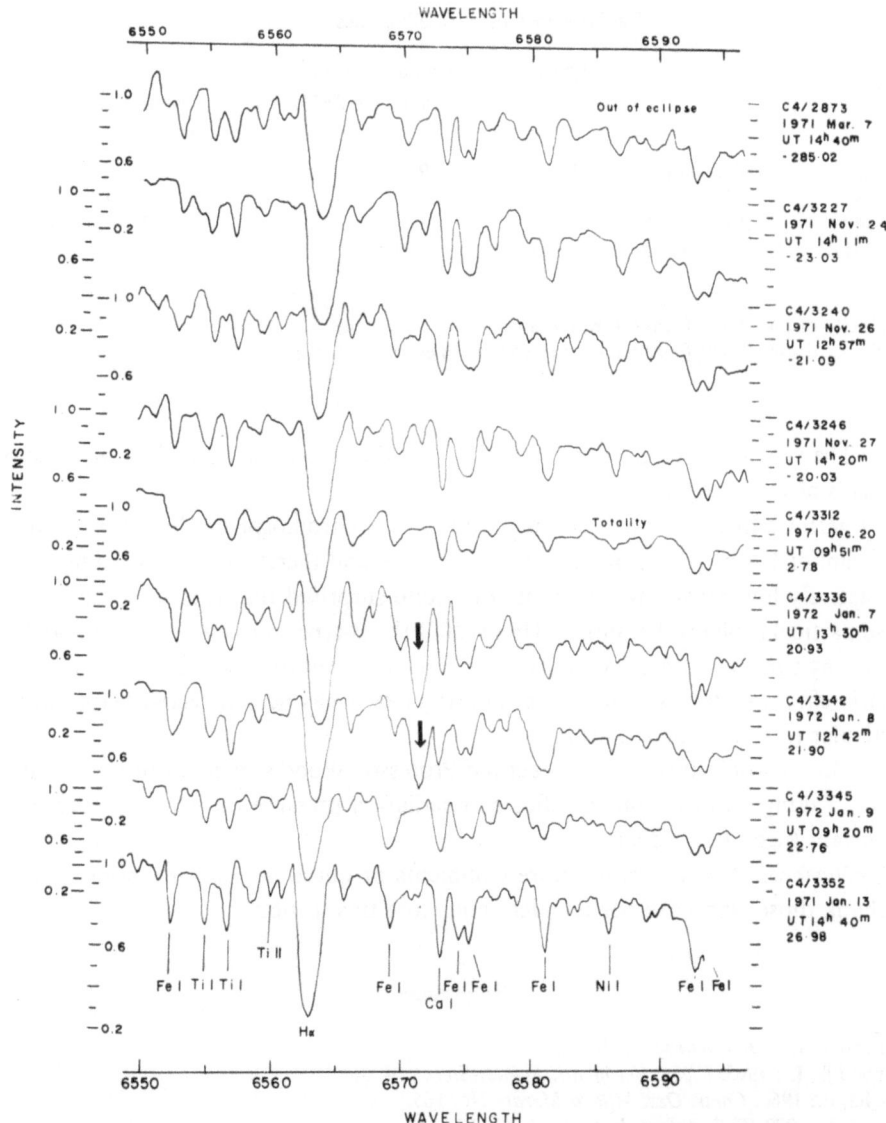

Fig. 5. Intensity tracings of spectrograms of ζ Aur obtained by M. Saito in the fall and winter of 1971–1972. The arrows point to the satellite lines of Ca I at $\lambda 6572$.

Simultaneously, spectroscopic observations were made with the 199-cm (74-inch) reflector at Okayama mainly by M. Saito, and 60 coudé spectrograms were obtained. Of these spectrograms, 20 were taken at the Hα region with a dispersion of 20 Å mm^{-1} and 40 in the blue and ultra-violet regions with the dispersion of 10 Å mm^{-1}. Figure 5 shows some interesting phenomenon that appeared at Ca I (λ6572), on the spectrograms taken around the fourth contact. The same thing can be seen at Ca I (λ4227) on the spectrograms taken on the same nights. These would indicate strong satellite

TABLE III

Radial velocities of satellite lines of Ca I

Pl. No.	Date	Phase	Radial velocities	
			Main Line λ6572[a])	Main line λ4226[b])
C4/3336	1072, Jan. 7	29$\overset{d}{.}$93	− 90 km s^{-1}	–
C4/3337	1972, Jan. 7	21$\overset{d}{.}$00	–	− 57 km s^{-1}
C4/3340	1972, Jan. 8	21$\overset{d}{.}$77	–	− 23 km s^{-1} [c])
C4/3341	1972, Jan. 8	21$\overset{d}{.}$90	− 74 km s^{-1}	–

[a] Velocity displacement from the K star's lines
[b] Velocity displacement from the chromospheric lines
[c] Faint

lines. The velocities are given in Table III. The detailed reduction of all the spectrograms is now in progress.

Wright: I'd like to make a comment about this chromospheric line of Ca I in the spectrum of ζ Aur. It's the intercombination line, and therefore it could be produced in relatively low-density populations. I'm quite surprised that it is so prominent on these particular plates, because at Hα (λ6500) the B-type star is very, very difficult to see at any time, and especially outside eclipse. This certainly seems to be a chromospheric effect. As far as I am aware, nothing like it has been detected before in this particular region of the spectrum.

Whelan: I would just like to underline Professor Wood's mention of Schuerman's paper. Taking into account the effects of radiation pressure on particles may be an important theoretical advance.

McNamara: If there are no more comments or questions this afternoon, I'd like to thank those who have participated, and close this session.

References

Binnendijk, L.: 1963, *Astron. J.* **68**, 22.
Binnendijk, L.: 1965, *Kleine Veröffentl. Remeis-Sternw.* **4**, 36.
Broglia, P.: 1960, *Contr. Oss. Milano-Merate* No. 165.
Grygar, J.: 1972, *Bull. Astron. Inst. Czech.* **23**, 175.
Günther, O.: 1959, *Astron. Nachr.* **285**, 97, 105.
Hall, D. S. and Garrison, L. M.: 1972, *Publ. Astron. Soc. Pacific* **84**, 552.

Hardie, R. H.: 1950, *Astrophys. J.* **112**, 542.
Hazlehurst, J.: 1970, *Monthly Notices Roy. Astron. Soc.* **149**, 129.
Hutchings, J. B. and Hill, G.: 1973, *Astrophys. J.* **179** (in press).
Hutchings, J. B. and Wright, K. O.: 1971, *Monthly Notices Roy. Astron. Soc.* **155**, 203.
Klinglesmith, D. A., Bernacca, P. L., and Frey, H.: 1971, *Veröffentl. Remeis-Sternw.* **9**, 205.
Lucy, L. B.: 1968, *Astrophys. J.* **153**, 877.
McLaughlin, D. B.: 1951, *Astrophys. J.* **114**, 47.
McNamara, D. H.: 1967, *Astrophys. J.* **149**, 723.
Mauder, H.: 1972, *Astron. Astrophys.* **17**, 1.
Miner, E. D.: 1966, *Astrophys. J.* **144**, 1101.
Molnar, M.: 1972, private communication to A. B. Underhill.
Moss, D. L. and Whelan, J. A. J.: 1970, *Monthly Notices Roy. Astron. Soc.* **149**, 147.
Mochnacki, S. W. and Doughty, N. A.: 1972, *Monthly Notices Roy. Astron. Soc.* **156**, 51, 243.
Norris, J.: 1971, *Astrophys. J. Suppl.* **23**, 235.
Peery, B. J.: 1966, *Astrophys. J.* **144**, 672.
Ruciński, S. M.: 1969, *Acta Astron.* **19**, 125, 245.
Struve, O.: 1944, *Astrophys. J.* **99**, 89.
Wright, K. O. and Larson, S. J.: 1969, in M. Hack (ed.), *Mass Loss from Stars*, Reidel-Dordrecht, p. 198.

EVOLUTIONARY ASPECTS OF CIRCUMSTELLAR MATTER
IN BINARY SYSTEMS

MIROSLAV PLAVEC

Dept. of Astronomy, Univ. of California, Los Angeles

Abstract. Several groups of close binary stars are considered in an attempt to explain their present state as a consequence of a large-scale mass transfer or mass loss in the past: Algol-like semidetached binaries, some shell stars (AX Mon), some binary X-ray sources (Cen X-3, Her X-1), the recurrent nova T CrB, helium-rich binaries υ Sgr and KS Per, and the symbiotic variables.

Algol-like binaries like U Sge cannot be products of a conservative case A of mass transfer; rather, mass loss from system and/or a temporary contact stage must be invoked. Nova T CrB as well as the symbiotic variables probably contain a mass-losing giant and a helium star, which again may be a product of a previous mass transfer of type B. Similarly, some of the X-ray sources may actually be binaries undergoing a second process of mass transfer. The systems υ Sgr and KS Per may contain helium stars expanding to the right of the helium main sequence, while the other component may be a rather inactive main-sequence star. Some shell stars may be products of mass transfer. Mass loss from convective envelopes is also discussed.

Loss of mass and of angular momentum from many binary systems must be anticipated. Behavior of the mass-accreting stars may often be decisive for the appearance and evolution of the system.

1. Introduction

A few years ago, I had a long discussion on close binary stars with Dr. Alan Batten. We could agree perfectly on only one conclusion: that the real hell for a sinful astronomer (if any such exists, of course) would be a round trip through the heavens. Knowing now how different were our ideas about Mars and Venus from what is revealed now by the space probes, can you imagine the mental torture of an astronomer who proposed models of β Lyrae, UX Monocerotis, ε Aurigae, HR 2142 etc., etc. – and suddenly sees the reality? And yet, he tried so hard to interpret all the available data... The trouble is that we know so little about masses, radii, luminosities, even about the evolutionary stage in which a particular system is just now.

You will understand that I am trying to mollify your attitude towards the paper I am going to present. For my task can be summarized as follows: Find the exact ways by which a given binary system has arrived just at its present stage, of which we know next to nothing! In his remarks on Sahade's paper at the La Plata Symposium on Stellar Evolution in 1960 – which happened to be one of the last meetings he attended – Otto Struve said (Sahade, 1962):

With regard to the discussion of the evolutionary characteristics I take little less interest in it. To me, much of this is still vague and perhaps subject to a great deal of doubt, and while I have nothing to criticize, I have the feeling that we know at present very little about what is going on in many of these binary systems. It is awfully hard to say anything about how such binary systems will evolve. There were times in my own experience when I thought that perhaps the existence of gas streams would have a very significant effect upon the evolutionary track of a star. At other times I felt that those phenomena were superficial and that they did not seriously modify the evolutionary track. And I think there is no consistent answer.

Batten (ed.), Extended Atmospheres and Circumstellar Matter in Spectroscopic Binary Systems, 216–259.

It is my intention to show that the propects of our understanding of evolution of binary stars may not be so gloomy today as they were twelve years ago. Before I attempt to show what has changed since Struve's times, let me answer briefly his dilemma whether gas streams are significant for stellar evolution or not. While we can no doubt find binary stars where circumstellar matter is only a superficial phenomenon, I think there is little doubt that in general circumstellar matter plays an essential role in the evolution of close binary stars. I have two simple arguments in favor of this point of view, both based on a statistical survey of eclipsing binaries.

First, it is interesting to realize that in order to be conspicuous, an eclipsing binary must violate the laws of single-star evolution. Photometrically conspicuous eclipsing binaries have the primary minimum deeper than, say, 1^m. This presupposes an occultation of a hot star by a cooler but larger companion. This cannot happen in a main-sequence pair. The Algol-like semi-detached binaries can display very deep minima, and in them a hotter main-sequence star is combined with a subgiant which is cooler, larger and of smaller mass – in clear violation of normal single-star evolution. Among the first 9 eclipsing binaries to be discovered, no less than 7 were of this type, according to a list compiled by A. H. Batten. Only one system of the nine is normal and detached, namely U Ophiuchi. The remaining star is β Lyrae. Although picked up as a bright variable, it is a foremost representative of eclipsing binaries which are spectroscopically conspicuous – due to the presence of circumstellar matter.

My second argument is based on two surveys of eclipsing binaries. Among 145 eclipsing binaries with somewhat reliable data and brighter than 8^m5 at maximum light, I found (Plavec, 1968) 59 systems with both components on the main sequence. Most of them are simple detached pairs with little interaction. About 20% of these main-sequence binaries, however, are contact or nearly contact, i.e. with strong interaction and probably with circumstellar matter in significant quantities. Circumstellar matter is present or must have been present in the past in most of the remaining 86 systems, since they are: 52 semidetached binaries, 14 systems of the W UMa type, 2 with a Wolf-Rayet secondary, and 18 eclipsing binaries with a giant or supergiant primary. Only among these 18 systems a few can be detached and simple, with little interaction.

More recently, I have surveyed, with the collaboration of R. S. Polidan, all eclipsing binaries contained in Batten's catalog of spectroscopic binary orbits (Batten, 1967) with periods longer than 5d. There are 60 such systems. Twenty-two among them can be classified as Algols or related to them. Three are of the AR Lacertae type, i.e. a pair of subgiants not filling the critical Roche lobes but with rather prominent H and K emission indicating possible extended envelopes. Fourteen systems cannot be classified easily but components are certainly interacting (β Lyrae, UX Monocerotis, VV Cephei and W Serpentis are included in this group). Sixteen systems appear to be detached and simple; some are main sequence systems (such as RR Lyncis), a few are detached giant and supergiant systems which may have evolved as single stars (ζ Aurigae is included here). Five systems are undecided. My estimate is that 70% of systems with periods longer than 5 d are either strongly interacting now or their evolution cannot be explained without large-scale mass transfer in the past.

Selection rules in both surveys eliminated eclipsing binaries with components below the main sequence, such as UX UMa or some U Gem stars and novae. These are clearly also unstable systems where circumstellar matter plays an important role. We can safely say that at least 60% and probably as much as 75% of all known eclipsing binaries are of the strongly interacting type, where circumstellar matter plays an important role.

Simple, well-behaving systems with little if any circumstellar matter can be found only on the main sequence and rather exceptionally among giants or even super-giants provided that such systems are sufficiently large to permit free expansion of the components. Pairs of very similar stars, quite frequent on the main sequence, are almost non-existent among the giant and supergiant systems.

In a recent article, Kopal (1971) used just this absence of similar pairs among giants as an allegedly strong argument "focusing attention on the weak points of the optimistic picture of the evolution of close binary systems based on the theory of large-scale mass loss or transfer". In his colorful language, he writes:

Thus the old charm that equally massive stars are equal also in size – so manifest on the main sequence – seems to lose its power soon after the star evolves away from the main sequence. It is almost as though some new factor – previously inactive – entered the control room of stellar evolution at this stage; but if so, we are as yet in the dark as to its identity (l.c., p. 529).

I think I should discuss this point here since Kopal repeatedly returns to it, saying that "This is an embarrassing question, and the answer to it still eludes us".

On the contrary, I think the explanation is quite simple. The evolution across the Hertzsprung gap and up the giant branch is very rapid, and is progressively much faster for stars of larger mass. Associated with this evolutionary phase is a large increase in radius. In Iben's models (Iben, 1967), the age of a star of mass m at the top of main sequence can be approximated by the linear formula (for $2 \leq m \leq 6$)

$$\log t^{\mathrm{I}} = 8.686 - 2.5\,(\log m - 0.352), \tag{1}$$

while for a red-giant tip we get similarly

$$\log t^{\mathrm{II}} = 8.771 - 2.66\,(\log m - 0.352). \tag{2}$$

Take now for example AR Aurigae, a system which Kopal mentions as one of the good examples of a main-sequence system with nearly identical stars. According to Popper (1970) the masses of the components are

$$\log m_1 = 0.41\,(m_1 = 2.57\ m_\odot),$$
$$\log m_2 = 0.36\,(m_2 = 2.29\ m_\odot).$$

Inserting into (1) and (2), we get $t_1^{\mathrm{II}} = 410 \times 10^6$ yrs, $t_2^{\mathrm{I}} = 460 \times 10^6$ yrs. This means:

the primary component develops into a helium-burning giant while the secondary star is still within the main sequence band; the primary reaches the red-giant tip 50 million yrs before the secondary ends its main-sequence evolution. Over a period of some 400 million yrs when both stars are relatively young, small difference of masses means small difference in radii. But once one of them crosses the Hertzsprung gap, its radius increases rapidly and considerably, and a small disparity in masses is sufficient to cause large disparity in radii, spectral types and so on. In fact, even a smaller difference in masses will have the same effect. Subtracting (2) from (1),

$$\log t_1^{II} - \log t_2^{I} = 2.50 \ \log m_2 - 2.66 \ \log m_1 + 0.14, \tag{3}$$

we find that a difference of 3 to 8% in masses is sufficient to bring the more massive star to the red-giant tip while its twin sister, almost identical but slightly lighter, is still a main-sequence star. The above figures hold good for stars between about $2M_\odot$ and 5 $M_\odot$, but are only insignificantly changed for other masses. Stars which are almost identical twins on the main-sequence become very different beyond it unless they are *exactly* identical. Naturally such systems are very rare, and among those enumerated by Kopal – β Aur, WW Aur, AR Aur, Y Cyg, YY Gem – probably only YY Gem fulfills the above condition of virtual equality of the components.

   While the absence of similar pairs among giants is in this way explained by the ordinary theory of single-star evolution, it is clear that a good deal of observed close binary systems cannot be explained without assuming large deviations from single-star evolution. I think a very fruitful fundamental concept is the assumption that each component in a close binary system has only a finite volume available for its expansion, and once it reaches the limit of dynamical stability, it begins to lose mass. Theoretical treatment shows that in many cases this mass loss proceeds on a time scale much shorter than the nuclear time scale of stellar evolution, and that large quantities of mass are lost before the star manages to restore secular equilibrium. Here is the theoretical foundation of the theory of large-scale mass loss or mass transfer.

   Actual calculations of these processes have been so far based on the assumption that the limit of dynamical stability is identical with the innermost contact surface of the Jacobian zero-velocity surfaces, or, as we briefly say, with the Roche limit. This may not be so if forces other than gravity are at play. How far this can affect the basic results of the theory of mass transfer is difficult to say. At times I am tempted to share some of Kopal's pessimism expressed in his article cited above. I will attempt today to apply current theory to various groups of binary stars. I hope this will stimulate further research, both theoretical and observational. If we eventually fail to explain the observed systems, we will have to change the theory.

2. Survey of Theoretical Results

The general picture of evolution of binary stars in the case of mass transfer will be given only very briefly here, since it was reviewed several times (Plavec, 1968, 1970a), most recently in a very brilliant exposé by Paczyński (1971a). The course and outcome of the process of mass transfer depend in the first place on the mass and evolutionary stage of the mass-losing component. Following Kippenhahn and Weigert (1967) it has become customary to distinguish cases – the term 'modes' would be more appropriate – A, B, and C. Case A occurs when the initially more massive component – I will call it consistently the principal component – reaches its limit of dynamical stability (henceforth simply called the Roche limit) while it still burns hydrogen in its core. When the principal component reaches the Roche limit later, at the stage when it is moving across the Hertzsprung gap and up the giant branch and burns hydrogen in a shell surrounding the hydrogen-depleted core, we speak of mode B. Mode C corresponds to cases where the Roche limit is reached when the star expands along the second giant branch, and as a rule possesses two energy-generating shells (hydrogen shell and helium shell). Which of these three modes will apply to a particular system depends on the masses of the components and their separation. A very convenient way of showing this dependence is to plot the period of the system against the mass of the principal component, as in Figure 1. Lines can be drawn to separate the modes, A, B, and C, respectively. The lines drawn in the Figure actually correspond to a mass ratio of 0.5. For other reasonable values of the mass ratio they would not be significantly different. A system of any mass ratio q can be plotted if instead of its actual period P we plot a 'reduced' or 'normalized' period P^* computed by the formula

$$\log P^* - \log P = -0.144 \log 2q . \tag{4}$$

(Cf. Horn et al., 1969; or Plavec, 1968). The lines in Figure 1 were drawn using the recent models of stellar evolution by Paczyński (1970). Plotted in the diagram as full circles are some model binaries for which evolutionary sequences have been computed in detail.

Since a star's radius increases only slowly as it traverses the main-sequence band, case A occurs only within a very limited range of short periods, typically about 1 d. In a rapid phase proceeding on a thermal time scale the principal component becomes typically a subgiant secondary of smaller mass. This subgiant evolves very slowly but eventually begins to rise up the giant branch, reaching the Roche limit for the second time. However, the separation of the components is always so small in case A that an expansion of the mass-accreting star may easily interrupt the evolution and lead to the formation of a contact system. This problem will be discussed in Chapter 3.

Models located in the region labeled *Br* are stars with radiative envelopes and hydrogen-burning shells. Mass loss proceeds again on a thermal time scale. Besides the instability on the surface due to the presence of the Roche limit, the star has another instability at its center since the hydrogen-poor core becomes thermally unstable when it exceeds the Schönberg-Chandrasekhar limit. This latter instability eventually predominates and mass transfer does not stop until helium is ignited at the center and the expansion of the envelope is halted. The remnant of the principal star – essentially only the helium-rich core – then crosses the H−R diagram and

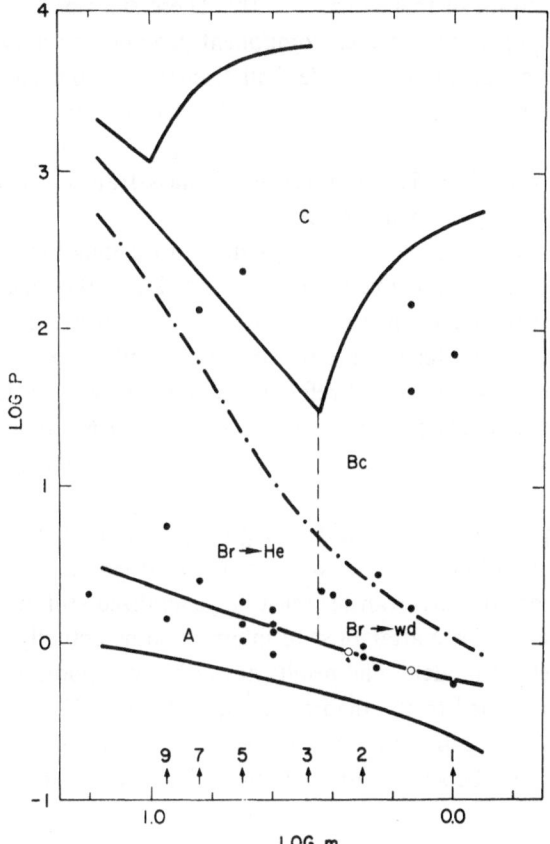

Fig. 1.   The cases (modes) A, B, C, are shown here on a plot of the orbital period P against the mass of the principal component m. Period is expressed in days, mass in solar units. Full circles indicate initial values for models that have been computed in detail. The two open circles indicate roughly the initial conditions at the beginning of the second phase of mass loss in case AB.

settles to a new equilibrium near the main sequence of the homogeneous helium stars.

Stars less massive than about 2.8 $M_\odot$, when trapped at the Roche limit soon after the end of the main sequence, have cores too small to ignite helium. The rapid phase of mass transfer gradually passes into a very slow phase where the contraction of the core and the associated expansion of the envelope are slowed down by increasing electron degeneracy in the core. Mass loss stops when practically no envelope above the hydrogen-burning shell is left. The remaining envelope virtually collapses on the star and the star becomes a white dwarf. In the stars of small mass ($<2.8\ M_\odot$) the mass of the core increases considerably as the star moves up the giant branch. Therefore, if such a star reaches the Roche limit at a later stage, it can still ignite helium and become a helium star rather than directly a white dwarf. Stars for which this can happen, i.e. those higher up on the giant branch, have by that time developed a deep convective envelope.

The region of deep convective envelopes is labeled Bc. Very few models have been computed so far, and all of them very recently. These stars pose a specific problem since the mass loss from the principal component proceeds on a scale much shorter than the thermal time scale of the star. We shall discuss them in Chapter 5.

Finally, one case was computed for mode C (Lauterborn, 1970). The outcome was a massive white dwarf composed of carbon. The region C will probably have more subdivisions with possibly different products of mass transfer. The upper limit of region C is given by the ignition of carbon.

It would be interesting to plot actual systems into a diagram such as Figure 1. For most systems this is impossible since we do not know the masses. Transformed systems which already underwent the process are by no means eliminated from the graph. A second mass transfer is quite possible and I think that certain systems can best be explained in this way. I will refer to this possiblity when talking about the Prendergast-Burbidge model of X-ray sources in Chapter 6, about the helium-rich binaries v Sagittarii and KS Persei in Chapter 7, and about the symbiotic variables in Chapter 8.

The variety of possible configurations is greatly increased if multiple processes of mass exchange are admitted. However, the observed systems display such a large variety that the theory of mass transfer must be generalized still in other respects. In the past several years, I have attempted to interpret some actually observed systems by means of theoretical models. The results are rather disappointing. Some binaries can be tentatively assigned to the theoretical types A, B or C of mass transfer. But the old Mephisto's saying proves again to be true: "Grau, teurer Freund, ist alle Theorie, und grün des Lebens goldner Baum". The real systems are enormously more complicated.

Interestingly, the star that is most frequently responsible for these complications is the mass-accreting component. No doubt the primary factor that brings complications into a binary system is the more massive component when it expands, becomes unstable and begins to lose mass. But it seems to play the role of an *éminence grise*, a grey shadow behind the scenes. The mass loss appears to proceed rather smoothly, although this may be an illusion due to the circumstance that the mass-losing star, typically a red giant, is seldom prominent spectroscopically or photometrically.

Nevertheless, really conspicuous phenomena – outbursts including nova outbursts, flares, X-ray emission, radio flares, emission lines, complex absorption profiles – most of these phenomena seem to be associated with the mass-accreting star. This star was almost completely neglected in the theoretical studies reviewed at the beginning of this chapter. This is certainly no longer possible, and recent identifications of binary systems with X-ray and radio sources make it quite plain.

Another recent development with immediate impact upon the problems of evolution of binary stars are the attempts to discover *collapsed objects* in binary systems. The presence of a black hole was suggested for a number of binaries – β Lyrae, ε Aurigae, BM Orionis, HDE 226868. In ε Aurigae there is probably enough space for a star to expand freely into the supergiant stage and thus to pass through all stages of normal stellar evolution. From this point of view, a collapsar in ε Aurigae would not be a surprise. All the other systems are much too small. If they indeed harbor a black hole, we must postulate a shortcut in stellar evolution in the sense that the progenitors of the black holes never passed through the stages of supergiants. The mass transfer of mode B or AB, in which heavy stars transform into helium stars, may be such a shortcut. Or else we must postulate a drastic reduction in the size of these systems, associated apparently with large mass loss and loss of angular momentum from the system.

It is still possible that the puzzles of β Lyrae and BM Orionis will be explained without the black hole hypothesis. However, the problem of loss of mass and angular momentum from binary systems will remain since it emerges also when we want to interpret the Algol systems, some X-ray sources, and other objects.

In his review article quoted above, Kopal puts forward another argument in favor of mass loss from the system (l.c., p.534): The total mass of some present-day Algol systems is so low that the initial principal component is unlikely to have had enough time to evolve to the point of instability within the lifetime of the Galaxy, unless of course it was originally much more massive than the present total mass of the system indicates. It would be possible to argue successfully about some of the masses Kopal uses, since they are based on absolute elements published in his catalog (Kopal and Shapley, 1956), which have in many cases been superseded by more recent investigations. Thus R Canis Majoris, for which he gives 0.6 $M_\odot$ as the total mass of the system, was found to have a total mass over 2 $M_\odot$ by Sato (1971) and to be much less anomalous than it was thought once. Another system mentioned by Kopal as having total mass less than 1 $M_\odot$, T Leonis Minoris, should not be used as an argument at all. Its spectroscopic elements (Struve, 1946) are based on three normal points each formed from a very scattered group of velocities corresponding nearly to the same phase. Struve says that they "are suitable only for a preliminary determination of the semi-amplitude of the velocity curve", and Batten (1967) justly assigns the lowest grade to this orbit. No mass determination should be based on such elements. Nevertheless, some of the Algol systems may represent an argument for mass loss from the system. However, I think we cannot go as far as to suggest that "... it is the mass loss, rather than a mere mass exchange, which seems to hold a clue to the actual situation" (l.c., p. 543). Even with Algols of low mass, the problem still remains that the less massive component appears more advanced in evolution being a subgiant – often considerably overluminous for its mass – while the primary is a bright main-sequence star. This paradox, I think, cannot be explained without admitting some mass transfer between the two stars. It seems, however, that the process often is not *conservative*, i.e. that the total mass of the system and total orbital angular momentum are not preserved.

Virtually all papers dealing with mass transfer assumed the conservative mode, although the authors knew quite well that this may not be the best assumption, as was mentioned in almost every paper and recently emphasized in the review by Paczyński (1971a). The conservative case may not even be significantly simpler to treat than a more generalized case: the formula giving the instantaneous Roche radius to be compared with the stellar radius would simply be different. I think the conservative case has been so attractive mainly because it is physically well defined. It appears difficult to formulate another physically consistent model without a cumbersome hydrodynamical treatment of the gas flow in the system. Paczyński and Ziólkowski (1967) formulated a nonconservative model by introducing two parameters, one representing the fraction of mass lost from the system, and the other representing the rate of loss of orbital angular momentum. Simplified model calculations showed an interesting variety of results, but the authors themselves admitted that the choice of the parameters had been rather arbitrary.

Weigert and Lauterborn in 1970 formulated a physically well-defined and consistent non-conservative model in which mass was streaming out from the system through the Lagrangian point L_2. Large loss of angular momentum greatly decreased the size of the system and its period. It is to be regretted that the model has not been published.

In spite of an annoying lack of observational data on mass loss, and lack of hydro-dynamical models of gas streaming, it is quite clear now that the study of non-conservative models is urgently needed. I hope to show it in the next two chapters.

3. The Semidetached Binaries of the Algol Type

By the Algol systems I will mean semidetached binaries in which the less massive component is apparently more advanced in evolution, i.e. it is located above the main-sequence band, usually in the region of subgiants of spectral types F through K. The more massive star is still on the main sequence and is of earlier spectral type.

(Incidentally, something should be done with the nomenclature on close binaries. It is very awkward to use the term "semidetached binaries of the Algol type", or any variant of it, to describe a class of eclipsing binaries so well defined that single word should describe it. However, I cannot use simply the term 'Algols' right from the beginning, since many authors still use the term 'eclipsing binary of the Algol type' to mean an eclipsing variable with constant light between minima. It has been argued several times (Kopal, 1955; Plavec, 1964) that this definition is ambiguous and does not specify a physically meaningful and homogeneous class of objects.

Case A of mass transfer – beginning when the primary component still burns hydrogen in its convective core – has often been invoked as a lucid example how a typical Algol system comes into being. Figure 2 is an example. The primary component reaches its Roche limit at *b*. As the rate of mass transfer increases quickly, the star moves down and to the right in the H–R diagram. Thermal equilibrium is not restored until at *e* when the rapid phase of mass transfer ends. By that time the original primary is already the less massive component but is situated well above the main sequence in the region of the subgiants. Subsequent evolution proceeds on a nuclear time scale, thus providing us with much better chances actually to observe the system. The subgiant continues to fill its critical Roche lobe, and loses mass at a low rate, of the

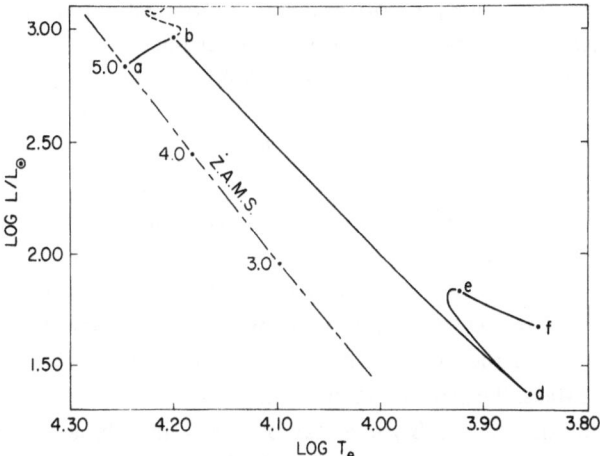

Fig. 2. Conservative mass transfer in case A. Mass outflow begins at *b*, both stars have equal mass at *d*, thermal equilibrium is restored at *e*, the subgiant begins to contract at *f*. Diagram drawn after Plavec *et al.*, 1969.

order of 10^{-7} $M_\odot y^{-1}$ or so. The secondary star is now a bright massive early-type main-sequence star.

This picture represents the Algol systems extremely well if we are satisfied with a general qualitative picture. We have believed that most of the Algols are in the slow phase of mass transfer, between *e* and *f* in Figure 2, but that for example U Cephei or RZ Scuti may be nearer to point *e* or even between *d* and *e* because of more prominent gas-stream phenomena and rapid rotation of the early-type component.

One difficulty in explaining the Algol systems by mass exchange in case A was noted almost at the beginning, however. Many Algols have relatively small total mass of the system, less than 4 $M_\odot$ – for example X Tri, RW Tau, TW Dra, RZ Cas and many others. Many of these systems are notable for large overluminosity of the subgiant – it lies well above the main sequence and is too luminous for its mass. Most striking are the cases of AS Eri, DN Ori and XZ Sgr, where the overluminosity amounts up to 5 to 10 mag. It is quite impossible to generate these luminous sub-giants by the conservative case A process. Extensive calculations (Plavec and Horn, 1969; Plavec *et al.*, 1969; Horn *et al.*, 1969; Ziółkowski, 1970) can be summarized into the following 'rule of thumb': *principal component tends to preserve the luminosity class in the process.* A main-sequence B star has luminosity class IV when it has evolved some distance away from the initial main sequence; if it fills the critical Roche lobe at that time, it becomes a subgiant, luminosity class IV. Mass transfer starting very shortly before the B star terminates its main sequence evolution actually gives giants, and indeed the B star at such stage would be classified as luminosity class III. Among the A and F stars only luminosity classes V and IV or IV–V correspond to the main-sequence stage; and indeed, the transformed stars are luminosity class IV–V at best. The remnants of principal components of small mass – say less than 2.5 $M_\odot$ – invariably remain quite close to the zero-age main sequence.

This is why it has been suggested (Weigert, 1968) that Algol systems of small mass are produced by mass exchange of mode B. In this case, the expanding primary reaches the Roche limit at the time when it burns hydrogen in a shell surrounding the helium-rich core where hydrogen has already been depleted. Thus at the beginning of the process the star is either traversing the Hertzsprung gap or climbing up the giant branch. These phases are much shorter than the main-sequence phase, but the star's radius increases so much that many more binaries should come under case B than under case A. However, our chances to observe such an object as a semidetached binary are good only if the mass of the system is sufficiently low, say below 4 $M_\odot$. For more massive stars, the process is too short.

Compare the modes A and B for a star of 5 $M_\odot$ in a binary system where the secondary component initially has 4 $M_\odot$. If the initial period is 1.27 d, the primary will reach its Roche limit when hydrogen at its center is depleted from $X_c = 0.60$ to 0.15 but the convective core still exists and produces all the energy of the star. The rapid phase of mass transfer lasts 640000 years and the principal component transfers 1.82 solar masses to the secondary. In the ensuing slow phase the primary loses another 0.66 $M_\odot$ but at a much slower rate since this phase lasts 20 million yrs. This is comparable with the main-sequence lifetime of the principal component, which was about 45 million yrs. Since the transformed system displays deeper eclipses, we believed that the phase of slow mass loss should be identified with observed Algol systems. (For a more detailed discussion of this case see Plavec *et al.*, 1969.)

If the sytem is slightly wider to begin with, and the initial period is about 2 d, the primary does not become unstable until after a hydrogen shell source is formed in its interior. This example of mode B was studied by Kříž (1969a). The rapid phase lasts 4.2×10^5 yrs, i.e. it is not much shorter than its counterpart in case A, but the star loses much more mass, namely 4.06$M_\odot$. The slow phase that follows lasts only 6.6×10^5 yrs, and ends when the primary – now only 0.69 $M_\odot$ – ignites helium in its core. The envelope contracts quickly and the star crosses the H-R diagram, changing from a red giant into a helium-burning subdwarf. The whole process lasts hardly longer than 1 million yrs as compared to the 21 million yrs in case A. Therefore it must be concluded that our chances are very low to catch such a system in the process of mass transfer.

Things are different, however, when the primary is initially less massive than about 2.8 $M_\odot$. When the hydrogen-rich envelope is stripped off, the remnant is not massive enough to ignite helium. Its final destination is the region of white dwarfs. But before it contracts to this stage, the primary climbs up the giant branch for a long time. The rapid contraction of the core that so accelerates the evolution in the preceding case of a star of moderate mass is now perceptibly slowed down by increasing electron degeneracy in the core. This makes the slow phase favorably long. For example, for a primary initially of 2 $M_\odot$ this slow phase when the system is semidetached lasts one quarter of the main-sequence lifetime (Kippenhahn *et al.*, 1967) while for the above discussed star of 5 $M_\odot$ the ratio was 1:80. Since the relative and absolute lifetimes are so favorable for systems of low mass, and since such binaries are more frequent than those of moderate mass (simply because stars of lower mass are more frequent), case B of mass transfer for systems of low mass appears a very adequate explanation of the existence of numerous Algol-type semi-detached binaries – except for one circumstance: The evolution is very slow towards the end of the process, when the primary is way up the giant branch. Therefore, among the observed Algols, those systems should be most frequent in which the late-type star is a real giant, very overluminous for its small mass. This does not appear to be the case. The above-mentioned extreme overluminosities found in DN Ori, AS Eri and a few other stars are exception rather than rule, and typical overluminosities do not exceed 4^m (Ziółkowski, 1969).

In order to remove this discrepancy, Ziółkowski (1969, 1970) suggested that many Algols are actually in the so-called mode AB of mass transfer. In discussing the case A for a star of 5 $M_\odot$, I assumed that the slow phase ends when the subgiant detaches itself from the Roche limit. Actually, this introduces only a short break in the process of mass loss. The slow phase of case A is a continuation of the main-sequence evolution of the primary star. It has now much smaller mass, has lost part of its envelope, but hydrogen continues to burn in its core. When it is eventually exhausted, the star shrinks (as the more massive stars typically do) at the end of the main-sequence phase. However, very soon it begins to expand again when the hydrogen burning shell is established around the hydrogen-depleted core. Thus, in our special case, within about 3×10^4 yrs the primary expands to the Roche limit again and a new phase of mass transfer sets in. (Stars of smaller mass do not have the contraction phase at all.) Since the star now burns hydrogen in a shell, we have case B of mass transfer. The process is soon dominated by the thermal instability at the star's center so that the outcome is the same as in the case B discussed above: when the mass-losing giant eventually ignites helium in its core, rapid contraction sets in and the star develops into a helium-burning subdwarf. The only significant difference between this case AB and the pure case B is the time scale. The process of mass transfer is much slower in case AB. This is understandable, since the principal component enters phase B as a star of only 2.5 $M_\odot$ so that the process must be considerably slower than for the 5 $M_\odot$ in pure case B. Horn (1971) studied the case AB for a star of 5 $M_\odot$. As usual in mode B, the process is very much the same independently of the initial mass ratio. Let us distinguish – somewhat schematically – between the subgiant phase (when the mass-losing star has luminosity class IV and is in the lower half of its giant evolutionary track) and the giant phase (when the star is on the upper part of its giant branch). The ratio of the corresponding lifetimes is 11:1 (18.5×10^6 as against 1.7×10^6 yrs) in favor of the phase of lower luminosity. The primary component spends most of the time as a subgiant, only moderately overluminous for its mass. This is different from case B discussed by Kříž where the ratio of lifetimes is roughly 3:1 in favor of the stages of higher luminosity.

Ziółkowski (1969, 1970), who discovered case AB independently of Horn, realized that the difference in lifetimes might be significant for the probability of discovery. Since typical subgiants in Algol-type binaries show only moderate overluminosity, he suggested that many of them are actually undergoing mass transfer of type AB. His histogram of the distribution of observed semi-detached binaries as a function of the total mass seems to indicate the existence of two separate groups: massive binaries with a total mass larger than 5 $M_\odot$, and low-mass systems with a total mass less than 4 $M_\odot$. Since in this latter case the initial mass of the principal component is likely to be below 2.8 $M_\odot$, the dichotomy appears rather natural. It was believed that the massive binaries are products of mode A, while the low-mass group were binaries observed in mass transfer of type B or AB; the ratio between these two groups remained rather uncertain.

The cause of the uncertainty was the problem how frequently can the case AB actually occur. In our survey of mass transfer of mode A from a star of 5 $M_\odot$ (Plavec et al., 1969; Plavec and Horn, 1969), we found that the slow phase is terminated either by an overall contraction of the mass-losing component, or by expansion of the mass-accreting star which reaches its own Roche limit. Clearly only the former alternative permits the primary star to enter stage B of mass loss. This happens only if at the end of the slow phase the system is relatively large, and/or the secondary is not very much more massive than the primary. These favorable conditions obtain only when initially there was not a large disparity in masses ($m_2/m_1 > 0.6$) and the primary star did not reach the Roche limit until shortly before the end of its main-sequence evolution (so that its central hydrogen content was already low to begin with, say $X_c < 0.25$). Even if the primary component is allowed to embark on case B of mass loss, the process is so slow that most likely the other component becomes unstable before it can be terminated. Thus larger overluminosities can be attained only for a much narrower range of original conditions ($X_c < 0.05$). These stringent conditions lead us to suspect that case AB will be rather rare. Indeed, from private communications it seems that Ziółkowski also has largely retracted his original optimistic opinion on the cosmic importance of case AB.

Although the basic problem of the massive Algol-type systems (with total mass above 5 $M_\odot$) appeared to be settled, I decided to take up the problem once more and from a different angle. Until now we have been satisfied with an overall, qualitative picture of the processes of mass transfer, and did not go beyond comparisons with statistical plots of certain quantities such as overluminosities of the subgiant components. I have now decided to compare theoretical evolutionary tracks directly with selected observed systems. This is not easy since observed characteristics of the Algol systems are as crude as our theoretical models. The semi-detached binaries are conspicuous photometrically and it is a serious problem to obtain reliable relative dimensions and inclination. Spectrographic observations are much more difficult since typically only one spectrum is observed and is often contaminated by gas streams. Batten spent a number of years trying to get better data for U Cephei. A similar effort was made in the case of U Sagittae – by Cester in Trieste photometrically (Cester and Pucillo, 1972) and by Grygar and myself – at Victoria in 1969/70 – spectroscopically. In neither case can the available results be considered final, yet some conclusions can be based on them with confidence.

I will show first that *U Sagittae cannot be explained by a conservative case A of mass transfer*. In the model by Cester and Pucillo, the total mass of the system is 5.7 $M_\odot$ of which the primary – now a subgiant – has 1.6 $M_\odot$ and the secondary has 4.1 $M_\odot$. Assume now that the total mass of the system (and also the total orbital angular momentum) has been preserved. The initial mass of the primary must have been $3\ M_\odot \lesssim m_1 < 5.5 M_\odot$. One plausible choice of initial parameters is $m_1 = 4 M_\odot$, $m_2 = 1.7\ M_\odot$. The configuration at the end of the rapid mass transfer can easily be predicted from the extensive tables by Horn *et al.* (1969). The outcome depends on three parameters, of which two – the initial masses – have already been fixed. Thus in Table I we can write down a family of solutions depending, for example, on the initial radius of the primary component (or its central hydrogen abundance):

TABLE I

Case A Models for U Sagittae

X_{c1}	R_1^0 $(R_\odot)$	m_1^e $(M_\odot)$	m_2^e $(M_\odot)$	P^e (days)	Sp 1	M_1	R_1 $(R_\odot)$
0.05	3.72	1.24	4.46	2.10	G1 IV	$+2^m0$	3.3
0.15	3.49	1.35	4.35	1.61	F6 IV	$+2^m1$	2.9
0.25	3.21	1.66	4.04	1.16	F0 IV	$+1^m6$	2.5
observed		1.6	4.1	3.38	G3 IV	$+2^m1$	4.9

None of the three models is satisfactory. Small initial X_c gives good spectral types but wrong masses; $X_c = 0.25$ gives good masses but wrong spectral type. In all three models, the period is too short and the size of the subgiant too small. Subsequent slow mass transfer would affect the parameters in the correct direction, but it can easily be shown without much calculation that good agreement cannot be achieved. The conservative case presupposes that the orbital angular momentum

$$J = G^{1/2} A^{1/2} \frac{m_1 \times m_2}{(m_1 + m_2)^{1/2}} \tag{5}$$

is preserved. Combined with the constant total mass this yields the condition that

$$C = A m_1^2 (m_t - m_1)^2 , \tag{6}$$

where m_t is the total mass, is a constant throughout the whole mass transfer process, rapid or slow. Using present masses $m_1 = 4.1\ M_\odot$, $m_2 = 1.6\ M_\odot$, and present separation $A = 16.9\ R_\odot$, we find $C = 728$. Inserting $m_1^0 = 4.0\ M_\odot$ as a possible initial value, we get the initial orbital radius $A_0 = 15.7\ R_\odot$. At the same time we know the initial mass ratio, therefore also the radius of the Roche lobe which the primary just filled.

In other words, we get the initial radius of the primary, and this comes out to be $R_1^0 = 7.2 \, R_\odot$. This is clearly too much since at the top of the Main Sequence, the star of 4 $M_\odot$ has a radius of only 3.73 $R_\odot$ (for initial hydrogen abundance $X=0.60$; for $X=0.70$ we get 4.2 $R_\odot$). Clearly the choice of the initial mass was wrong. But if we pick up any other initial mass within the permitted range 3 $M_\odot$ to about 5.5 $M_\odot$, we invariably find that the initial radius was too large for a main sequence star. A possible way out might be that the observed parameters of U Sagittae are wrong and do not represent any real system. I therefore tried a different set of parameters, based on Irwin's photometry and my rediscussion of Joy's spectroscopic observations of the subgiant and on McNamara's orbit for the bright component (cf. Plavec, 1967): total mass 7.8 $M_\odot$, present masses 5.8 $M_\odot$ and 1.9 $M_\odot$, respectively, present orbital radius 18.8 $R_\odot$. The differences with respect to Cester's newer elements show how poorly known are the characteristics of prominent Algol systems. This change introduced another dependence between initial radius and initial mass of the primary, as shown in Figure 3. However, the main conclusion is not changed at all: again, no main-sequence binary can generate the present system of U Sagittae by conservative mass transfer in case A. Figure 3 shows that the same is true of Z Vul, RS Vul, V 3 56

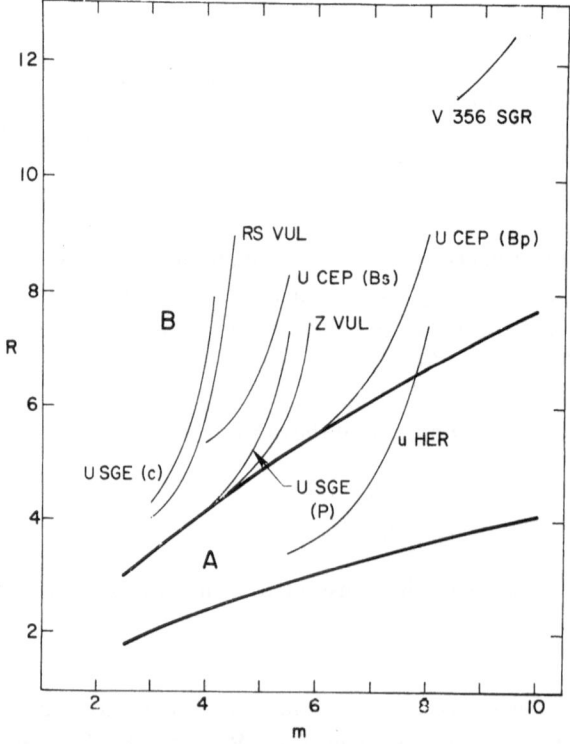

Fig. 3.   Suppose the Algol system shown in this diagram originated by conservative case of mass transfer, mode A. Then the parental star must fulfill the indicated relation between its mass m (in solar units) and radius R (in solar units) at the beginning of mass transfer. We see that – except for u Her – the parental star could not be a main-sequence star, hence conservative mode A is ruled out.

Sgr and U Cep (where again the uncertainty is indicated by plotting two sets of elements, both contemplated by Batten – one based rather on spectroscopy, the other on photometry). Only u Herculis can probably be a product of case A, but this is not a very typical Algol system. Many systems of intermediate mass are not sufficiently well studied, but I have little doubt now: I think that *conservative case A cannot explain the Algol systems.* I shall discuss more direct physical arguments supporting this contention later in the text.

The problem is now: What shall we drop? The words 'case A' or the adjective 'conservative'? Let us consider the alternatives in turn. It is more difficult to review all possible intermediate stages in case B since no grid of models exist that would adequately describe the process. Fortunately, Harmanec (1970a) studied this case for a star of initially 4 $M_\odot$ with companions of initially 3.2 $M_\odot$ and 1.6 $M_\odot$, respectively. The latter combination gives a total mass almost equal to that of Cester's model of U Sagittae. Let me quote values for two adjacent models from Harmanec's table (see Table II).

TABLE II

Case B Models for U Sagittae

Age (10^6 y)	m_1 ($M_\odot$)	m_2 ($M_\odot$)	P (days)	Sp 1	M_1	R_1 ($R_\odot$)
0.259	1.32	4.28	2.39	G1 IV	$+1^{\mathrm{m}}9$	3.7
0.349	1.06	4.54	3.82	G2 IV	$+1^{\mathrm{m}}6$	4.6
observed	1.6	4.1	3.38	G3 IV	$+2^{\mathrm{m}}1$	4.9

A complete agreement would be very surprising in view of quite arbitrary initial conditions. I feel that in general this model in case B represents the observed system better than anything we could find in case A. Harmanec's model starts mass loss at the time when the primary component is not far from the Main Sequence: the initial radius, 4.78 $R_\odot$, is only by 1 $R_\odot$ larger than the maximum main-sequence radius. This is quite consistent with Figure 3 which shows that, if U Sagittae is an outcome of a conservative process of mass transfer, then the initial primary was probably 4 $M_\odot$ or somewhat less and reached its critical radius not far from the main sequence. The same is apparently true of U Cephei, RS and Z Vulpeculae, and even for V 356 Sagittarii, although in this last case the parental star would be displaced somewhat farther away from the main sequence. The hypothesis that these Algol systems are products of mass transfer of the 'early B' type is favorably supported by the statistical survey of main-sequence detached systems I made in 1968 (Plavec, 1968). Among 45 eclipsing binaries of this type, 11 are so close that the principal component will eventually lose mass in type A process, while the majority, 34, or 75%, correspond to case B, many of them to 'early mode B'.

However, even without detailed model calculations, arguments can be presented against interpreting the Algol systems as products of a conservative case B. One is the predicted rate of mass loss. If U Sagittae is approximately represented by Harmanec's models, the current rate of mass transfer would be about $3 \times 10^{-6} \, M_\odot y^{-1}$, and the orbital period should secularly increase at a rate of nearly 1 s yr^{-1}. This is definitely refuted by the observations. Minima of U Sagittae have been observed since the end of the past century, and analyzed for period changes by Svechnikov (1955) and by Cester and Pucillo (1972). The period seems to be shorter than it was most of the time in the past. We should therefore look for models which give secular increase of period about 2 orders of magnitude lower.

Another argument against accepting conservative case B is the statistics of the Algol systems. The rapid phase of mass loss in case B is followed by a longer phase of slow mass transfer during which the mass-losing star is already a luminous giant, not a subgiant. For each binary like U Sagittae we should expect at least five binaries with luminous giants, bolometric magnitude between $+1^M$ and -2^M, spectral type G4 (for the fainter ones) to A7 (for the brighter giants). These systems would have periods between 10 and 25 days. Selection effects would be somewhat adverse to these binaries, since the bright main-sequence components have small fractional radii (0.12 to 0.05), hence eclipses would be less frequent, and also less deep, since the difference in surface brightness is smaller. The well-known dislike of the photometrists for longer periods (with the notable exception of Dr. Douglas Hall) would also disfavor such objects. There exist Algol systems with longer period, S Cancri, RS Cephei, RW Persei and others. Nevertheless I think we must conclude that we cannot find enough systems with bright giants to comply with the theoretical prediction.

A third argument against conservative case B is the short duration of the process of mass transfer. The process is eventually dominated by the thermal instability of the helium-rich core, which develops almost independently of the conditions in the envelope. Therefore the mass loss takes about the same time as the normal evolution of the parental star up to the giant tip. Then why don't we observe any eclipsing binary systems in which the more massive component fills, or nearly fills, its critical Roche lobe? Our argument has always been that this phase is very much shorter than the phase of mass loss. This would not be true if the mass transfer were of type B.

We must conclude that the Algol systems like U Sagittae are products of non-conservative processes. Some material must be lost from the system, and carry away a part of the orbital angular momentum. But we need a definite physical picture how this happens in order to compute corresponding model sequence. Let us therefore consider a star of 5 $M_\odot$ which fills its critical Roche lobe near the end of its main-sequence evolution. The star can be classified as about B5 IV, and has a radiative envelope How does it lose mass? Struve always spoke about prominences on its surface. I think solar physicists do not like this terminology. Solar prominences are not vehicles to carry away significant amount of material. The picture may be of some value for late-type components. We have no evidence about prominences on B stars. Kruszewski (1966) advocated the idea that the principal component rotates asynchronously and the material flies off along the tangent when it approaches L_1. I like more the idea that the Jendrzejec type of laminar flow applies here as soon as the star exceeds the

Roche limit. There is probably no serious difference between the latter two alternatives. Material streams out at fairly low velocities, of the order of several km s^{-1}. Then it flows from L_1 towards the other component. In case A, we always have a small orbital radius, so that the fractional radius of the other component is large. Benson (1970) studied the combination 5 $M_\odot$ +2.5 $M_\odot$, initial period 1 d, which makes the orbital radius equal to 8.24 $R_\odot$. Then the fractional radius of the mass-accreting component is 0.21. So large a star will capture directly all in-coming material. Since the stream falls obliquely with a positive velocity component in the direction of rotation, the rotation velocity of the star will be accelerated. The impact area is heated. The crucial problem is how much of the kinetic energy of the stream is converted into heat and how fast the extra energy – in any form – is distributed through the star. Benson concluded that the radiative envelope is heated and assumes such a structure (deviating from radiative equilibrium) that as the accreted material is deposited, the star's radius increases rapidly. A transfer of only 0.1 $M_\odot$ suffices in this case for the secondary to fill its critical Roche lobe, too, so that a contact system is formed. Dr. Ulrich at UCLA pointed out to me that Benson's solution is not necessarily the only one or the most likely one. It is conceivable that the falling stream invokes shear turbulence which transports a considerable amount of the acquired energy rapidly into the interior so that the radiative envelope is much less heated. Another factor probably acting in the same sense is the decrease in rate of mass loss from the principal component as soon as the secondary component does no longer increase its mass. I think that such a drop in mass outflow rate is bound to come. For it is not a rapid expansion of the principal star that drives the mass outflow – in fact, the expansion is quite negligible – but rather the increasing radius excess over the Roche radius due to a rapid shrinking of the latter. But when the material is not deposited on the secondary component, the Roche limit will decrease much less rapidly. These 'delaying actions' may well take place but it seems that in most cases A eventually a contact system will be formed.

In this case, the future course of events may be quite different from the course chartered by computations describing the pure conservative case A. Very little is known about the structure and evolution of contact binaries, but recent activity has been very promising. From the brief reports published so far (Hazlehurst and Meyer-Hoffmeister, 1971; Biermann and Thomas, 1971) it is hardly possible to assess fully the impact of their work on the problem of the Algol systems – most attention is paid to the W UMa systems. From a paper by Moss (1971) and from a discussion with Dr. Whelan I gather that in contact systems the important problem is not only transfer of mass but also transfer of energy between the two stars. At a certain stage mass transfer may be reversed, go from the less massive star to the more massive one, and make the system detached again. Quite possibly this process must be studied if we want to understand the Algol systems.

On the other hand, we do not have any direct observational evidence in favor of one of our basic assumptions, namely that the mass outflow from the principal component is a quiet laminar outflow at low velocities, essentially sonic or subsonic.

UX Monocerotis is a binary system not unlike the Algol systems, yet ejections of gas were observed from the late-type component with velocities of the order of 150 km s^{-1} to 250 km s^{-1} (Struve and Huang, 1958). If this is a rule rather than an exception, radical changes in our fundamental concepts will be necessary. Mass loss from the system will be an important factor.

One possible way of getting some insight into such a non-conservative process is to accept the formalism of Paczyński and Ziółkowski (1967) and introduce two parameters, f and g, to describe the process. Suppose an amount of mater Δm_1 leaves the principal component. Then we assume that $f\Delta m_1$ escapes into space while $(1-f)\,\Delta m_1$ is intercepted by the secondary component. Clearly $0 \leqslant f \leqslant 1$, and $f=0$ is the conservative case. The escaping material carries away a certain amount of angular momentum ΔJ. Let us assume that

$$\frac{\Delta J}{f\,\Delta m_1} = g\,\frac{J_0}{M_0},\tag{7}$$

where M_0 is the initial total mass of the system and J_0 the initial orbital angular momentum (at the beginning of mass transfer). If $g=1$, then 1 g of escaping material

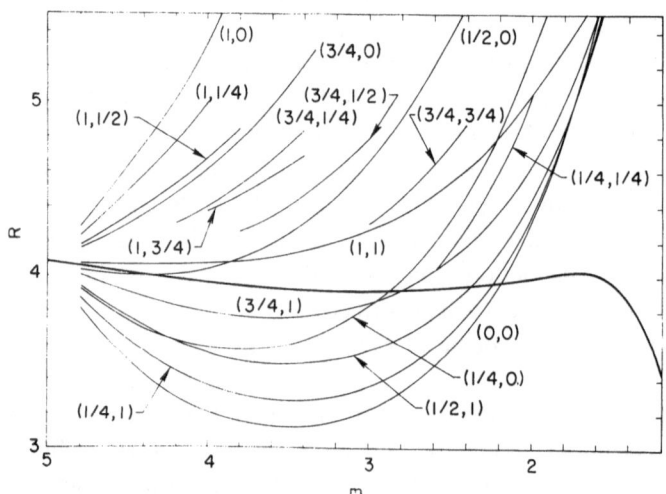

Fig. 4.   The outcome of the rapid phase of mass loss of a non-conservative case A can be found always on the cross-section of the locus of thermal equilibrium models (heavy line) with the curve (f, g) representing the variation of the Roche radius with mass.

carries away the initial average angular orbital momentum per 1 g. The difficulty is with assigning a physically meaningful value to the parameter g. Clearly $g=0$ is unacceptable but $g>1$ is possible at least at early stages of the process. Most likely the parameter g cannot be constant in real systems.

It is well known that configuration of the system at the end of the *rapid phase* of mass transfer of mode A can be obtained if we find the model for which the thermal equilibrium radius is equal to the Roche radius. This method was widely used by Paczyński and others, and many models similar to those we are going to use here were calculated by Plavec and Horn (1969), by Plavec et al. (1969), and by Horn et al. (1969). Figure 4 shows a diagram of the case when in a system initially consisting of a star of 5 $M_\odot$ and a star of 3 $M_\odot$, the principal component reaches the Roche limit when its central hydrogen content has dropped to $X_c=0.15$. The heavy curve is the locus of models in thermal equilibrium, with decreasing mass but fixed state of nuclear evolution in the interior. The arcs show the change of Roche radius of the principal component with its mass. They are labeled (f, g) to indicate the dependence on the two parameters defined above. The instantaneous total orbital angular momentum J is related to the instantaneous orbital radius and to the masses by

$$A = \frac{J^2 (m_1 + m_2)}{G \quad m_1^2 m_2^2} , \tag{8}$$

but by the definition of g,

$$J = J_0 - \Delta J = J_0 (1 - gf \, \Delta m_1/M_0), \tag{9}$$

and by the definition of f,

$$M = m_1 + m_2 = M_0 - f \Delta m_1, \tag{10}$$

and

$$m_1 = m_1^0 - \Delta m_1, \qquad m_2 = m_2^0 + (1 - f) \Delta m_1. \tag{11}$$

Subscripts and superscripts '0' refer to the initial parameters at the beginning of the mass loss. The instantaneous Roche radius of the mass-losing component is then

$$R_1^* = A r_1^* (m_1/m_2), \tag{12}$$

where the fractional Roche radius r_1^* was calculated more accurately but can be approximated by

$$r_1^* = 0.38 + 0.2 \log (m_1/m_2). \tag{13}$$

Figure 4 shows how the intersections of the family of curves representing the Roche limit with the locus of the equilibrium models determines the end of the rapid phase. Plotted are curves for $g \leqslant 1$. Among them, the conservative case gives the largest mass transfer: the system is transformed from 5 $M_\odot$+3 $M_\odot$ to 2.18 $M_\odot$+5.82 $M_\odot$. Also, the rate of mass outflow will be largest since the excess of the equilibrium radius over Roche radius is a maximum for the conservative case.

However, modes with moderate mass loss from the system ($f=\frac{1}{4}$ or $\frac{1}{2}$) and heavy loss of angular momentum ($g=1$) come very close to the same final mass of the mass-losing star. Indeed, for $g>1$, larger mass transfer and somewhat l tained. To represent the observed system of U Sagittae, we need m

$A = 16.9$, $P = 3.38$ d, $R_1 = 4.9$ $R_\odot$. The value of the radius shows immediately that none of our equilibrium models for 5 $M_\odot$ can represent the present state of the system. Either the parental star had more than 5 $M_\odot$, or the principal star has evolved rather far in the slow phase of mass transfer. Both alternatives may be true at the same time. We have not yet been fully successful in finding a model that would represent U Sagittae satisfactorily. Also, even if we do succeed, the model will be acceptable only if the adopted values of f and g can be physically interpreted. However, the conclusion seems quite plausible now that U Sagittae, and probably many other Algol systems, are generated by processes of mode A or quite possibly AB in which mass loss from the system is not negligible, and in which phases of contact configurations may play an important role.

4. Duplicity of Shell Stars

Mass transfer in case B proceeds at a high rate, in particular in massive stars and in stars with convective envelopes. However, the outcome is rather modest. Take for example the case studied by Harmanec (1970a). The initial configuration is two B type stars, with masses 4 $M_\odot$ and 3.2 $M_\odot$, respectively, revolving in an orbital period of 1.8 d. On the main sequence, such a system represents a well-observable spectroscopic double-line binary provided the orbital inclination is not too low. Eclipsing binaries U Ophiuchi and ζ Phoenicis are not much different from our model. Over a period of some 90×10^6 yrs it is transformed into a very unequal couple where the primary component (provided it accretes all the mass coming from its mate) is an early B-type star with 6.7 $M_\odot$ while the remnant of the originally more massive star has only 0.5 $M_\odot$ and a radius of only 0.2 $R_\odot$. Spectroscopically and photometrically it would be difficult to detect that the B star is not single. The predicted rate of mass transfer reaches a peak value of nearly 10^{-5} $M_\odot$ yr^{-1} for our star of 4 $M_\odot$, and can be much higher for more massive stars or for stars with convective envelopes. Moreover, spectroscopic binaries with B-type components and periods of the order of days or tens of days are rather commonplace. Spectroscopic phenomena associated with high-rate mass transfer or mass loss must be quite conspicuous. Relatively high frequency of systems that must be at this stage partly compensates for the short duration of the process. Therefore at least a few binary systems should be observed just at the stage when mass is streaming in the system. Where are they?

One would expect that the gas streams are most likely to be observed when they form a disk or an extended atmosphere around the mass-accreting component. The mass-losing star is a late-type giant with low surface brightness, rather difficult to observe if the system is not eclipsing. Also the gaseous streams between the two stars will in general not project on the disk of the bright component when the system is not eclipsing. The accumulation of material around the accreting star should be much easier to observe. It was this line of reasoning that led me in 1968 to suggest that *some* shell stars may actually be binary stars of this type. The peculiar eclipsing binaries β Lyrae and V 367 Cygni have spectra which bear many resemblances to shell stars.

A recent paper (Harmanec *et al.*, 1972) seems to promote a hypothesis that *all* shell stars may be binaries. I think such a hypothesis is more difficult to defend. Physically the advantage of the binary model over a rapidly rotating equatorially unstable single star may be that the mechanism that supports the extended atmospheres is simply the continuous supply of material with excess angular momentum, coming from the other component. Observational evidence, however, lends little support. As a sample of well or fairly well observed shell stars let us take the list of 19 stars published by Underhill (1966). This list contains stars brighter than 7^m. For the sake of completeness, let us add to it 4 Herculis, 88 Herculis, 14 Comae, and HR 2142, thus increasing the number to 23. Among these 23 stars, 2 are certainly known to be binaries (17 Leporis, AX Monocerotis), and 6 others (*o* Andromedae, 88 Herculis, β Monocerotis, ϕ Persei, ζ Tauri and HR 2142) are suspected of being binaries. Even if all of them eventually prove to be binaries, this would still be only 35% of the total number. It is estimated that at least half of all early-type stars are spectroscopic binaries. Thus the duplicity of these shell stars still might be quite accidental. Presence of a nearby companion may easily affect the shell phenomenon, but this still is not what we are looking for. The problem is whether at least *in some shell stars the existence of the extended envelope is* not only affected but *directly caused by a companion of certain type.*

Let us survey briefly the suspected systems. That of ζ Tauri displays radial-velocity changes that may be periodic with a period of 133 d. However, according to Underhill (1952) the range is only 9 km s^{-1} so that the mass function is very small, about 0.01. The star ζ Tauri is an early B-type star, and most likely its mass will not be very much different from 10 $M_\odot$. The mass of the companion then might be 1 $M_\odot$ to 2 $M_\odot$ if the inclination is not too small. In a large system this companion will hardly play a significant role, and there is no evidence of its presence or influence. Delplace (1970) suggests that the radial-velocity variations are due to oscillations in the extended envelope. Underhill admits that the radial-velocity curve is based on the sharp shell lines rather than on the broad and shallow underlying photospheric lines, but points out that at the time when the star was studied the shell was stable and quiet. In any case, the shell is much more likely an intrinsic phenomenon.

Kříž (1969b) pointed out the agreement in the spectral type of the primary and in the orbital period between ζ Tauri and his model of mass transfer of mode B. In this case ζ Tauri would be interpreted as a system which recently finished mass transfer, so that one component is already a small helium star while the other has not yet accomodated all the material that the mass-losing component had supplied. There is no fresh supply now, however, and I suspect that this stage must be very short if it exists at all. So far no evidence has been furnished that would make this model more attractive than that of a single, rapidly rotating star.

Even less favorable for the binary hypothesis is the case of β Monocerotis. Possible duplicity was recently suggested by Cowley and Gugula (1972). The period indicated by radial-velocity variations and certain other periodic or cyclic phenomena is 12.5 yrs. In this case the binary system is wide, i.e. no mass transfer can occur. Moreover, the mass function is very large, 12.5, contrary to the case of ζ Tauri where it was very small. Cowley and Gugula find it very difficult to assign acceptable masses. Most likely the radial-velocity variations are not due to binary motion.

According to Hynek (1940) and Hickok (1971), ϕ Persei shows a repetitive radial-velocity curve which is, however, rather seriously and systematically distorted. Both investigators believe that lines of the secondary component can be seen at the helium lines $\lambda\lambda 4026$ and 4471; thus the secondary would also be a B star. Hickok's model, not very different from Hynek's, suggests two stars with masses 24 $M_\odot$ and 12 $M_\odot$, separated by 350 $R_\odot$. Although the presence of such a companion will no doubt affect the distribution of mass in the system, the dimensions of the system are too large and the spectral types do not indicate that the extended atmosphere is formed by material coming from a thermally unstable component.

No radial-velocity curve exists yet for HR 2142. Suspicion of duplicity is based on surprisingly regular recurrence of a short-lived shell (Peters, 1972). The shell appears in cycles of about 81 d, although with variable intensity and with somewhat variable duration, which is in each case very short, probably less than 10% of the period. Even if duplicity is demonstrated, the recurrent shell may be a consequence of modulation of the conditions in the extended atmosphere of the primary component by the companion, rather than to be due to gas streaming between the components.

The star o Andromedae is a very controversial object. The shell appears sporadically with no conspicuous periodicity. Spectral lines are so broad that radial velocity variations found by different authors disagree. Schmidt (1959) concluded from his photoelectric observations that the star is actually an eclipsing binary with very distorted components very nearly in contact; the period indicated by his observations was 1.6 d. Such a configuration could be a result of case A mass transfer leading to contact systems, and the envelope producing the shell spectrum would surround the whole system. Short-period radial-velocity variations were reported for EW Lacertae (HD 217050), for which also cyclic fluctuations of light were discovered by Walker (1953), and for 48 Librae (cf. a remark by Sahade, 1971). While unstable contact systems cannot be ruled out, atmospheric oscillations are a more likely explanation in all three systems including o Andromedae for which eclipses have so far not been confirmed.

The star 88 Herculis was recently studied by Harmanec et al. (1972). They find that the radial velocities oscillate with a range of 10 km s^{-1} and a period of 87 d. The mass function is small, $f(m)=0.008\ M_\odot$. Since the spectral type is B6 IV–V, a mass of some 6 $M_\odot$ is suggested for the visible component. Then the invisible star could have about 1 $M_\odot$ and the separation would be 150 $R_\odot$. A model of mass transfer of mode B studied by Harmanec (1970a) yields similar elements. The idea is that, as in the case of ζ Tauri, the secondary component is losing mass or was losing it relatively recently, and that this material now forms a ring around the blue component. So far there is no other observational evidence. I searched for the invisible star in the red and infrared spectral regions with the Lick 120-inch telescope, but the result was negative. Harmanec and his associates remark quite correctly that atmospheric oscillations may be an alternative explanation. The radial velocities were determined from the sharp shell lines, since the photospheric lines are too broad.

We come to the two shell stars in which duplicity has been firmly established. They are 17 Leporis and AX Monocerotis (HD 45910). These two systems are rather similar, since they combine a B star with a red giant. The question is whether this combination is accidental. I think that it is not, and that the instability of the red giant leads to gas streams which in turn produce the extended atmosphere around the B star – either directly, forming a ring, or indirectly, by interacting with the B star which then ejects matter as a secondary effect. Let us examine the evidence for this hypothesis. 17 Leporis was recently studied by Cowley (1967). The primary component is probably about B9, the secondary is about gM1 and its lines show up in the near infrared. Only the red star gives a satisfactory radial velocity curve. The relatively small value of the mass function, $f(m) = 0.24$, indicates that the orbital inclination will probably be small. From a high-dispersion study of the Mg II line which is not directly affected by the shell but otherwise rather poor, Wright (1957) and Widing (1966) suggested that the mass ratio is about 4:1 or more, the blue component being the more massive. Cowley finds the following very tentative model as most satisfactory: $m(B) = 5.6 \; M_\odot$, $m(M) = 1.4$, $i = 24°$, $R(B) = 2.6 \; R_\odot$, $R(M) = 75 \; R_\odot$, $A = 325 \; R_\odot$. In this model, the red giant becomes unstable at the time it approaches periastron. The mass outflow would then tend to have a pronounced periodical character since the eccentricity is fairly large, $e = 0.13$.

Many shell lines occasionally develop double or multiple components, a phenomenon which was called 'outburst' by Struve. Major outbursts, according to Cowley, tend to occur between a quarter and a half of an orbital revolution past the time of periastron passage. The period is 260 d, so the delay is between 65 and 130 d. Our graduate student R. Crawford calculated the travel time of particles ejected from the red star at periastron. This travel time depends very strongly on the ejection velocity. For velocities about 5 km s^{-1}, the particles reach the vicinity of the blue star within about 45 d. For velocities about 25 km s^{-1} the travel time is only about 25 d. I think the result for the low velocities is encouraging and may indicate that the activity of the blue star is triggered by particles ejected from the red giant, perhaps as a consequence of convective overshoot. The greatest problem is how the streaming mass interacts with the blue star or with the material that already surrounds it. Specific to 17 Leporis are two difficulties: the masses are very poorly known, and the orbit is eccentric. For these reasons, AX Monocerotis appears to be easier to study, and our work on it will be reported in the next section.

5. Convective Envelopes. AX Monocerotis and T CrB

AX Monocerotis is included in lists of shell stars and Be stars, but its spectrum varies in a peculiar way, not typical for either group. The object is actually a binary system, in which an early B star and a K giant revolve in a period of 232.5 d. The system was studied in detail by A. P. Cowley (1964), and also by Boyarchuk and Pronik (1967) and by Péton (1971). The early-type spectrum is that of a rapidly rotating B3:IV: star, but in particular the hydrogen lines are very complex and variable. As a rule they show emissions with P-Cygni profiles, but at times the absorption components are double or multiple. The early-type component is evidently surrounded by an extended variable atmosphere. Material is probably occasionally ejected by the star since for example Hα shows a nova-like profile, although the ejection velocities must be lower by a factor of three to four than in a real nova (Cowley).

At times, the star develops a shell spectrum with sharp absorption lines of ionized metals. Cowley found that this shell spectrum reappears periodically, although with variable intensity and duration. (Péton showed that the peak intensity of the shell lines varies rather smoothly with time and has been lower in recent years than in 1960). Shell spectrum is observed within the interval of phases $0^P.6$ to $1^P.0$ or at most $1^P.1$, the phases being reckoned in units of the period with zero epoch at the time of conjunction with the K star in front (this would correspond to primary minimum if the binary were eclipsing). Boyarchuk and Pronik confirmed this periodicity and suggested that the extended atmosphere has the shape of a tail through which gas is streaming out of the Be star. However, the radial velocities of the shell lines do not lend much support to this interpretation since the difference with respect to the Be star, 10 km s^{-1}, observed by Boyarchuk and Pronik is probably not significant and in any case cannot lead to the formation of the tail. I think the explanation advanced by Cowley is more promising, namely that material is streaming from the K star into the vicinity of the B star. Negligible difference in radial velocities (actually observed only at a certain phase) may mean that the stream moves there approximately at right angles to the line of sight. This is in qualitative agreement with the general pattern of gas streams as indicated by the calculation of trajectories.

If we accept this hypothesis, then the K star plays an important role as a source of circumstellar material. Cowley suggested, much in accord with the ideas of Struve, that the K star is the seat of huge prominences. This may well be the case but I would like to point out that the present configuration in AX Monocerotis can hardly be explained without assuming a large-scale mass transfer in the past. The B star appears to lie near the main sequence, while the K star is a bright giant, perhaps K2:II:. The radial-velocity curve of the K star is fairly reliably established. For the Be star, one must rely on the assumption that periodic displacements of the hydrogen emission lines represent the motion of the B star. While the mass ratio $m(B):m(K) = 2.5$ is only a crude estimate, it would be difficult to disprove the conclusion that the K star, although apparently more advanced in evolution, is the less massive component. In other words, AX Monocerotis can be considered as a large Algol system. It is tempting to complete this picture by assuming that the red giant actually fills the critical Roche lobe.

In the model I have considered (originally with Harmanec), the B star has 8.3 $M_\odot$, the K star 3.2 $M_\odot$, the orbital radius of the circular orbit is 360 $R_\odot$ and the inclination is 66°. Adopting again a conservative case of mass transfer, we have a choice as to the initial mass of the principal component which is now the K star. The range is about $6M_\odot$ to 10 $M_\odot$, and we adopted 7 $M_\odot$. This arbitrary choice may naturally mean that our models will turn out to be rather far from the actually observed system, but one general feature of the problem will not be affected. Namely, with the long period of the system, it is inevitable that the principal component should begin to lose mass when it has *a deep outer convective zone*.

This brings about very serious difficulties in studying the process of mass transfer. In predominantly radiative atmospheres, the expansion of the outer layers due to reduced pressure absorbs so much energy that the radius decreases considerably. Therefore the condition can be fulfilled that the star always just fills the critical Roche lobe – in numerical model calculations we achieve this by proper choice of the rate of mass transfer. This is impossible for convective envelopes (Paczyński, 1965; Paczyński *et al.*, 1969; Refsdal and Weigert, 1969; Paczyński and Sienkiewicz, 1972; Lauterborn and Weigert, 1972). Even if the mass of the star is decreased considerably, it is impossible to reduce the radius of the adiabatic envelope without somehow radiating away considerable part of its heat content. As soon as mass transfer to the other component starts, the Roche radius will decrease while the stars's radius will remain nearly unchanged. The star will overflow its critical Roche lobe. We believe that a laminar flow will build up under the pressure of the gas, and that gas will begin to stream through the nozzle at L_1 towards the other component. Jendrzejec (see Paczyński and Sienkiewicz, 1972) derived a formula where the rate of mass outflow depends, in the case of an adiabatic envelope, on the third power of the radius excess over the Roche limit. Since the Roche limit decreases rapidly while the star's radius remains nearly constant, soon after the beginning an extremely high rate of mass transfer is established. Paczyński and Sienkiewicz, using a rather simple but very ingenious method, found that in the case of a red giant originally of 1 $M_\odot$ and 36 $R_\odot$, the rate becomes as high as 0.13 $M_\odot$ y^{-1}!

Mass transfer in the model of AX Mon has been studied by R. S. Polidan and myself at UCLA. Knowing that such a high rate of mass transfer may not be supported by any observational evidence, we wanted to see if it can be cut down by introducing two additional modes of mass loss. Dr. R. K. Ulrich, at UCLA, derived a formula for convective overshoot based on his theory of convection in red giants. When a convective blob arrives into the vicinity of L_1, it may escape from the star because of its outward velocity. We believed that this process may be effective when the star just reaches the Roche limit but the radius excess is still too small for the Jendrzejec laminar flow to be important. In addition, mass loss was assumed to occur from the radiative atmosphere above the convective envelope. This thin atmosphere is nearly isothermal, and a modification of Jendrzejec's formula was applied here. A laminar outflow of material from the isothermal atmosphere is assumed to occur as soon as the atmospheric (surface) radius of the star exceeds the Roche radius, i.e. before the convective envelope begins to lose mass.

These two 'safety valves' proved quite inefficient. Figure 5 shows the evolution of the mass-losing star on the H–R diagram. The 7 $M_\odot$ star begins to lose mass when its radius (from now on I will mean by this term the 'effective' or 'photospheric' radius)

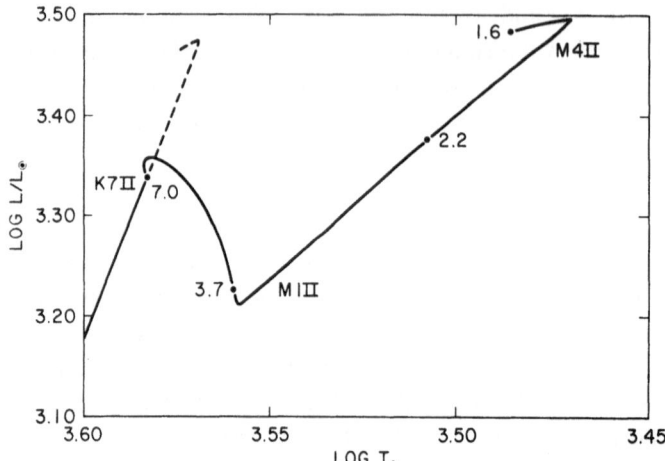

Fig. 5.   Theoretical Hertzsprung-Russell diagram for the mass-losing component in the model of AX Monocerotis.

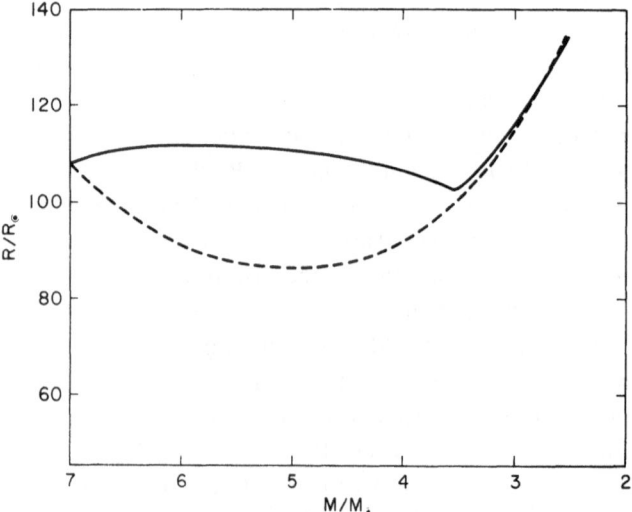

Fig. 6.   Mass loss from the convective envelope of the principal component in the model of AX Monocerotis: Dashed line shows the variation of the critical Roche radius, while the full line shows the actual (photospheric) radius of the star. Calculations were made with a modified Paczyński's evolutionary code and cannot represent events occurring on a dynamical time scale.

is 108 $R_\odot$, its luminosity 2228 $L_\odot$, and $\log T_e = 3.583$. By this time, the convective envelope begins 2.32 $M_\odot$ outwards from the center, and reaches right up to the photosphere.

At the beginning, mass is being lost only from the radiative atmosphere. The rate is so low, not exceeding $10^{-7} M_\odot y^{-1}$, that the star continues to move up along the giant branch, increasing its radius and luminosity for a while. In this way its radius soon exceeds the Roche limit and the rate of mass loss accelerates. This halts the normal undisturbed evolution, the radius ceases to increase but does not decrease either, see Figure 6. Since fairly large amount of material is now already being transferred to the secondary star (we assume the conservative case), the Roche limit shrinks, the radius excess increases, the process is self-accelerating. Within 1000 yrs, the rate of mass loss reaches a maximum of nearly 0.2 solar masses per year. Computations are very difficult at this stage and represent an extremely crude approximation only. Convergence was achieved only when the time step between two consecutive models was cut down to 5 yrs. The rate of mass loss remained higher than $10^{-2}\ M_\odot\ y^{-1}$ for about 20 yrs. The maximum rate of mass loss indicates a time scale of the order of 15 to 20 yrs, which is very much shorter than the thermal time scale but still longer than the dynamical (pulsational) time scale. The thermal time scale is 6.3×10^3 yrs, the dynamical time scale is 2.2×10^{-2} yrs, so that the time scale of the process at its peak happens to be just about the geometric mean between the two. These results are in full agreement with those obtained by Paczyński and Sienkiewicz (1972) for a giant of 1 $M_\odot$ and 36 $R_\odot$.

The rate of mass loss begins to drop when the principal component becomes the less massive star, since then the Roche radius of the mass-losing star increases again and the excess over the Roche limit diminishes. Some 1900 yrs after the beginning of mass loss, the Roche radius becomes again almost equal to the star's radius, and the surface instability is removed. However, the star remains thermally unstable since its core contracts and envelope expands, similarly as before the mass loss. The process is slower now since the mass of the star is only half the original mass. It continues to very nearly fill its critical Roche lobe, losing mass first at a rate of $7 \times 10^{-4}\ M_\odot\ y^{-1}$, but slowing down. Mass loss from the convective envelope ceases when the mass is 2.6 $M_\odot$, some 4000 yrs after the beginning of mass loss. Radius and luminosity increase steadily, but effective temperature decreases more than it would in undisturbed evolution. This phase lasts 10000 yrs, and at its end helium burning becomes so significant in the core that expansion is halted. The star reaches its maximum radius, 214 $R_\odot$, and luminosity, 3130 $L_\odot$, as an early M bright giant with mass 1.67 $M_\odot$. It is brighter than a 7 $M_\odot$ giant although it has only 24% of its mass. There are practically no losses of luminous flux by absorption since the star is in thermal equilibrium, and surface layers are already enriched in helium while the hydrogen abundance has dropped from $X=0.70$ to 0.58. Mass loss from the radiative atmosphere continues for some time but at a negligibly low rate. Our computations have so far been carried up the point where the star's mass is 1.65 $M_\odot$. The radius is already decreasing, the star moves to the left in the H–R diagram, and we expect that it will move rapidly over to the region of helium stars.

In our model, the present configuration of AX Monocerotis, with the red star of 3.2 $M_\odot$, corresponds to the relatively slow phase of mass transfer when the radius and luminosity increase again, and the rate is $4 \times 10^{-4}\ M_\odot\ y^{-1}$. This material flows to the other star which is assumed to have a mass of 8.3 $M_\odot$. If it is in thermal equilibrium, its radius is only 0.04 of the orbital radius. Therefore the gas stream does not fall on it but bends around and forms a ring. From the qualitative description of the gas ring model developed by Prendergast and Burbidge (1968) it transpires that part of the material of the ring will probably eventually be lost from the system while part is eventually accreted on the B star. Since the B star is so very much smaller than the K star, one must assume that the gas stream, although it tends to flow in the orbital plane, produces a ring or disk sufficiently thick perpendicular to the orbital plane to make the accretion roughly spherically symmetrical. Assuming that half of the material falls on the star, we must ask now whether a steady influx of some 2×10^{-4} $M_{\odot} \, y^{-1}$ can produce the phenomena observed in the spectrum of the B component. Much more critical is the problem whether the time interval of some 3000 yrs has been sufficient for the B star to accomodate the 3.8 $M_{\odot}$ of material transferred to it during the rapid phase. Loss of material from the disk means, of course, that the process cannot be strictly conservative, but this probably will not reduce the rate of mass outflow profoundly.

There is no doubt that the majority of close binary stars undergo this type of mass transfer since the increase in radius of the principal component is so large at the time it has a deep outer convective zone that it will most likely reach the Roche limit then. With the high rate of mass outflow the duration of the process is extremely short. However, the rate of mass transfer is so high that very conspicuous phenomena should occur and be occasionally glimpsed.

I suggest that the recurrent nova T Coronae Borealis may be an example of the rapid process. Kraft (1958) showed that T CrB is a binary consisting of a gM3 star and blue object which he classified as Q. The most remarkable thing is that the period of the system is $227^{d}\!.6$, nearly identical with AX Monocerotis. If the source of gas outflow is to be the M star, then it must be a real giant, and possess a deep outer convective envelope. On the other hand, the blue object is probably not a degenerate star. Kraft found that the lower limits on the masses are 3.7 $M_{\odot}$ for the red giant and 2.6 $M_{\odot}$ for the blue object. This lower limit follows from the fact that the stars do not eclipse, so that the orbital inclination can hardly be larger than 70°. Paczyński (1965b) revised Kraft's masses downward and obtained as lower limits 2.6 $M_{\odot}$ for the red giant and 1.9 $M_{\odot}$ for the blue star. But there is no compelling reason why the inclination should be as high as 70°. For $i = 45°$, for example, the total mass of the system and its size would be closely similar to our (and Cowley's) model of AX Monocerotis.

It is unlikely that the mass of the blue star could be as low as the Chandraskehar limit for white dwarfs. On the other hand, the red component is the more massive star in T CrB. If the system is at the stage of mass loss from the red giant, then it is near the beginning when the mass-losing star is still the more massive component. We must then expect a higher rate of mass outflow. Is the difference in behavior between the nova T CrB and the peculiar Be star AX Mon due to this difference in mass influx? Or is it primarily caused by a different character of the mass-accreting stars? In AX Mon, the blue component may be more or less normal main-sequence star. In T CrB, Kraft estimates the color temperature of the blue star at 25000 K and derives a value of 0.2 $R_{\odot}$ for the radius. In other words, the blue component would in this case be a helium star, and what we observe would be a second process of mass transfer. A more meaningful analogy would then exist between T CrB and AG Peg. But again, the difference between a nova-like object such as AG Pegasi and the recurrent nova T CrB might then again be in the rate of mass transfer.

Even if the proposed models for T CrB and AX Mon are basically correct, I still have very grave doubts about the reality of the very high rates of mass transfer we get from theoretical calculations. I think rates smaller by some two orders of magnitude would still be sufficient for the models proposed here. But then – how to dispose of the theoretical predictions? One can easily argue that the model calculations are extremely crude. They represent an effort to carry the usual model-sequence calculations over a period when the star is unstable on a very short time scale – so short indeed that the computational codes used cannot reproduce the star's behavior properly. Interestingly different is the approach by Bath (1969, 1972) who found that the ionization zones in convective envelopes cause instabilities on the dynamical time scale. While his approach is quite different, his results are similar – again he calls for a sudden outflow of material at a high rate. Since the outburst is accompanied by a sudden increase of luminosity, he predicts U Geminorum-type outbursts of the cool component. Naturally, a subsequent violent reaction of the mass-accreting star is not precluded. The outburst is short-lived so that not much material is transferred in spite of the large rate; moreover, a good deal of the ejected material will probably be lost from the system. One such outburst does not solve the evolutionary dilemma of a giant trapped at the Roche limit. Our calculations reported above were made on the assumption that while our computing program cannot treat these dynamical instabilites, it may still give a good overall picture of the evolution of the system. Observationally, Bath's results also pose some embarrasing questions. Why don't we observe some of those outbursts in binaries of longer periods? Perhaps T CrB is one example, and some symbiotic stars may be other similar cases. But in them, the blue component seems to erupt while the cool star probably does not change significantly. The problem of mass loss from convective envelopes is very puzzling.

6. Case B and the X-Ray Sources

Mass transfer in case B invariably produces a star of small mass and radius – either a helium star or a white dwarf. In the conservative case, it leads to relatively large orbital radii and orbital periods. For example, the calculations by Kippenhahn et al. (1967) show that a system originally $2 M_\odot + 1 M_\odot$ moving in an orbit with orbital period 1^d15 is transformed into a main-sequence A star with $2.74 M_\odot$ and a low mass white dwarf of $0.26 M_\odot$, with an orbital period of 24 d. Such a system will be very difficult to detect – spectroscopically or photometrically.

It is therefore very exciting to realize that a second mass transfer in such a system could produce an X-ray source. The problem was recently reviewed by Kraft (1972). The idea is that the kinetic energy of a gas stream falling on the mass-accreting component can be under suitable conditions converted into X-radiation of sufficient intensity. The *maximum* temperature that can be reached is

$$T = 1.5 \times 10^7 \, \frac{m}{R},$$ (15)

where both the mass m and radius R of the star are in solar units. This formula is very crude. The right-hand side should be multiplied by a factor easily as low as 10^{-2} since the efficiency of energy conversion is certainly very low. The UHURU satellite records radiation between 2 and 6 keV, so that the required temperature of such a hot spot is about 10^7 to 10^8 K; in other words, the ratio m/R should be roughly 10 to 10^3 or so. This may still include neutron stars for which the ratio is about 10^4. Also, mass transfer in case A or B is ruled out since the recipient is a main sequence star, but a second mass transfer in case B may be what we are looking for. The mass-accreting star is then a white dwarf or a helium star. For the white dwarf obtained by Kippenhahn *et al.* (1967) the ratio $m/R=6$. The white-dwarf models by Hubbard and Wagner (1970) give the same value for a white dwarf of 0.20 $M_\odot$, while for 1.05 $M_\odot$ the ratio is about 13. The helium star obtained by Harmanec (1970) gives $m/R=2.75$. A homogeneous helium star on the helium main sequence (Paczyński, 1971b) with a mass of 1 $M_\odot$ has $m/R=5$ and this ratio increases with mass, reaching 10 for 8 $M_\odot$. Massive helium stars seem to be almost equally promising as massive white dwarfs. Systems of low mass are naturally more common and Kraft points out that quite a significant fraction of observed X-ray sources are associated with the region of the galactic center, suggesting that they may be binaries of small mass.

Suitable temperature is of course not the only condition for a detectable X-ray source. The problem was studied in greatest detail by Prendergast and Burbidge (1968). In their model, the material is not accreted by the star directly. Rather, it first forms a ring around it, as it indeed must because of its excess of angular momentum. But viscosity tends to equalize the angular momentum across the ring. Therefore the outer particles will be accelerated and eventually some of them may excape from the system. The inner particles are slowed down and fall on the star. Prendergast and Burbidge were unable to investigate this final process. However, they find that that X-rays are generated in the inner ring where the gas is optically thin. The required rate of mass influx is 2×10^{19} gm s^{-1}, or 3×10^{-7} $M_\odot$ y^{-1}. If the rate were one order less, the mechanism would not work. Kraft explains that in U Geminorum stars and old novae the mass transfer is probably driven by emission of gravitational waves (Faulkner, 1971) rather than by thermal instability in the interior, since the masses are too small for significantly rapid nuclear processes. Hence the transfer rate is perhaps three orders of magnitude lower, and these systems cannot be expected to be X-ray sources.

The rate of mass transfer is quite sufficient in case B when the mass-losing star is a giant of several solar masses. Indeed, I fear that the rate is in many cases somewhat too high. Paczyński (1971a) found that the maximum rate of mass transfer actually obtained in model calculations is very well reproduced if we simply divide the initial mass of the mass-losing star by its thermal time scale. This ratio can be written

$$\frac{dm}{dt} (M_\odot y^{-1}) = 3.2 \times 10^{-8} \frac{R_0 L_0}{m_0},$$ (16)

where m_0, L_0 and R_0 are the initial mass, luminosity and radius of the mass-losing star. This simple formula gives good orientation values but cannot be exact since at the beginning of mass transfer its rate is affected by the rate at which the Roche limit shrinks, and this of course depends on the mass ratio and on the size of the system. Harmanec (1970a) found, for the same initial values of the principal component of 4 $M_\odot$, a maximum rate of 9×10^{-6} $M_\odot y^{-1}$ when the secondary has 3.2 $M_\odot$, and 3×10^{-5} when the secondary has 1.6 $M_\odot$. The above formula gives 2.2×10^{-5}. Generally speaking, systems of low mass give values just about equal to the rate 3×10^{-7} $M_\odot y^{-1}$ postulated by Prendergast and Burbidge, or somewhat lower. In the system $2M_\odot + 1$ $M_\odot$ studied by Kippenhahn, Kohl and Weigert, the maximum rate is 1.3×10^{-6} but the average rate is much lower, 2×10^{-7}; Paczyński's formula gives 1×10^{-6} for the maximum rate.

More massive systems will invariably give rates higher than 3×10^{-7} $M_\odot y^{-1}$. The average rate in case of the system 9 $M_\odot + 3.1$ $M_\odot$ (Kippenhahn and Weigert, 1967) is 2×10^{-4}. Paczyński's formula gives 6×10^{-5} $M_\odot y^{-1}$ for the maximum rate. This is two orders of magnitude more than postulated by Prendergast and Burbidge. They assumed that the mass-accreting star has half a solar mass and 0.2 solar radius. They postulated this star to produce 10^{35} ergs of X-rays per second at a temperature of about 40×10^6 K, and concluded that for this case the influx rate must be 2×10^{19} g s^{-1}, which is 3×10^{-7} $M_\odot y^{-1}$ (and not 3×10^{-6} as they, and also Kraft, say). We may conclude that a helium star of 0.5 or 1 $M_\odot$, combined with a star of several solar masses which expanded to the Roche limit and is losing mass by way of mode B, fully satisfies the conditions of the Prendergast-Burbidge model and should be an X-ray source. If a crude extrapolation is permitted, then a similar but more massive system appears even more promising.

Sco X-1 and Cyg X-2 may be systems built on this model but the complexity of observed phenomena makes it very uncertain. Very interesting is the periodic source Cen X-3, which is an eclipsing binary according to Schreier *et al.* (1972). The X-ray source is eclipsed periodically in a period of 2^d087. Short period (4.8 s) X-ray pulses are modulated periodically with the same period. If this modulation is interpreted as the Doppler effect, the system can be treated as a single-spectrum binary, and the mass function is 15.4 $M_\odot$, very large indeed. Attempts to establish limiting elements of the system were made by van den Heuvel and Heise (1972), by Wilson (1972), and by McCluskey and Kondo (1972). Unless we make some assumption as to the nature of the system, the limits are too wide and they reveal very little. Our interest here is narrower; we ask whether Cen X-3 can be explained by the Prendergast-Burbidge model. The X-ray source must be a white dwarf or helium star, while the other component must fill its critical Roche lobe. In addition, we can assume that the radius of the X-component is very small compared to the radius of the occulting star. Van den Heuvel and Heise found an upper mass limit 0.70 $M_\odot$ for the pulsar, and 17 $M_\odot$ for the unseen companion provided $i = 90°$. If i is less than this, both masses are smaller. Wilson found that the upper limit on the mass of the pulsar can be put at 0.23 $M_\odot$. Mass ratios and hence masses derived via the Roche limit depend very sensitively on the value of the Roche radius and I do not think that the arguments in favor of a very small mass are really compelling. There is little doubt however, that the mass of the unseen star is large. According to McCluskey and Kondo, if the X-ray source has less than 3 $M_\odot$, the other star must have at least 15 $M_\odot$. Such a system no doubt fulfills the conditions of the Prendergast-Burbidge model, but is rather acutely embarrassing to people who calculte evolution of close binaries. For a second process of mode B the period of $2^d.087$ is uncomfortably short.

Van den Heuvel and Heise attempted to evade this difficulty by constructing a rather elaborate scheme. After the first mass transfer, a helium star of 4 $M_\odot$ is formed, which subsequently explodes as a supernova and leaves a neutron star with about 0.5 $M_\odot$. This star is now receiving mass from its mate which has about 15 $M_\odot$. The regular 4.84 s pulses of the X-ray source are interpreted here as due to rotation of the neutron star with an embedded dipole magnetic field. As already pointed out by Kraft, there is a simpler way out, preserving the assumed white-dwarf or helium-star character of the X-component. We must assume that the first process of mass transfer was not conservative, that some mass escaped from the system and carried away part of the orbital angular momentum, hereby diminishing the orbital period.

Let me remark here that the optical identification of Cen X-3 with the eclipsing binary LR Centauri can now be practically definitely discarded. The periods, $2^d.087$ and $2^d.095$ respectively, are firmly established and incompatible. The error box defining the uncertainty in the position of Cen X-3 has now been considerably reduced and no longer includes LR Cen. The spectrum of LR Centauri was studied by M. S. Bessell (1972) who claims that the star that occults the X-component during the eclipses observed by UHURU is indeed peculiar and thus LR Cen should be seriously considered as the optical counterpart of Cen X-3, in spite of the discrepant position and period. However, Bessell's arguments are not convincing. He found a B8.5 III star with a rotational velocity $v \sin i = 300$ km s^{-1}. He assumes that the star is rotating synchronously with orbital revolution, and obtains then large radius, large mass, high luminosity and concludes that the star is filling the critical Roche lobe. However, the assumption of synchronism is hardly warranted in this case. In Algol systems, the primary components tend to rotate faster than synchronism requires, at times very much so (Plavec, 1970b; van den Heuvel, 1970): U Cephei is a system with similar period and similar high velocity of rotation for the primary, which is, however, much smaller than the Roche limit and does not rotate in synchronism. I think that LR Cen may be a normal Algol-type system; there is nothing in the spectrum as described by Bessell that would indicate a really peculiar system.

Another periodic and probably eclipsing X-ray source is Her X-1. According to Tananbaum et al. (1972), in this case the period is $1^d.700$, and the mass function

$f(m)=0.85\ M_\odot$. Since the system is eclipsing, the inclination cannot be too low, so that the small value of the mass function must be primarily due to much smaller mass of the unseen component, compared to Cen X-3 .The variable star HZ Herculis appears to be the optical counterpart of Her X-1 (Liller, 1972; Lamb and Sorvari, 1972; J. and N. Bahcall, 1972). Photometric and spectroscopic evidence begins to build up that HZ Her is an eclipsing variable. If so, the Prendergast-Burbidge model may well apply. The difficulty with too short a period is repeated here, although less strikingly because the masses are considerably smaller.

Finally, we have the case of Cyg X-1 which has been tentatively identified with HDE 226868, and with the radio source that 'turned on' in April 1971. HD 226868 is a spectroscopic binary with period $5^d.6$. Its spectrum is that of a B0 Ib supergiant, rather surprising for a binary with such a relatively short period. The mass function has been found to be $f(m)=0.12$ by Webster and Murdin (1972) or 0.16 by Bolton (1972a). Most recently, Bolton (1972b) and Brucato and Kristian (1972) announced that the emission line of He II at $\lambda 4686$ shifts in antiphase to the absorption lines of the supergiant. If this line comes from the region surrounding the X-ray component, then the mass ratio is about 1.4 or 1.5. If the B0 Ib star is a normal supergiant with mass about 20 $M_\odot$, then the secondary must be about 13 $M_\odot$. While the X-ray source may be a region heated by falling material which streams from the supergiant, the secondary component cannot be a white dwarf or a neutron star. Bolton and others suggested that it has many properties of a black hole.

If so, then the star that collapsed into a black hole could not have passed through the normal supergiant stage of stellar evolution since the supergiant component observed today is too close, only some 43 $R_\odot$ away. A star stripped first of its hydrogen envelope in mass transfer of type B can evolve without ever becoming a red supergiant. Again, this first mass transfer and the supernova outburst that may be anticipated during further evolution create some problems in accounting for the total mass of the system. Evolution with loss of mass and angular momentum is almost inevitable. However, observations are still to unreliable to guarantee the masses upon which all the speculation is built.

7. The Helium – Rich Binaries

Two stars with pronounced hydrogen defficiency in their atmospheres are single-spectrum spectroscopic binaries. They are υ Sagittarii and KS Persei (the latter is perhaps better known as HD 30353 or Bidelman's star). Since the stars show signs of circumstellar mass, it is tempting to explain their photospheric hydrogen deficiency by large-scale mass loss. Indeed, a process such as mass transfer in case B appears to be the easiest way to produce a helium-rich and hydrogen-poor star. However, case B cannot explain the observed degree of hydrogen depletion. While a considerable part of the hydrogen-rich envelope is stripped off, the process never reaches deep enough. Kippenhahn et al. (1967) and Kříž (1969a) showed that the mass transfer stops when the envelope has been stripped off down to the layers where the hydrogen abundance is about half the initial value. Such a mild hydrogen deficiency cannot be recognized spectroscopically. In υ Sagittarii, hydrogen abundance is down by a factor of about 10^2 while KS Persei is even more extreme and the factor is about 10^4 (Hack, 1967).

The simple case B is therefore ruled out. Much less is known about case C, but the one example studied by Lauterborn (1970) also does not lead to an extreme hydrogen deficiency. Towards the end of the mass transfer, a very tenuous envelope (in terms of mass) with mild hydrogen underabundance is left above the hydrogen-burning shell source. But this shell is an effective barrier for convection, so that the really hydrogen-poor material from deeper layers cannot be mixed into the atmosphere unless the hydrogen-burning shell dies off. Lauterborn did not follow the process into these crucial phases. Cases B studied by Harmanec (1970a) and Horn (1971) seem to offer some clues. There the shell was still active when the envelope above it collapsed and the star was transformed into a helium star near the helium main sequence. Horn points out that the outermost envelope may become rotationally unstable after the collapse. Since no additional energy seems to be available, it is unlikely that the envelope could be dissipated in this way. One may surmise that the surface hydrogen depletion will become much more complete by convective mixing in the subsequent evolution when the helium star expands again and returns into the giant region of the H–R diagram.

No model was followed completely through the stage of a helium star. Moreover, the helium stars obtained by Harmanec and Horn have small masses (0.5 $M_\odot$) and cannot be expected to move into the giant region again. Paczyński (1971b) found that homogeneous helium stars move into the red giant region (provided the universal Fermi interactions exist) if their masses lie between about 0.9 $M_\odot$ or 1.0 $M_\odot$ on one side and 2 $M_\odot$ on the other side. Biermann and Kippenhahn (1971a, 1971b) found a wider range of masses for such models. If the models have a degenerate core and a helium-burning shell, the range is $0.6 < m/M_\odot < 2.5$; if carbon burns in the core, the range is 1.0 $M_\odot$ to 3.6 $M_\odot$. Their models are simplified stationary models, while Paczyński's represent a true time-dependent evolutionary sequence. Neither set is entirely adequate for our purpose since it is assumed that even the outermost layers are completely devoid of hydrogen. However, it is very tempting to see whether υ Sagittarii and KS Persei could be roughly interpreted as evolved helium stars.

We do not have enough reliable data on these two stars, but a crude model will be attempted here.

The radial-velocity curve of υ Sagittarii is apparently reliable (Seydel, 1929), although a modern revision is desirable. The mass function is $f(m) = 1.58$. Orbital inclination is much more uncertain. The star was reported by Gaposchkin (1945) to display partial eclipses. The eclipses are very shallow; the depths found by Gaposchkin are $0^m\!.15$ and $0^m\!.08$ respectively, while the probable error of his photographic estimates was $\pm 0^m\!.20$. Obviously no elements can be based on such observations. In particular, no weight should be given to Gaposchkin's estimated inclination $i = 47°$. This was based on the assumption that both components are of equal size. The period of υ Sagittarii is 138 d, and systems of such long periods tend to have components of very unequal radii, which is quite understandable if we recall our discussion of differential stellar evolution. Limited photoelectric observations reported by Eggen *et al.* (1951) confirmed the existence of eclipses and the fact that the *invisible* component is eclipsed at primary minimum. The invisible star is therefore hotter, and it is reasonable to assume that it is much smaller than the visible star. The mere existence of eclipses then indicates that the orbital inclination cannot be very far from 90°, and I think $i = 80°$ is quite a plausible assumption.

Keeping in mind that our working hypothesis postulates the mass of the helium star to be about 1 $M_\odot$ to 1.5 $M_\odot$, we get the set of acceptable masses shown in Table III.

TABLE III

Possible Masses of υ Sagittarii

m_{invis}/m_{vis}	m_{vis} ($M_\odot$)	m_{invis} ($M_\odot$)	A ($R_\odot$)
2.0	1.98	3.95	62.0
2.5	1.31	3.29	67.5
3.0	0.99	2.98	70.9
4.0	0.66	2.62	75.6

From Table III it follows that the invisible component must be the more massive star. This dilemma is well known in cases such as β Lyrae or ε Aurigae and led some astronomers to the suggestion that the invisible star is a collapsed object, a black hole. I do not think that any such hypothesis is needed here. At effective temperatures near 10000 K, Paczyński's models have bolometric magnitudes near -5^M or -6^M. If we assume that the other component is a main sequence star of about 3 $M_\odot$, then it should be some five magnitudes fainter and yet its surface brightness can still be somewhat higher than that of the helium star.

I assume that the helium star is a remnant of an originally much more massive star which lost its hydrogen-rich envelope by the process of mode B. This type of process is suggested by the present period of υ Sagittarii. A star initially of 7 $M_\odot$ gives in this case a helium-rich remnant of 1.13 $M_\odot$ (Harmanec, 1970b). The process is not hampered by the 3 $M_\odot$ companion which evolves much more slowly: it takes 250 million yrs to cross the main sequence band, while the 7 $M_\odot$ star can develop into a giant, lose all its envelope, shrink into a helium main sequence star and expand again into a helium rich star, all together within not more than some 50 million yrs.

This picture, however, implies that almost all the material of nearly 6 $M_\odot$ shed by the 7 $M_\odot$ giant was dissipated into space rather than transferred onto the companion – for that star has today only 3 $M_\odot$ and certainly started its life with not much less than that.

If my hypothesis is correct, we are witnessing a second mass loss from the same star, and this time it is the less massive component – a rather unusual event. An emission feature at Hα, accompanied by a violet-displaced absorption, and other peculiarities in the spectrum indicate that mass loss is going on. Therefore the star should not be far from the Roche limit. The adopted mass ratio 0.37 and size of the system, 69 $R_\odot$, put the Roche limit around the helium star at 20 $R_\odot$. Its actual radius is unknown within rather wide limits. One source of great uncertainty is the effective temperature. The spectral type is usually given as A0p, but cannot well be used to derive effective temperature since the star's atmosphere has abnormally low opacity which enhances the metallic lines. This difficult problem was studied for the other helium-rich star, KS Persei, by Danziger, Wallerstein, and Böhm-Vitense (1967). They found $T_{\rm eff} = 10000$ K and $M_v = -3\overset{M}{.}2$. KS Persei is usually classified as A5p, and is probably somewhat cooler than υ Sagittarii. If we take $T_{\rm eff} = 10000$ K and $M = -5\overset{M}{.}3$ for υ Sagittarii, its radius will be 33 $R_\odot$, much larger than the Roche limit. However, $T_{\rm eff} = 13000$ K, a more likely value, gives a radius of 17 $R_\odot$. The model is crude but does not involve an open contradiction.

I think that complex observations of υ Sagittarii can verify and develop the model. Photometric observations can determine the difference in effective temperatures, even perhaps yield a crude estimate of the ratio of radii. Revision of the estimates of absolute magnitude and effective temperature is desirable and feasible with modern techniques.

Thanks to the effort by Wallerstein and his associates, KS Persei is actually better known although the star is fainter and circumstances less favorable. The system does not eclipse, so that orbital inclination is uncertain within wide limits. However, with very unequal components moving in a large orbit (the period is 360 d), the inclination can still be fairly near 90°. I adopted tentatively 70°. For inclinations much smaller than this the masses are too large for almost any mass ratio. Table IV

TABLE IV

Possible Masses of KS Persei

$m_{\rm invis}/m_{\rm vis}$	$m_{\rm vis}$ ($M_\odot$)	$m_{\rm invis}$ ($M_\odot$)	A ($R_\odot$)
3.0	2.61	7.84	97.4
4.0	1.72	6.89	103.9
5.0	1.27	6.35	108.2
6.0	1.00	6.00	111.3
7.0	0.82	5.76	113.6

shows that for acceptable masses of the helium star, the mass ratio must be 5:1 or 6:1 in favor of the unseen companion. With a mass about 6 $M_\odot$ a main-sequence star will have visual absolute magnitude near -1^M, while the visual absolute magnitude of the helium star was estimated at $-3\overset{M}{.}2 \pm 0\overset{M}{.}7$. Our explanation why the spectrum of the companion is not visible may still work. The bolometric absolute magnitude of the helium star is about $-3\overset{M}{.}5$, or about $1\overset{M}{.}5$ too faint for Paczyński's models. Combined with $T_{\rm eff} = 10000$ K this gives a radius of 14.5 $R_\odot$. In this model, the orbital radius is 111 $R_\odot$, and with the mass ratio $\frac{1}{6}$ the radius of the critical Roche lobe of the helium star is 25 $R_\odot$. Higher luminosity of the helium star would give better agreement with theoretical models as well as a star more nearly filling the Roche critical lobe. Again, the helium star can be considered as a remnant of a star originally of 7 $M_\odot$. Since the companion is about 6 $M_\odot$, the process might have been more nearly conservative, and a good deal of the mass lost by the principal component was probably captured by its companion. The main-sequence lifetime of a star of 6 $M_\odot$ is, however, only about 35×10^6 yrs, while we estimated the age of the helium star at 50×10^6 yrs.

In spite of this discrepancy – which can probably be removed by adjusting the model slightly – the hypothesis that KS Persei is a helium star undergoing second phase of mass loss remains very attractive, in particular through the work of Wallerstein, Greene and Tomley (1967). They in fact proposed a similar hypothesis already five years ago. Studying the abundances of light elements in KS Persei, they found the ratio $H/He = 10^{-4}$ in good agreement with the previous work of Nariai (1963). High abundance of nitrogen is suggestive of carbon cycling, but the two ratios found, $C/N = 10^{-3}$ and $O/N = 10^3$, are not easily reconciled. Wallerstein *et al.* prefer an evolution in which the central temperature originally was 24×10^6 K but subsequently dropped to $10–15 \times 10^6$ K, with both stages lasting relatively long. Such a process is indeed conceivable, since this is exactly what happens in case A. The masses would be too low should we take the term 'central temperature' literally. Since Wallerstein *et al.* favor an initial mass about 5 $M_\odot$, we should perhaps interpret the values given above as average temperatures in the convective core. A star of 5 $M_\odot$ can spend some 30 to 40 million yrs on the main sequence, then reach the Roche limit and transform rapidly into a star of 2 $M_\odot$ or even less. A long phase can follow in which the evolution is slow again. Since we postulate an ultimate loss of the whole envelope, the process must eventually pass over into the B mode; in short, we have a case AB.

A great trouble of this picture is the long period, 360 d. A very short initial period is required for case A to occur. Assuming for a while the conservative case, we can for example conjecture that the present configuration 1 $M_\odot$ + 6 $M_\odot$ was originally 5 $M_\odot$ + 2 $M_\odot$ or even 4 $M_\odot$ + 3 $M_\odot$. But in this case the period could have increased by a factor of 8 while we need a factor of 360. Even if we admit mass loss from system, it is difficult to find a physical model that would remove this discrepancy.

The two systems with helium-rich stars, υ Sagittarii and KS Persei, seem to be a rarity. This is not surprising since helium stars can develop into giants only if their masses lie within a very narrow range, perhaps narrower than 1 $M_\odot$. On the hypothesis put forward here this also means a fairly narrow range in the masses of the parental stars.

The two helium-rich stars need not fill the critical Roche lobe at the present time and yet a non-negligible mass outflow from them is possible. As explained in the review paper by Dr. Böhm, helium stars have such a strong output of acoustical energy that very dense coronae are likely to be generated, and the mass outflow may come from these coronae. This idea has already been proposed by Nariai (1967).

One last remark: β Lyrae was reported to be hydrogen-poor, too (Boyarchuk, 1959; Hack and Job, 1965). According to Hack, hydrogen abundance is down by a factor of 10. In this case the B8 II component of β Lyrae would be rather similar to υ Sagittarii, and one could speculate whether or not the systems are in fact similar. But before we do so, the abundance determination for β Lyrae should be re-examined.

8. Symbiotic Variables

Helium stars smaller than about 1 $M_\odot$ or larger than about 2 $M_\odot$ always remain to the left of the hydrogen main sequence. They have small radii and can hardly be expected to fill ever again their critical Roche lobes. In such systems, instability leading to mass transfer or mass loss can occur rather as a consequence of the secular expansion of the other component. Provided the helium star is a remnant of the initial principal component, we have a second process of mass transfer in the same system. This time, however, the one star is losing mass that originally accreted it. This may be considered as a kind of cosmic justice, only it is questionable whether the star that was originally robbed is capable of accepting the material back. From the available evidence one would conclude that the helium star does not accept the on-flowing material quietly, rather the interaction leads to a series of flares or minor outbursts which may occasionally attain a nova-like character. This, I think, is the basic explanation of the symbiotic variables.

A survey paper on the symbiotic variables was recently published by Boyarchuk (1969) who lists 21 objects which satisfy the following criteria: present in the spectrum must be absorption lines of a late-type component, as well as emission lines of He II, O III or similar ions; these emission lines should not be wider than about 100 km s^{-1}; the light of the object can vary with an amplitude up to 3 mag. and in cycles with several years duration.

The symbiotic stars are rather faint objects, and while photometric data are incomplete, spectroscopic observations are scarce.

Nevertheless, we probably can sketch a general picture of behavior of the symbiotic variables. The late-type component changes very little, while the blue component undergoes occasional flares. While its visual and photographic light changes considerably, it seems that the bolometric luminosity is nearly constant. A flare of Z Andromedae in 1939 and again in 1961 completely suppressed the late-type absorption features as well as the highly excited emission lines. Instead, a strong blue continuum dominated the spectrum and displayed absorption lines typical for early A stars.

More revealing is the history of AG Pegasi (Boyarchuk, 1967). The star brightened up as a very slow nova between 1855 and 1885, increasing its visual brightness by about 3 mag. The decline has been even slower and today the star is about as bright as it was before the flare. Between 1893 and 1920, its spectrum was that of a Be star with P-Cygni type emissions and numerous absorptions. By 1921, the absorption spectrum weakened a great deal, while the emission lines strengthened. Bright lines showed gradually increasing degree of excitation: today He II λ 4686 and N IV λ

4057 lines are among the dominant features, and also the forbidden [O III] and [Ne III] lines are strong. The B-type absorption spectrum has disappeared. Instead, first traces of a late-type spectrum were noted in 1930, and today many features of an M-type spectrum are easily visible. Thus AG Pegasi has behaved on the whole similarly to the other symbiotic objects, only the flare has been much slower.

What is the cause of this behavior? There is little doubt today that AG Pegasi is a binary star. Mean variation of radial velocites, as well as of various spectral features, indicates a period of about 800 days (Boyarchuk, 1967, 1969; Hutchings and Redman, 1972). The complex character of many lines did not permit in the past to determine the mass ratio, but the flare disturbance has settled down sufficiently to make the prospects much better in the near future.

The spectral changes of AG Pegasi can be interpreted as the formation of a shell around the hot star. The shell was originally of small size and optically thick, but as it expanded it gradually became optically thin. Today it has the form of a nebula of such a low density and large size that the strong ultraviolet radiation from the hot component can penetrate it to a great distance and produce emission lines of high excitation. If this picture is basically correct, what was the cause of the flare? If this flaring up is an intrinsic property of the hot component, then why is it always accompanied, in symbiotic variables, by a red star? It is more natural to assume that the outbursts of the blue star are triggered by material flowing to it from the unstable late-type component. The process of mass ejection may well be intermittent or fluctuating on the time scale given by our observations. Indeed, convection in the envelopes of red giants itself must lead to statistical fluctuations in the rate of mass loss by convective overshoot. However, I think that the most important cause of flaring is that conditions of instability build up gradually in the blue star. I think that symbiotic variables may not be basically too different from novae and U Geminorum stars; the term 'nova-like variables' is probably more meaningful than one might think at first sight.

What is the nature of the components in a symbiotic binary such as AG Pegasi? The cool component probably is typically an M giant. Hutchings and Redman (1972) found that the late-type spectrum is not quite normal. Perhaps we should not expect a fully normal spectrum in an atmosphere which is losing mass at a high rate, and moreover is illuminated by a hot star. It was very encouraging to see that our calculations of mass loss for the model of AX Monocerotis described in Chapter 5 (Plavec and Polidan, to be published) gave effective temperatures corresponding to an M-type giant at the phases of mass loss – both rapid and slow. At later stages of mass loss, the abundances are no longer normal.

What is the nature of the blue component? It is much hotter than a main sequence star of spectral type B. Boyarchuk (1966, 1967, 1969) suggests that the blue component of AG Pegasi is a Wolf-Rayet star, WN 6. Perhaps the term 'Wolf-Rayet star' should not be taken too literally in this context. By Boyarchuk's own definition of the symbiotic variables, the emission lines are relatively narrow, less than 100 km s^{-1}. This is much less than the typical width of the emission bands in Wolf-Rayet stars which corresponds to 500 km s^{-1} or 1000 km s^{-1}. Moreover, blue-shifted absorption components are typical for the W–R stars but not for the symbiotic variables.

I suggest that the blue components are helium stars near the helium main sequence. According to Boyarchuk (1969), in Z Andromedae the radius of the blue component varies between about 0.25 and 0.75 solar radii, and the visual absolute magnitude between $+1^M$ and $+3^M$. These values certainly do not indicate W–R character. Rather, they remind us of Paczyński's models of pure helium stars of about $2M_\odot$ (Paczyński, 1971b). The hot component of AG Pegasi may have a similar mass. Since the radial velocity curves indicate a mass ratio of about 5:1 in favor of the red component, this star may be a giant of about 10 $M_\odot$. With a period of 800 d, the orbital radius would then be 820 $R_\odot$, and if the red giant fills its critical Roche lobe, its mean radius is 430 $R_\odot$. A star of 10 $M_\odot$ reaches this radius near the time of carbon ignition. If we cut the masses of both components to one half, which is probably still acceptable, the orbital radius will be 650 $R_\odot$ and the radius of the giant about 340 $R_\odot$. Again, a 5 $M_\odot$ expands to this size on the second giant branch. Long periods of other symbiotic viables indicate a similar picture. Out of five stars sufficiently well studied, four show pariodic variations of radial velocities. Besides AG Pegasi, they are: BF Cygni (period 750 d), RW Hydrae (370 d), and R Aquarii (740 d) (Boyarchuk, 1969). These periods indicate that at least some sybiotic stars, including AG Pegasi, may be examples of case C of mass transfer.

It would be, of course, already a second process of mass transfer in the system, because we must postulate a previous one, probably of mode B, to account for the presence of the helium star. As was shown in Chapter 7 when we discussed the helium-rich binaries, lifetimes of the two components must be seriously considered before we can decide whether the model does not contain contradictions already in its crudest form. With the great uncertainty as to the masses, and as to the possible mass loss from the system, such a discussion appears premature. What is badly needed now are better data on at least one system. AG Pegasi appears to be in a suitable phase now, and an effort to derive the mass ratio should certainly be made.

9. Conclusion

I have attempted to offer an evolutionary interpretation of several groups of close binary stars. In doing this, I acted contrary to Dr. Huang's wise advice "not to stick one's neck out". I think precisely this must be done. I hope that my remarks will stimulate new research, both observational and theoretical, on the problems discussed here. If this happens, most of the models I have proposed will probably be changed or discarded entirely. Yet it is necessary to distinguish between two possible ways in which this can happen. Either it will be found possible to maintain the fundamental assumptions, and the theory of a large-scale mass transfer and/or mass loss will be further developed. Or it will be found that the principles are wrong as well, and we will have to start right from the beginning again. I don't know which alternative is more exciting.

I have neglected in my talk such important classes of binary stars like the novae, stars of the U Geminorum type, supergiant systems like VV Cephei, the W Ursae Maioris stars, and of course systems like β Lyrae, ε Aurigae and others, suspected by some of harboring black holes. These classes of binary stars are very popular now, and very good review papers already exist or again none can be written since almost every week there is a new model proposed. The clue to the mystery of these very popular systems may well be found in a patient study of the less conspicuous but simpler ones. The basic configurations will probably not be too many. Therefore, whichever system you pick up, you are contributing to one of the most exciting areas of contemporary astrophysics.

Acknowledgements

I wish to thank Mr. R. S. Polidan for his help with computations and diagrams, and Miss J. Kuebler for careful typing.

The work on this paper and the original investigations reported in it have been supported by grant GP-32886X from the National Science Foundation.

References

Bahcall, J. N. and Bahcall, N. A.: 1972, *IAU Circ.*, No. 2427.
Bath, G. T.: 1969, *Astrophys. J.* **158**, 571.
Bath, G. T.: 1972, *Astrophys. J.* **173**, 121.
Batten, A. H.: 1967, *Publ. Dominion Astroph. Obs.* **13**, No. 8.
Benson, R. S.: 1970, *Bull. Am. Astron. Soc.* **2**, 295.
Bessell, M. S.: 1972, *Astrophys. J.* **175**, L133.
Biermann, P. and Kippenhahn, R.: 1971a, *Veröffentl. Remeis-Sternw., Bamberg* **9**, 54.
Biermann, P. and Kippenhahn, R.: 1971b, *Astron. Astrophys.* **14**, 32.
Biermann, P. and Thomas, H. C.: 1971, *Veröffentl. Remeis-Sternw., Bamberg* **9**, 285.
Bolton, C. T.: 1972a, *Nature* **235**, 271.
Bolton, C. T.: 1972b, *IAU Circ.*, No. 2424.
Boyarchuk, A. A.: 1966, *Soviet Astron.* **10**, 783.
Boyarchuk, A. A.: 1967, *Soviet Astron.* **11**, 8.
Boyarchuk, A. A. 1959, *Soviet Astron.* **3**, 748.
Boyarchuk, A. A.: 1969, in L. Detre (ed.), *Non-Periodic Phenomena in Variable Stars*, D. Reidel Publ. Co., p. 395.
Boyarchuk, A. A. and Pronik, I. I.: 1967, *Izv. Krymsk. Astrofiz. Obs.* **37**, 236.
Brucato, R. J. and Kristian, J.: 1972, *IAU Circ.*, No. 2421.
Cester, B. and Pucillo, M.: 1972, preprint.
Cowley, A. P.: 1964, *Astrophys. J.* **139**, 817.
Cowley, A. P.: 1967, *Astrophys. J.* **147**, 609.
Cowley, A. P. and Gugula, E.: 1972, preprint.
Danziger, I. J., Wallerstein, G., and Böhm-Vitense, E.: 1967, *Astrophys. J.* **150**, 239.
Delplace, A. M.: 1970, *Astron. Astrophys.* **7**, 459.
Eggen, O. J., Kron, G. E., and Greenstein, J. L.: 1951, *Publ. Astron. Soc. Pacific* **62**, 171.
Faulkner, J.: 1971, *Astrophys. J.* **170**, L99.
Gaposchkin, S.: 1945, *Astron. J.* **51**, 109.
Hack, M.: 1967, in M. Hack (ed.), *Modern Astrophysics*, 163.
Hack, M. and Job, F.: 1965, *Z. Astrophys.* **62**, 203.
Harmanec, P.: 1970a, *Bull. Astron. Inst. Czech.* **21**, 113.
Harmanec, P.: 1970b, *Bull. Astron. Inst. Czech.* **21**, 316.

Harmanec, P., Koubský, P., and Krpata, J.: 1972, *Astrophys. Letters* **11**, 119.

Hazlehurst, J. and Meyer-Hoffmeister, E.: 1971, *Veröffentl. Remeis-Sternw., Bamberg* **9**, 289

Hickock, J.: 1971, preprint.

Horn, J.: 1971, *Bull. Astron. Inst. Czech.* **22**, 37.

Horn, J., Kříž, S., and Plavec, M.: 1969, *Bull. Astron. Inst. Czech.* **20**, 193.

Horn, J., Kříž, S., and Plavec, M.: 1970, *Bull. Astron. Inst. Czech.* **21**, 45.

Hubbard, W. B., and Wagner, R. L.: 1970, *Astrophys. J.* **159**, 93.

Hutchings, J. B. and Redman, R. O.: 1972, *Publ. Astron. Soc. Pacific* **84**, 240.

Hynek, A. J.: 1940, *Contr. Perkins Obs.*, No. 14.

Iben, I.: 1967, *Ann. Rev. Astron. Astrophys.* **5**, 571.

Kippenhahn, R. and Weigert, A.: 1967, *Z. Astrophys.* **65**, 251.

Kippenhahn, R., Kohl, K., and Weigert, A.: 1967, *Z. Astrophys.* **66**, 58.

Kopal, Z.: 1955, *Ann. Astrophys.* **18**, 379.

Kopal, Z.: 1971, *Publ. Astron. Soc. Pacific* **83**, 521.

Kopal, Z. and Shapley, M. B.: 1956, *Jodrell Bank Ann.* **1**, 141.

Kraft, R. P.: 1958, *Astrophys. J.* **127**, 625.

Kraft, R. P.: 1972, *Contr. Lick Obs.*, No. 368.

Kříž, S.: 1969a, *Bull. Astron. Inst. Czech.* **20**, 127.

Kříž, S.: 1969b, in M. Hack (ed.), *Mass Loss from Stars*, D. Reidel Publ. Co., p. 257.

Kruszewski, A.: 1966, *Adv. Astron. Astrophys.* **4**, 233.

Lamb, D. Q. and Sorvari, J. M.: 1972, *IAU Circ.* No. 2422.

Lauterborn, D.: 1970, *Astron. Astrophys.* **7**, 150.

Lauterborn, D. and Weigert, A.: 1972, *Astron. Astrophys.* **18**, 294.

Liller, W.: 1972, *IAU Circ.*, No. 2415.

McCluskey, G. E. and Kondo, Y.: 1972, preprint.

Moss, D. L.: 1971, *Monthly Notices Roy. Astron. Soc.* **153**, 41.

Nariai, K.: 1963, *Publ. Astron. Soc. Japan* **15**, 7.

Nariai, K.: 1967, *Publ. Astron. Soc. Japan* **19**, 564.

Paczyński, B.: 1965, *Acta Astron.* **15**, 89.

Paczyński, B.: 1956b, *Acta Astron.* **15**, 197.

Paczyński, B.: 1970, *Acta Astron.* **20**, 47.

Paczyński, B.: 1971a, *Ann. Rev. Astron. Astrophys.* **9**, 183.

Paczyński, B.: 1971b, *Acta Astron.* **21**, 1.

Paczyński, B. and Ziółkowski, J.: 1967, *Acta Astron.* **17**, 7.

Paczyński, B., Ziółkowski, J., and Żytkow, A.: 1969, in M. Hack (ed.), *Mass Loss from Stars*, D. Reidel Publ. Co., p. 237.

Paczyński, B. and Sienkiewicz, R.: 1972, preprint.

Peters, G. J.: 1972, *Publ. Astron. Soc. Pacific* **84**, 334.

Péton, A.: 1971, *Compt. Rend. Acad. Sci. Paris*, B **273**, 1062.

Prendergast, K. H. and Burbidge, G. J.: 1968, *Astrophys. J.* **151**, L83.

Plavec, M.: 1964, *Bull. Astron. Inst. Czech.* **15**, 156.

Plavec, M.: 1967, *Bull. Astron. Inst. Czech.* **18**, 334.

Plavec, M.: 1968, *Adv. Astron. Astrophys.* **6**, 201.

Plavec, M.: 1970a, *Publ. Astron. Soc. Pacific* **82**, 957.

Plavec, M.: 1970b, in A. Slettebak (ed.), *Stellar Rotation*, D. Reidel Publ. Co., p. 133.

Plavec, M. and Horn, J.: 1969, in M. Hack (ed.), *Mass Loss from Stars*, D. Reidel Publ. Co., p. 242.

Plavec, M., Kříž, S., and Horn, J.: 1969, *Bull. Astron. Inst. Czech.* **20**, 41.

Popper, D. M.: 1970, in K. Gyldenkerne and R. West (eds.), *Mass Loss and Evolution in Close Binaries*, p. 20.

Refsdal, S. and Weigert, A.: 1969, *Astron. Astrophys.* **1**, 167.

Sahade, J. (ed.): 1962, *Symposium on Stellar Evolution, La Plata*, p. 225.

Sahade, J.: 1971, *Veröffentl. Remeis-Sternw. Bamberg* **9**, 285.

Sato, K.: 1971, *Publ. Astron. Soc. Japan* **23**, 335.

Schmidt, H.: 1959, *Z. Astrophys.* **48**, 249.

Schreier, E., Levinson, R., Gursky, H., Kellogg, E., Tananbaum, H., and Giacconi, R.: 1972, *Astrophys. J.* **172**, L79.

Seydel, F. L.: 1929, *Publ. Am. Astron. Soc.* **6**, 278.

Struve, O.: 1946, *Astrophys. J.* **104**, 253.
Struve, O. and Huang, S.-S.: 1958, in S. Flügge (ed.), *Handbuch der Physik*, Springer Verlag, **50**, p. 269.
Svechnikov, M. A.: 1955, *Peremennyje Zvezdy* **10**, 252.
Tananbaum, H., Gursky, H., Kellogg, E. M., Levinson, R., Schreier, E., and Giacconi, R.: 1972, *Astrophys. J.* **174**, L143.
Underhill, A. B.: 1952, *Publ. Dominion Astrophys. Obs.* **9**, 129.
Underhill, A. B.: 1966, *Early-Type Stars*, D. Reidel Publ. Co., p. 233.
Van den Heuvel, E. P. J.: 1970, in A. Slettebak (ed.), *Stellar Rotation*, D. Reidel Publ. Co., p. 178.
Van den Heuvel, E. P. J. and Heise, J.: 1972, *Nature Phys. Sci.* **239**, 67.
Walker, M. F.: 1953, *Astrophys. J.* **118**, 481.
Wallerstein, G., Greene, T. F., and Tomley, L. J.: 1967, *Astrophys. J.* **245**, 150.
Webster, B. L. and Murdin, P.: 1972, *Nature* **235**, 37.
Weigert, A.: 1968, in L. Perek (ed.), *Highlights of Astronomy*, D. Reidel Publ. Co., p. 414.
Widing, K. G.: 1966, *Astrophys. J.* **143**, 121.
Wilson, R. E.: 1972, preprint.
Wright, K. O.: 1957, *Publ. Astron. Soc. Pacific* **69**, 552.
Ziółkowski, J.: 1969, *Astrophys. Space Sci.* **3**, 14.
Ziółkowski, J.: 1970, *Acta Astron.* **20**, 213.

EIGHTH DISCUSSION SESSION

(Monday morning; 11 September, 1972)

(Following Sections 1–3 of review paper by Plavec)

Chairman: T. J. HERCZEG

Herczeg: Now the discussion is opened for the first part of this review paper: first, Dr. Smak.

Smak: I would like to make several comments and ask a few questions. Possible loss of mass and momentum from the system actually affects rather little the final properties of the mass-losing component, these being determined primarily by the properties of its core. The effect is more important, of course, for the dimensions of the orbit, mass-ratio, etc. and – most of all – with regard to the properties of the mass-collecting component, particularly in the rapid phase. We really badly need a good description of what is going on around this star before trying to answer the question of how much mass and how much momentum should leave the system at that phase.

Secondly, I want to mention the case of very-low-mass secondaries, for which masses of the order of, or even less than, $0.2\ M_{\odot}$ have been determined (examples being AS Eri and DN Ori). These masses are definitely too small, as compared with the theoretically obtained data, and it should be pointed out that the theoretical limit is rather well determined by the maximum possible mass of the helium core. It is my feeling that in some cases at least we should be aware that the accuracy of these observed masses may be too low for us to claim any significant discrepancy here.

Finally, I wish to discuss the problem of undersize subgiants. These appear to be quite common in Algol-type systems and statistically, as far as their physical properties are concerned, do not differ from the contact subgiants. Circumstellar matter is known to be present in systems which contain undersize subgiants as well. It may even be surprising to note that the emission lines originating in the disk are observed more often in those systems than in the semi-detached ones.

The most convincing and self-consistent interpretation of undersize subgiants seems to be that given recently by Paczyński (1971). It assumes that at an earlier epoch, when the subgiant was in contact with its Roche lobe and was capable of losing its mass, a large amount of mass and, particularly, of the angular momentum was stored in the disk surrounding the primary component. After this phase ended, i.e. when the secondary subgiant restored its thermal equilibrium, we had a semi-detached Algol-type system. During the subsequent slow phase of evolution, however, the angular momentum from the disk has been returned to the orbital momentum leading to the increase of the orbital radius and to the expansion of the Roche lobes. It must be noted that this explanation accounts for nearly all observed facts.

A few important conclusions can be drawn from the Paczyński's hypothesis concerning the properties of disks in such systems. First, if we accept that the amount of

Batten (ed.), Extended Atmospheres and Circumstellar Matter in Spectroscopic Binary Systems, 260–262.

angular momentum stored temporarily in the disk was so large as later to change the orbital dimensions in an observable way, then it means also that the amount of mass stored in the disk was also large. Crude, order-of-magnitude estimates give that it must have been larger than about 1% of the mass of the subgiant. Second, since the disks are still observed in many such systems, it appears that the time-scale of disruption of such disks is rather long, definitely longer than the time-scale of processes responsible for the exchange of angular momentum.

I must admit, though, that there is still the possibility that the existence of detached subgiants is a spurious effect and, I understand, Dr. Hall may wish to comment on that.

Biermann: Dr. Thomas in Munich and I are doing calculations very similar to those Plavec has mentioned with the two parameters f and g. We had a very simple picture for the angular momentum and mass exchange. We argued about the potential difference that gives energy to the mass flow in the following way: in the fast phase of mass exchange, all the volume around the mass-receiving star is filled with turbulent matter, because the energy gained by the matter flowing down the potential well is very high and not easily dissipated. If this is true, then you can easily set up a differential equation for the parameter f by saying that any matter that is travelling down the potential well gained energy that is dissipated by sending matter out of the system over the potential difference L_1 to L_2. We encountered the same problem as you did: that the parameter g, which is essentially the effectiveness of loss of angular momentum, is guiding the whole thing. It can very easily happen that the mass-receiving star fills its lobe. We began with the system BD $16°$ 516 which seems to show very strong evidence of high mass loss from the system, because it contains a white dwarf and a low-mass main-sequence star, and yet it seems to be a relatively young system. We have not yet been able to reproduce that kind of a system.

Bath: I was just wondering how dependent do you think the mass-transfer process is on the way in which you treat the envelope structure of the star, and particularly on the specific boundary condition that is imposed to simulate mass loss. It seems to me that the way in which the latter is formulated completely determines the detailed results you have been examining. This is particularly the case for stars with ionization zones and associated convective envelopes, which, according to certain treatments seem to be dynamically unstable to mass transfer. Could you comment on this point?

Plavec: I have treated only case A in this part of my paper. This case of mass transfer involves relatively massive main-sequence stars, which have radiative envelopes. I agree that mass loss from stars with convective envelopes is very high.

Bath: Does this model move during mass transfer into a region of the H–R diagram where it has a convective envelope?

Plavec: Yes.

Bath: Isn't it possible that this might affect the later evolution of the system?

Plavec: The model has a convective envelope only at the later stages of slow mass loss. Even convective mass loss is slow then, because the radius of the Roche lobe is already increasing.

Hall: Dr. Plavec said that in deducing the original mass of the initially more massive component, he found the need to assume considerable mass loss from the system. But couldn't you just as well assume that a significant fraction of the angular momentum coming over in the stream goes into angular momentum of rotation, stored in the outer layers of the hotter star? This seems to me just as reasonable a way out of the problem, because we do know that at least some angular momentum does go into rotation. Furthermore, although rotational angular momentum can be redistributed rather quickly throughout the hotter star, downwards and upwards, we know it may take much longer for rotational angular momentum to be removed from the star, if it ever is. The question is simply how large a fraction of the angular momentum is stored semi-permanently in rotation – whether the fraction is negligible or whether it is large enough to get you out of the difficulty.

Plavec: Prendergast and Burbidge (1968) published a short article telling us that a very good program had been built for computing angular-momentum transfer and radiative transfer in a disk. Because of viscosity the part of the disk closest to the star eventually slows down; the outer part, on the other hand, is accelerating and some mass is lost from the system. This seems to be very significant progress, but, unfortunately, the program has never been published. Everyone is waiting for it and hoping it will be published soon!

Popper: A couple of additional observations on Kopal's concern about not finding the giant successors of main-sequence systems with equal detached components: In the first place, there is not enough room in most detached systems for the components to evolve to the giant stage. There are systems with nearly equal components such as ZZ Boo, El Cep, and DM Vir in which the components are evolved well through the main-sequence band. And finally there is SZ Cen, a pair of A5 giants, with masses of early A main-sequence stars, that almost certainly lie outside the main-sequence band of core hydrogen burning. It seems remarkable to me that even one system would be found with masses so nearly identical that the components still have nearly equal radii at this stage of evolution.

References

Paczyński, B.: 1971, *Ann. Rev. Astr. Astrophys.* **9**, 183.
Prendergast, K. H. and Burbidge, G. R.: 1968, *Astrophys. J. Letters* **151**, L83.

NINTH DISCUSSION SESSION

(Monday afternoon; 11 September, 1972)

(Following Sections 4 and 5, of Plavec's review paper)

Chairman: D. M. POPPER

Hall: If AX Mon is losing mass at such a very high rate, even though it is relatively difficult to determine an orbital period accurately from spectroscopic observations, you might be able to put an upper limit on the amount of mass exchange by looking for a change in the period. Is there any known variation in the spectroscopically determined period?

Plavec: No, we thought of that. You get a fairly large change of period that should be detectable over fifty years; but for spectroscopic binaries period determination is much less accurate than for eclipsing binaries. When AX Mon was rediscussed by Anne Cowley, the change introduced by the improved observations was larger than the real change in the orbital period.

Cowley: On the other hand, the original orbit for AX Mon had a period of 235^d and I looked at those observations rather carefully, to see if they could be fitted to the revised orbital elements. I decided that they could be.

Smak: One thing bothers me, namely, is it possible – when we are talking about rapid mass transfer – to have the mass-collecting component look like a normal star, say a B-type star? Let me quote a crude estimate concerning the amount of matter to fall on the surface of such a star. Assuming that the rate is $5 \times 10^{-5} M_{\odot} y^{-1}$, and that the star has a radius of $4 R_{\odot}$, that only 1% of the mass transferred falls on the star, and that this matter is distributed uniformly over its surface, we get a flux of about 10^{20} incoming particles $s^{-1} cm^{-2}$ of the stellar surface. This seems quite a lot, even without mentioning the problem of energy. I haven't made any estimates on that, but I believe that many of us here feel quite strongly that in the case of such major infall of material, its radiation must obliterate completely the spectrum of the central star.

Concerning T CrB, however, I feel that Dr. Plavec's interpretation could be the correct one. It may be noted that the theoretically predicted rate of mass transfer could probably agree (within the uncertainties involved) with the amounts of mass ejected by the star during its repeating nova-outbursts. I also have two minor comments to make on this system. First, according to a rediscussion by Paczyński (1965a) the limit on masses are $1.9 M_{\odot}$ and $2.6 M_{\odot}$. Second, if I remember, Kraft's estimates of the parameters of the 'blue' star were – at least partly – based on the assumption that the emission lines from the disk are due to the ionization of the disk by the central star, which may not be true.

Plavec: I know about Paczyński's paper and his values are included in my manuscript; they are not significantly different because the most important thing which has

Batten (ed.), Extended Atmospheres and Circumstellar Matter in Spectroscopic Binary Systems, 263–276.

been established in both cases is that the blue component is most likely not a degenerate star. As to the behaviour of the mass-receiving star, I am equally puzzled as you, Smak. One thing is just to study the mass-losing star and come to this conclusion from which somehow I would like to escape, but I don't exactly know how. This is why I presented the case of terrible mass loss from the convective atmosphere. The other thing is whether this is a possible explanation for AX Mon. There certainly is something happening to the mass-accreting star, but not so terribly much. If this discussion stimulates some further investigation into the problem of the reaction of the mass-receiving star, then it will have accomplished something.

Popper: Dr. Bath, you may give your summary.

Bath: I intend to discuss some work I've been involved with for the past four years relating to this problem of dynamical instabilities in semi-detached binary systems. The approach I have been using is somewhat different from that used by Dr. Plavec, and others, who have been using programs which, designed specifically to investigate longer time-scale thermal instabilities, employ the approximation that acceleration terms are negligible. Because of this difference of approach it may be that we are talking about somewhat different phenomena and using the same term, dynamical instability, to describe it.

The point of departure of my own approach has been exactly that point which Dr. Smak mentioned before lunch. That is, that dynamical instabilities in semi-detached systems will, if they occur, be totally analogous to those in single stars and that one must therefore use the same methods as are employed in pulsation studies to analyse them. All the physical terms that arise in the single star problem will be relevant to the case of dynamical mass transfer in semi-detached systems. The only distinction between the two cases is the boundary condition. A single star is free to expand and contract at will, whereas the contact component in a semi-detached system is constrained by the potential fields to remain within some fixed volume defined by the Roche lobe. To simulate this constraint one must use a boundary condition which acts at the Roche lobe surface, and not on some fixed mass layer at the surface of the star. The major problem is to define exactly what this new boundary condition is and to apply it to stability tests of detailed model envelopes.

This brings in the problem of treating an essentially asymmetric system in which the constraining boundary acts in a simple way at only one point, the inner Lagrangian point, by employing spherically symmetric stellar models with an associated spherically symmetric boundary condition. This is a particular difficulty in the case of dynamical instabilities, in which hydrodynamic terms must be included.

In the case of the thermal instabilities described in Dr. Plavec's review, no hydrodynamic boundary condition need be employed. The physical boundary of the star is assumed to be the photospheric surface and mass is removed at such a rate as to keep the photosphere in contact with a spherical surface whose volume is equal to the Roche lobe at all times. We have seen that in some cases, in particular in stars with deep convective envelopes, it is impossible to remove mass at such a rate as to fulfil this condition. It appears that the faster mass is removed, the faster the photosphere expands. In the model described by Dr. Plavec the star goes through a phase when the photosphere overflows the lobe by something like 10%–20%. Any attempt to remove mass at an even faster rate leads to numerical instabilities.

If one includes all the hydrodynamic terms, such an approach cannot be employed. For then an additional boundary condition must be introduced to control the acceleration equation at the Roche surface. This boundary condition affects the behaviour of the star only when overflowing the lobe. Rather than arbitrarily controlling the mass loss by the behaviour of the photosphere, which, after all, reflects only the position of a region where the optical depth has some specified value, which is really irrelevant in the hydrodynamic situation, one must control the mass loss by a boundary condition that reflects the flow properties of the real system.

In an attempt to simulate this in spherical models I have employed the boundary condition,

$$\frac{\partial P}{\partial t} = 0,$$

That is, the pressure is forced to be constant at all times, at some fixed radius, equivalent to the 'effective' radius of the Roche lobe. The justification for this is based on the argument that the equipotential surfaces, including the limiting surface of the Roche lobe, must be constant pressure surfaces in the static situation. Once an instability insues material will stream towards the second component, to be accreted, probably, as a disc. So long as the companion is well detached, and the amount of mass transfered is small, no increase in back pressure across the Lagrangian point will occur. Thus the outer critical equipotential will remain at constant pressure. This sort of boundary condition receives some support from more detailed treatments of the flow based on the similarity of the potentials in the vicinity of the Lagrangian point to a de Laval nozzle.

With this condition of constant pressure at a fixed surface one can use all the techniques and methods of stability theory in single stars but, since the boundary condition is Eulerian, the equations must be transformed to an Eulerian coordinate system. Some work based on this point of view has been completed using both linear and non-linear, and both adiabatic and non-adiabatic methods.

The linear adiabatic solutions indicate that, if the photosphere is initially in contact with the Roche lobe, then dynamical instabilities will occur in all stars with convective envelopes, that is, with any extent of ionisation zones in the surface. The growth times of the instabilities suggest an extremely rapid phenomena. A non-linear, but adiabatic, time dependent code has confirmed this, and shows that large temperature changes, and associated luminosity variations can be expected. These arise as a consequence of an increase in the temperature gradient in the superadiabatic region of the evnelope. For, as the outward flow builds up, material in the hydrogen and helium I zones recombines. It is just this recombination energy which heats the gas and drives the instability. Adiabatic cooling is not sufficient to keep the temperature gradients down to their initial values within the original static model. The regime where the adiabatic approximation must break down only occurs after considerable luminosity increase has occurred at the surface.

All these conclusions have recently been confirmed by a fully non-linear, non-adiabatic program, similar to that used by Christy in pulsation studies. Only the early stages have so far been followed, but a rapid rise in surface luminosity again occurs. This is true with both the boundary condition I have described, and a somewhat different, less constrained, condition.

How these instabilities relate to the observational problems we have been discussing remains to some extent obscure. That they may relate to the U Gem phenomena, as originally suggested by Paczyński (1965b), is still possible from a theoretical point of view. Certainly the energy requirement is available in the alternate removal and re-establishment of the hydrogen and helium ionisation zones. Whether novae are in any way related to dynamical mass transfer, as Dr. Plavec mentioned, is still open to question also. Furthermore it is not clear how the instabilities I have been describing, driven by ionisation zones in the outer envelope, relate to the more 'structure' dependent instabilities that Dr. Plavec and others have found. A more rigorous analysis of the relevant boundary conditions that theoreticians are using to simulate an asymmetric situation with spherical models is required on the one hand, and a concerted effort by observers to find out exactly which components are changing in variable and explosive semi-detached binaries on the other.

R. E. Wilson: All of these results are consequences of your boundary condition, so *that* should be examined very closely. In a formal sense it's not surprising that the pressure gradient is zero at the surface, since the pressure itself is zero there.

Bath: No, not if you've got a reasonable initial model; not if you've got a detailed treatment of the photosphere. In the spherically symmetric case $P = g/k$ at the surface. The pressure is very small, but not zero.

Whelan: I think you've got to emphasize that the pressure is not zero at the surface of the star. The density is about $10^{-8} \mathrm{gm\ cm}^{-3}$.

Bath: The pressure cannot be zero in the spherical case. As long as the pressure there is constant at a fixed radius, even if it is zero, this boundary condition is very, very important! And the difference between it and the free boundary condition is crucial.

R. E. Wilson: Yes, but aren't you effectively imposing a formal condition and then extrapolating it from regions of very low-energy processes down into regions of very high-energy processes and thus dictating the behaviour of much of the outer part of the star?

Bath: No, because all I'm doing in the non-linear models is putting an initial perturbation in the outer three or four zones where there are no energy sources at all. These outer zones communicate their motion, in accordance with the hydrodynamical equations, to regions which then drive the instability. They are driven by recombination in exactly the same way as in the model of planetary nebulae proposed by Paczyński and Ziołkowski (1968), and independently by Lucy (1967). Their model is just the same sort of dynamical instability but of single stars. Dynamical instabilities occur in single stars of very high luminosity, just because the energy from the hydrogen and helium ionization zones can, as they penetrate inwards, be converted into kinetic motion of the envelope.

R. E. Wilson: Can you demonstrate some direct way that your boundary condition is going to put important constraints, in a physical sense, on the value of the pressure gradient at some level deeper in the star?

Bath: Just by doing a linear analysis, that's all. Then you're tying a boundary condition into the solution of a linear differential equation, which thus determines the behaviour of the solution at the boundary.

Popper: I'm afraid this exchange is likely to get out of hand. I don't want to minimize its importance but I think clearly there are some difficulties which people working on this program will have to clear up.

Whelan: The important thing about using rotating polytropes is that you leave out some physics. It is more important to include the physics and leave out the rotation terms. I think the importance of Dr. Bath's work is that it shows that you cannot have these rapid mass transfers without worrying about the dynamical terms.

Popper: I think perhaps we should move on. Obviously this is a very pertinent field and maybe there will be some occasions to come back and have some more discussions on it. I think we'll get Dr. Plavec to go on with the next block of his subject.

Plavec: Let me make one final comment on that last discussion. It's no doubt that it's much better if analysis is done so that dynamical terms are taken into account and then you can discuss what happens on time scale much shorter than you can do if you use the normal evolutionary code.

(Following Sections 7 and 8 of Plavec' review paper)

Devinney: The light curve of υ Sgr is quite peculiar for a long-period star ($P = 138^{\rm d}$) because the light curve is characterized by a general light variation throughout the cycle. If my memory is serving me right, the light and radial-velocity curves are inconsistent in that the deepest minimum occurs when the radial velocity is increasing.

Plavec: Gaposchkin's (1945) photographic observations are still the only ones available, except for some reported in a short paper by Eggen *et al.* (1951) that essentially confirms Gaposchkin's picture. There is nothing else. The depths of minima are $0^{\rm m}15$ and $0^{\rm m}08$, but the probable error of a single determination is $\pm 0^{\rm m}20$. It's true that the deeper minimum is the eclipse of the star we don't see. This other star is $3M_{\odot}$ or so, sits on the main-sequence, and it is actually the star which has the larger surface brightness. So, if the eclipse is grazing, the helium giant of about $15R_{\odot}$ or more eclipses, at the time of primary minimum, a small star which is probably of a higher effective temperature – namely a B-type star. So there is no inconsistency, but of course the light curve should be reobserved.

Thackeray: If one examines Boyarchuk's list of 19 symbiotic stars, for the mean galactic latitude we find $|b| = 17°$, while for 13 of these objects with known radial velocities, 9 have mean radial velocities $|V| > 50$ km s^{-1}. This is to be contrasted with the low latitude of the supergiant VV Cep stars. Thus we should regard the symbiotic stars as an older population than the VV Cep stars, which is perhaps relevant to Dr. Plavec's discussion.

R. E. Wilson: I have proposed a model for Cyg X-2 which involved a white dwarf accreting material from the G-type primary in this system (Wilson, 1970). The spectrum of Cyg X-2 has broad absorption lines, which are very interesting, because absorption lines give information about physical conditions in the vicinity of the stellar component near which they are produced, whereas emission lines may come from gas which is far from both components. These absorption lines are very broad, and indicate mass motions of the order of 1000 km s^{-1}. One can eliminate broadening mechanisms other than mass motions. Since there must be a continuum against which we see the absorption lines, and the G-star is unlikely to be responsible, my model postulated an optically thick region of accreting material in the form of a disk or cloud around a white dwarf. It is interesting that Dr. Plavec mentioned a helium star as an accretion source because a helium star could be bright enough to serve as a continuum source for the absorption lines, yet faint enough to escape spectroscopic detection through its own lines. Therefore, a variation of the above-mentioned model involving a helium star might also be interesting to consider. It is very important to continue observing Cyg X-2, since Kraft's interesting observations were discontinued after only five or six night's work. Furthermore, the optical identification, which was only at the fifty percent confidence level several years ago, is now virtually certain.

Smak: Regarding various models for binary X-ray sources, it seems to me that in some cases it might easily be that the two parameters, namely the rate of infall of matter and the temperature, could be independent. If so, the shape of the X-ray spectrum could tell us about the mechanism involved, while the intensity could tell us about the rate of mass transfer. In the case of the Prendergast-Burbidge model, or – more generally – whenever circumstellar matter is involved, the two parameters are closely connected because of the essential role being played by the disk.

Regarding the absorption lines, mentioned by Dr. Wilson, I wish to recall that broad, shallow absorption lines of hydrogen are often observed in the spectra of U Gem stars at their outbursts. To explain such cases, I feel that a continuous emission from optically thick circumstellar material may produce the background against which the absorption lines produced in the outermost parts are seen.

Bolton: I would like to make a few general remarks on X-ray sources. For almost all of the sources the X-ray spectra are not well enough known to allow us to determine the mechanism that produces the X-rays. This is because of several things including poor observations, lack of observations, and source variability. It is now known, however, that the binary sources, of which eight or nine are known, have generally flatter spectra than other sources. Attempts to fit the spectra of a variety of X-ray sources with simple source models, such as synchrotron radiation, optically thin thermal bremsstrahlung, or thermal black-body radiation, have generally not been very satisfactory. Usually at least two models fit the spectra equally well, at least in the lower-energy region from 2 keV to 20 keV. For only a few X-ray sources is there enough high-energy data so that anything can be said about their spectra above

20 keV. Let me repeat that the spectral situation is very complicated. In Cyg X-1 it may be that there are two or three different source regions with different kinds of spectra and variability. Sorting this all out will require much better X-ray data than are currently abailable. I would like to make some remarks on Cyg X-1 in particular. I think in light of Dr. Hutchings earlier remarks on the antiphase motion of He II $\lambda 4686$ that my interpretation of that motion is very much open to question. This would not, however, change the basic interpretation of the system. The mass function of the B0 Ib star requires that $M_2 > 4.8 M_\odot$ if $M_{B0Ib} = 20 M_\odot$. The observational requirement that there be no eclipse of the B0 Ib star raises the lower limit to $M_2 > 5.4 M_\odot$ for $M_1 = 20 M_\odot$, $\log g = 3.3$, and a distance of 2 kpc. Both of these lower limits are comfortably above the upper limits on the masses of neutron stars and white dwarfs.

Finally, I would like to point out that the derived eccentricity of 0.09 is quite large for a binary system if this period and primary star of this spectral type. I have checked carefully to see of the eccentricity could be spurious, but I now believe that it is not. This eccentricity is consistent with the behaviour of the emission lines, and it is possibly telling us something about the evolutionary state of the system.

Bopp: If I might make a few comments on Cyg X-2, as I recall from a paper by Kraft and DeMoulin (1967), the radial velocities did not yield a consistent period. The object certainly did not seem to be a double-lined spectroscopic binary, as several authors had hoped, with the He II emission being produced by one component and the hydrogen absorption lines by the other. The final comments of Kraft and De Moulin concerning binary motion showed them to be dubious regarding it.

We have obtained some observations of Cyg X-2 in the past year, and the behaviour of the system seems to be quite different from what it was at its discovery in 1967. Photometrically there are no variations, in contrast to the results of Kristian *et al.* (1967). Recent spectra obtained at McDonald show that C III – N III emission at $\lambda\lambda 4640$–50 has suddenly appeared. The He II line has developed a P Cyg profile, and the radial velocity of the He II emission seems to be constant.

Popper: What is the optical apparent magnitude of Cyg X-2?

Bopp: About $14^m.5$ – and the optical identification is quite certain.

Bath: Do you suppose that the orbital eccentricity arises from the mass loss? Is it continuous mass loss?

Bolton: I think the eccentricity is a result of mass loss, but this would have to be on a time scale comparable to or less than the period of the binary system.

R. E. Wilson: I would like to mention that there is another binary for which a black hole might be indicated. I have done some work on BM Ori and the model in this case is very similar to that of ε Aur. It involves the transit of an inclined disk with a transparent central opening. The arguments are not identical in the two cases but again I come to the conclusion that there is likely to be a black hole in this system. Now in the case of ε Aur it is difficult to make a definitive comparison between the predicted and the observed optical light curves because the primary is intrinsically variable and there are fluctuations of the order of $0^m.2$ or almost that large, in the F0 supergiant primary. The light curve of BM Ori (Hall and Garrison, 1969) shows a similar scatter, but in this case it arises from the difficulty of observation. It is very difficult to do photometry of this system because nebular emission from the trapezium affects measures made with broad-band filters very severely. It is important now to obtain a very precise light curve for the primary eclipse in which one goes to the greatest extremes possible to eliminate the background light of the nebula.

Plavec: The case of BM Ori of course is the most difficult problem. It is an eclipsing binary and is the faintest star in the Trapezium. The period is 6d, so you have two serious objections to your picture. First the period is a bit terribly short for producing a black hole; second, the Trapezium is a classic example of a very young, young as possible, multiple system. Now you say that this very latest stage of evolution is present there. Again, it's possible, but I would say that probably it would be wise to try to find other explanations. This really is a good model but these two objections from the point of view of stellar evolution, I consider serious.

R. E. Wilson: Well, I've listed these two objections, which are also objections of my own, in the paper, but of course we have similar problems with Cyg X-1 and I personally find it very difficult not to believe the case for Cyg X-1. By implication, therefore, this kind of objection is to be treated very carefully at this stage.

Smak: If I understand properly, there are at least two crucial points involved. First, we are looking for a binary with sufficiently well determined masses or limits on masses which are large enough to exclude the possibility of the invisible object being a white dwarf or another low-luminosity star. Secondly, however, we are looking for some details in the light curve which definitely cannot be explained by all other effects we have been talking about during the last few days, but could be explained by some effects due to the black hole. I am sorry to be pessimistic and sceptical at this point, but it seems to me that if a really convincing evidence for a black hole is ever to be presented at a symposium on binaries, the title of this symposium will more likely be: 'Binaries Without Circumstellar Matter'!

R. E. Wilson: The light curve of ε Aur has a flat bottom, which is usually taken as the fingerprint of a total eclipse. There cannot be a total eclipse in this system, and I have tried to show that the light curve can only be produced by a thin disk, highly inclined (inclination 89°) with a central opening through which the F0 supergiant can shine (Wilson, 1971). This model will reproduce the light curve and the spectroscopic features. No other model but Dr. Huang's will do so – and that meets certain difficulties. The radial velocities of certain satellite lines are in very good agreement with the model (Kuiper *et al.* 1937). The secondary mass, which holds the ring together is not veiled, as in earlier models, but should be in full view. Since it has not yet been detected at any wavelength, its luminosity must be far less than normal for its mass. Finally, since that mass is of the order of 12 $M_\odot$, it cannot be a white dwarf or a neutron star, and seems, therfore, to be a black hole.

Herczeg: Do you think that your model has definitely shown the existence of a black hole in ε Aur?

R. E. Wilson: In effect, I say so in my paper.

Huang: Dr. Wilson mentioned my model of ε Aur. My philosophy is to allow a great amount of freedom of thought in the interpretation of observations. Everyone can propose a model from his own perspective. We should produce as many models as we possibly can, and then, after twenty years, we can combine the best parts of these models together, and form a better one. I think every model of ε Aur we have now has some merits – and some defects – we should let time take care of them. After all, stellar events took millions of years – we can afford to be patient.

Bolton: It is my opinion that given what we know of the physics it will never be possible to point to a specific system and say that that system contains a black hole. As I understand the theory a black hole has three characteristics: (1) a certain mass, (2) a radius defined by the mass (3) no luminosity. The only way that a black hole will call itself to your attention is through material falling into it, and that is only likely to be important in a binary system. Mass accretion by a black hole in a binary system may produce X-rays, but the accretion process will probably produce a disk around the black hole. This disk will yield a spurious radius in any eclipse measurement, and if seen edge on, may block the X-ray emission. Thus about all we can hope to do is to detect a number of massive underluminous objects and hope that this collection of objects will show some regularities that can be theoretically explained.

Underhill: What kind of mass do you require in a black hole?

Bolton: Well, if you can believe the theories of such people as Cameron and Ostriker, then neither a neutron star nor a rotating white dwarf can be more massive than about $3 M_\odot$. Under very special and unlikely conditions a rotating white dwarf could be a bit more massive than this.

R. E. Wilson: I would like to point out that our faith in the law of conservation of energy is based on the fact that we have never seen it violated. It is not based on any original principle that someone found engraved on a stone. The hypothesis was made that energy is conserved; people have made experiments and failed to find that it is not conserved. Therefore, we believe energy is conserved. I say all this to emphasize that ideas of this kind are accepted eventually because of the weight of overwhelming auxiliary evidence, which accumulates over many decades and centuries.

Biermann: You might be able to determine the existence of a black hole from the high-energy X-ray spectrum.

Batten: I sometimes find myself wondering if we shouldn't draw a lesson from quite recent history. Neutron stars were talked about for about thirty years before pulsars were discovered, and when pulsars were discovered they were such a surprise, we didn't believe at first that they were neutron stars. If black holes ever are discovered, might they not be discovered by some completely unpredicted property? I think that if and when they are discovered, it will be beyond reasonable doubt what they are, and as long as there is reasonable doubt, I prefer to adopt the principle: entities should not be multiplied unnecessarily.

Popper: Well, since you are in a philosophical vein, let me put it another way. Nature in general is more imaginative than human beings!

Devinney: The question was raised: will we ever find binaries that are well behaved in which we may find some evidence for black holes? There are systems like δ Gem, which has a large mass function although the visible component is an apparently normal A-type star. Perhaps amongst systems of this kind, a black hole may be found.

Popper: Any further comments on this topic of crazy stellar objects? If not then I think we should open up the floor for general comments on Dr. Plavec's presentation.

Oliver: I have a question for Dr. Plavec that does not involve black holes or X-ray sources. There exists a group of binary systems that have in the past been called Algol-like systems with undersized subgiants. We've talked about them already in this Symposium, and we hope to give more details tomorrow.* I would like to hear your comments now on their evolutionary status. The mass ratios in these systems seem to be near unity. The secondary star, usually like a K0 subgiant, may be either more or less massive than the hotter star. The secondary does not fill its Roche lobe, and there are rather too many of them with mass ratio close to one. Can you say anything about the mass transfer that seems to be going on in these systems?

Plavec: No, I can't. I, too, am puzzled. Many years ago I studied period changes in eclipsing binaries and I found that if one component of the system is off the main-sequence, there are period changes. If so, and if we accept that period changes always indicate mass transfer, then we must conclude that all giants are losing mass – which would be horrible to accept in general.

Popper: Now we can proceed with some of the listed topics. Dr. Walter, would you like to present your remarks on the period-eccentricity relation?

Walter: Among the mutual relations between the components of binaries from which we may get important information one should not forget the relation between periods of revolution and eccentricities of the orbits. It is a widespread opinion that this relation is only a statistical one in the sense that in general for systems with close components the orbits are circular and, for larger periods of the systems, the orbits pass over to elliptic ones, and that only the mean values of the eccentricities, for a large number of systems, are larger, the longer the periods.

But this is a rather superficial view, and I must criticize people who worked on the problem in recent years or reported on it for not having studied older literature. They are not aware of the fact that, with large probability, for main-sequence stars of spectral types B9 and later and within distinct regions of periods, the relation between periods and eccentricities is very close. The observations show that, depending on the physical properties of the components of the systems, the period-eccentricity relation has a quite characteristic appearance.

The details of the relations between period and eccentricity for systems containing stars within certain spectral-type ranges are shown in the illustrations of my three papers (Walter, 1950, 1951, 1954). Recently I repeated some of this work, using Batten's (1967) catalogue. The number of systems available for these investigations increased only modestly in the two decades since the older Lick catalogue (Moore and Neubauer, 1948), but the relations suggested in my earlier work no longer seem valid for systems containing main-sequence stars of spectral types B8 and earlier.

---

* See p. 279.

Spectroscopic binaries containing primary components of spectral type F8V to K5V and showing only one spectrum have orbits that are circular, or nearly so, if $P < 10d$. If $10d < P < 25d$, there is a close correlation between period and eccentricity, which breaks down for $P > 25d$.

Binaries containing main-sequence components in the range of spectral types B9–A7 show a more complicated relationship. Those containing stars of nearly equal masses can be divided into two groups, a and b, which differ in the period at which circular orbits give way to elliptical ones. Systems with very small mass functions seem to have the same maximum value of the eccentricity as do those containing equal stars. Moreover, features in the (P, e) relation for these stars appear to be related to the rotational velocities of the A-type components. In some systems, the A-type component cannot be forced into synchronous rotation. Systems with intermediate mass functions also show a division into two groups. It is of interest that α CrB, which belongs to group a in this class does indeed rotate much faster than synchronism would require. It is located in the (P, e) diagram by the component of smaller mass.

Here I would like to touch only shortly on the theoretical aspects. My earlier statement that, because of the effects of tidal oscillations of the components, elliptic orbits should have a tendency to be transformed to circular ones, was independently put forward and improved by Zahn (1966) by means of detailed model computations. But, disregarding the existence of the above mentioned (P, e) relation, this author did not realize the fact that as well an opposite tendency, one by which the eccentricities should grow with time, must be assumed in order that, as a result of an equilibrium between the two opposite influences, orbits may exist with the observed eccentricities. In one of my cited papers I have shown that differential rotation of binary components like that known for the Sun could increase orbital eccentricity. In the meantime it has come to seem very probable that magnetic fields exist in Algol systems connecting the components as I mentioned earlier in this Symposium.* Therefore one could imagine that besides gravitational forces, magnetic fields are also important for the transfer of rotational energy from the components into the orbit and that the bifurcation in the (P, e) relations found for A-type stars might be caused by different combinations of the signs of the magnetic fields on the two components.

In any case one should think that observations of the orbits of spectroscopic binaries may give us information which are important for the study of the mutual connections between the components and the properties of main-sequence stars.

If one looks at the diagrams in my papers, one may still have some doubts of the statistical significance of the (P, e) relation. I believe it would be very important to enlarge the observational material of spectroscopic orbits of main-sequence systems not only in order to ensure the reality of that relation but also to make an observational basis for detailed studies. It is a task which may be done by spectroscopists without outstanding instrumental means, since there are many stars of apparent

* See p. 74.

magnitude between 7^m and 8^m, among them over thousand A-type stars which have not yet been investigated for their binary characteristics. According to the probability of frequency of spectroscopic binaries, there should be many systems of binaries among them which could contribute to a comprehensive filling up of the (P, e) diagrams. It is conceivable that, without an actual object and physical purpose in mind, spectroscopists during the last years and decades had little interest in further enlarging the number of spectroscopic systems. I would like to encourage them to continue work which I believe will be urgently needed in future by theoreticians.

Bolton: While I would number myself among the doubters as far as the reality of the effects you have described, I would like to echo and amplify upon your comments about the need for getting more statistical data. For the past two years, I've been using the 74-inch telescope of the David Dunlap Observatory during bad weather to look at bright early-type stars. I have found or confirmed literally dozens of B-type and A-type spectroscopic binaries, between second and fifth magnitude, which have no orbits.

Hill: We are doing the same thing at the Dominion Astrophysical Observatory, and can confirm Bolton's statement.

Wright: At noon, I received a letter from Dr. Aller. He was to have been present here but was prevented at the last minute. He would like some of his views to be presented to you. He particularly wants to stress the limitations imposed by kinetic theory on the behaviour of the gas in rings, shells, jets, etc. around binary systems. He cannot follow at all the justification of particle dynamics which, it seems to him, must fail by many orders of magnitude. The circumstellar matter should have high thermal and electrical conductivity, and he emphasizes the possible role of magnetic fields. Streams, he believes, would be very viscous. Over limited ranges of the possible conditions in gas streams, the particle-dynamics results may approximately represent the true flow – but this must be thoroughly tested in each case.

Popper: Dr. Huang is anxious to comment on this.

Huang: Dr. Aller mentioned things that were discussed in my paper, so I would like to make a few comments. The version of my paper that was circulated, which Dr. Aller refers to, is my first draft. The final version is quite a bit different. I have discussed the hydrodynamical nature of the problem, but there is the great difficulty of how to impose the boundary conditions as I explain in my review paper (final version). You have to assume a certain physical state in order to derive some approximate solutions, because of this difficulty. You have seen the results of Dr. Kitamura who has produced such a solution by the hydrodynamic approach. If you compare his result with my result derived from the three-body particle problem, you find that they agree very well. This shows that in this case particle dynamics is not a bad approximation at all, because the flow is laminar – that point was not dealt with in my final manuscript. Thus, particle dynamics has the advantage of being very simple, while hydrodynamics is the correct approach to the problem. But, I emphasize that hydrodynamics gives no satisfactory solution at present, because of the difficulty with boundary conditions.

Biermann: I was asked to comment on this letter, and I would say I agree with almost all of it, except the part on viscosity. It's not important to know the Reynolds' number involved. In my paper (1971) I discussed that point briefly. It's very evident that the Reynolds' number is very large (many powers of 10) and I would say the flow is turbulent. Turbulence is very difficult to treat, and, of course, has to be neglected in nearly all the approximations we can make at the moment. The same reasoning makes the thermal conductivity probably unimportant.

Chen: I agree with Dr. Biermann about the large value of Reynolds' number that one would find for circumstellar flow. Hence, as in the case of flow around Be stars, the viscous terms are many orders of magnitude smaller than the inertial term.

Kitamura: I would like to comment on two points. First, in treating the problem of gaseous rings in close binary systems the use of particle dynamics may not be best. Even so, for practical application it would perhaps be alright as the first approximation, because as I showed from hydrodynamic equations*, the pressure-term of the form $\partial P/\partial \xi$ that appears in the solution is found to vanish at the maximum gas density within the ring. Secondly, in the hydrodynamic approach to the present problem it would be important to find a device by which the mathematical complexity can be reduced as far as possible. What I would like to mention is that the Roche coordinates would be a most convenient means for mathematical treatment in this sense. By using these coordinates, the surface of a parent star with an emission ring in a close binary can be best approximated by one of the zero-velocity surfaces on which the boundary conditions can be set in much more tractable form.

Biermann: I did some unpublished work on two-dimensional subsonic flow, and it turned out that the numerical complexity involved in expressing the boundary conditions nicely is so great that my impression, at the time, was that the advantage of the proper coordinates is offset by the complexity. It was almost impossible to check the computer code.

Popper: It's exactly 5.30 p.m., the time set for the termination of this session. Perhaps it would be a good time to finish.

References

Batten, A. H.: 1967, *Publ. Dominion Astrophys. Obs.* **13**, 119.
Biermann, P.: 1971, *Astron. Astrophys.* **10**, 205.
Eggen, O. J., Kron, G. E., and Greenstein, J. L.: 1951, *Publ. Astron. Soc. Pacific* **62**, 171.
Gaposchkin, S.: 1951, *Astron. J.* **51**, 109.
Hall, D. S. and Garrison, L. M.: 1969, *Publ. Astron. Soc. Pacific* **81**, 771.
Kraft, R. P. and Demoulin, M.-H.: 1967, *Astrophys. J. Letters* **150**, L183.
Kristian, J., Sandage, A. R., and Westphal, J. A.: 1967, *Astrophys. J. Letters* **150**, L99.
Kuiper, G. P., Struve, O., and Strömagren, B.: 1937, *Astrophys. J.* **86**, 570.
Lucy, L. B.: 1967, *Astron. J.* **72**, 813.
Moore, J. H. and Neubauer, F. J.: 1948, *Lick Obs. Bull.* **20**, 1.
Paczyński, B.: 1965a, *Acta Astron.* **15**, 198.
Paczyński, B.: 1965b, *Acta Astron.* **15**, 89.

* See p. 107.

Paczyński, B. and Ziołkowski, J.: 1968, *Acta Astron.* **18**, 255.
Walter, K.: 1950, *Astron. Nachr.* **279**, 1.
Walter, K.: 1951, *Astron. Nachr.* **280**, 149.
Walter, K.: 1954, *Astron. Nachr.* **282**, 122.
Wilson, R. E.: 1970, in L. Gratton (ed.), 'Non-Solar X- and Gamma-Ray Astronomy', *IAU Symp.* **37**, D. Reidel, Dordrecht, p. 242.
Wilson, R. E.: 1971, *Astrophys. J.* **170**, 529.
Zahn, J.-P.: 1966, *Ann. Astrophys.* **29**, 313, 489, 565.

TENTH DISCUSSION SESSION

(Tuesday morning; 12 September, 1972)

Chairman: D. M. POPPER

Popper: As you know, there is going to be a brief summary of the highlights of our proceedings by Dr. Sahade. There is still time for anyone who wishes to make comments to do so, but I thought we'd get some kind of organization in the proceedings if we kept at first to a list of speakers. So I'll call on Dr. Lloyd Evans for his remarks.

Lloyd Evans: Plavec used evolutionary tracks by Iben to estimate the critical orbital period for which a star fills its Roche lobe at various evolutionary stages. The values for a star of 5 $M_\odot$ are 10 d and 100 d at the Cepheid stage and the red giant tip, respectively. It is generally accepted that most Cepheids are in a post-red giant stage of evolution. The shortest known orbital period, 506 d for S Mus ($P=9^d\!.7$), was found in an extensive survey of Cepheid radial velocities carried out at the Radcliffe Observatory. Given the considerable theoretical uncertainties, especially in red giant radii, and the small size of the sample (5 known periods, though another ~ 5 binaries almost certainly all have longer periods), agreement with the theoretical minimum $P \sim 100$ d is satisfactory.

Plavec: The period of 100 d for a star of 5 $M_\odot$ at the red-giant tip is the maximum period for mass exchange to take place before the star gets there. Any period above 100 d makes the system sufficiently wide for the stars to evolve independently. This is what you need to obtain a Cepheid.

Andersen: What is the amplitude of the orbital motion relative to that of the pulsations?

Lloyd Evans: They are roughly the same.

Batten: What is the latest estimate for binary frequency amongst Cepheids?

Lloyd Evans: Still about 15%.

Hummer: Although the N III lines $\lambda\lambda 4634$, 40, 41 are seen in emission in the spectra of many types of objects including Of stars, for which they are classification lines, the mechanism for their formation has remained obscure. Because Otto Struve himself was one of the first people to discuss the problem of these lines (Swings and Struve, 1940) and because they are of some importance for the extended atmospheres of certain of the objects being discussed at this Symposium (for example, Cyg X-1)*, I would like to say a few words about some recent work explaining how these lines are formed. A full account of this work will appear shortly (Hummer and Mihalas, 1973). The multiplet in question is $3d^2D-3p^2P^0$. The subsequent transition $3p^2P^0-3s^2S$ is always seen in absorption, at least in O-type stars. The Bowen mechanism has sometimes been invoked as an explanation, but fails on several counts. Swings (1948) proposed that $3d$ is pumped from the ground state $2s^2 2p^2 P^0$ by continuum radiation

---

* See p. 110.

Batten (ed.), Extended Atmospheres and Circumstellar Matter in Spectroscopic Binary Systems, 277–285.
All Rights Reserved. Copyright © 1973 by the IAU.

which for Of stars peaks near the energy of the $2p$–$3d$ transition; $3p$ is then drained by some unspecified mechanism. By solving the statistical equilibrium and radiative transfer equations for a model N III ion using Mihalas' recent non-LTE plane-parallel model atmospheres (NCAR Technical Note, NCAR-TN/STR-76; Boulder, Colorado), we showed that $3d$ was populated primarily by dielectronic recombination and that $3p$ was drained significantly by two-electron transitions to $2s2p^2$ in addition to the other dipole-allowed transitions. Not only did the $\lambda4640$ multiplet appear in emission for the values of effective temperature and surface gravity for which the emission is actually observed, but the equivalent widths were in accord with those measured by Peter Conti for Of stars in which $\lambda4686$ was in absorption or neutralized, i.e. those designated as O((f)) or O(f) by Walborn (1971). This comparison has already been published by Mihalas *et al.* (1972).

On the basis of our work, I would like to make three points. The first of these concerns the atmospheric structure of Of stars. We find that the N III lines, which are weak and therefore formed deep within the atmosphere, can be understood on the basis of a static, plane-parallel model, at least for those Of stars with $\lambda4686$ in absorption. Yet for one of these objects, ξ Per, Morton, Jenkins and Macy, (1972) observe strong ultra-violet lines that show the outer part of the atmosphere to be rapidly expanding. Our current picture is that as $\lambda4686$ goes from absorption into emission and $\lambda4640$ strengthens, (the progression O((f)) → Of), the fraction of the atmosphere that is expanding rapidly increases, that is, the sonic point lies at increasingly greater depths in the atmosphere. Although in our calculations the mechanism proposed by Swings was not effective because the $2p$–$3d$ transition is saturated, we expect that as more of the atmosphere expands rapidly this transition will become desaturated and pumping by the continuum will add to the strength of the $\lambda4640$ emission. Further work along these lines is in progress in Boulder by Castor, Mihalas and myself.

The second point is that our detailed understanding of the N III ion makes the lines of this spectrum available for use as temperature and density indicators for a wide variety of atmospheric conditions. More atomic parameters are now known for this ion than for any other (although our group at JILA is now producing atomic data to the same degree of completeness for other ions of C, N, O and Si). Although a substantial amount of work is required to interpret the N III spectrum, our program could be used for any atmospheric model.

Finally, I would like to stress the danger of making inferences about the structure of an atmosphere from the appearance of emission lines without understanding the underlying physical mechanisms. The $\lambda4640$ multiplet is seen in emission in Of stars only because of certain atomic properties of that ion, i.e. it is an intrinsic emission line, and does not imply mechanical heating, outward temperature rises, extended atmospheres or any of the other atmospheric features that are commonly associated with the appearance of emission lines.

Bolton: I agree with what you say about making errors if incorrect physics is used in the interpretation. In the case of Cyg X-1 I think that you must invoke something different from the particular non-LTE phenomena you have described here, even though you find some of the same lines found in the spectra of the Of stars. The primary star in Cyg X-1 lies outside the region of the H−R diagram where the Of phenomena are found. The Cyg X-1 emission lines may arise from a similar process to that in the Of stars, but something else must be happening as well.

Hummer: I'm pretty sure that our proposed mechanism holds for all of these situations, although perhaps in the supergiants the Swings mechanism plays a more important role. I can see no way for the Bowen mechanism to work under these conditions.

Hutchings: Can you say whether these emissions will be enhanced in extended atmospheres?

Hummer: As the atmosphere becomes more and more extended, the streaming velocity of the gas increases and the $2p$–$3d$ transition is desaturated, so that the Swings mechanism almost certainly becomes effective. Of course the population of $3d$ via dielectronic recombination continues to be effective until the temperature becomes too low.

Popper: But presumably both lines would be in emission in that case?

Hummer: We know of no cases of $\lambda 4097$ appearing in emission in the spectra of Of stars. In some of the objects in whose spectra it does appear, we know that the Bowen mechanism is operative.

Oliver: We have several times referred to the RS CVn class of binaries during this Symposium. I thought it might be useful to list the characteristics of the class. They are:

(i) Eclipsing binaries containing late-type undersized subgiant components.

(ii) They show H and K emission in the spectra of the secondary, or both, components. Six of these stars were first singled out by this characteristic (Struve, 1946). A larger list was published by Hiltner (1947).

(iii) They show two spectra. The mass ratios q (m_{cool}/m_{hot}) range from about 0.8–1.3. Eggen (1955) called attention to the group and Popper has repeatedly mentioned them in recent years (e.g. Popper, 1970).

(iv) The secondary component is usually of spectral type K0–K1 IV, the primary may range from F0–G5 IV or V.

(v) The light curves show both short-term and long-term variations.

(vi) There are fairly large and frequent period changes $\Delta P/P$ up to 10^{-3} or 10^{-4}.

(vii) The secondary stars do not fill their Roche lobes (Plavec, 1967).

(viii) They generally show a wave-like distortion of the light curve. At some epochs this may make the maxima appear asymmetric, but the wave moves slowly toward earlier phases so as eventually to reverse the asymmetry. This was first seen in RS CVn itself as a result of long series of careful photometric observations at Catania (Catalano and Rodonò, 1967), but a similar feature is observed in the light curves of most, if not all, of the members of the class.

In Table I, I list some probable members of the class. Dr. Popper has kindly provided provisional values for the masses. The components are distinguished by the subscripts h and c – for hot and cool.

Smak: I believe that these subgiants are not over-luminous.

Oliver: They are certainly not. It's hard to talk about the ratio of the luminosities without a good model for the 0^m25 (wave-like) variation. If the secondary component is the source of the variation, it must be varying by considerably more than 0^m25 – perhaps as much as 0^m75.

Popper: I think the fact that these are double-line binaries shows that the luminosities of the components are not very different; the masses are not very different – so obviously they are not very over-luminous.

TABLE I

Some Probable RS CVn Binaries

System	Period (d)	$m_h + m_c$ ($M_\odot$)	$q = m_c/m_h$
SS Boo	7.6	1.75	0.92
RS CVn	4.8	2.75	1.04
WW Dra	4.6	2.6	1.00
Z Her	3.99	2.3	0.90
AR Lac	1.98	2.6	0.99
SZ Psc	3.97	3.0	1.24
RW UMa	7.33	2.2	1.00

Hall: I want to talk about the evolutionary status of these RS CVn binaries and about some of the problems which arise in understanding their physical nature. I have already said (p. 143) that I think these binaries are in some sort of pre-main-sequence evolution (Hall, 1972). Although this is difficult to prove conclusively (the evidence must be regarded as somewhat circumstantial) it is much more tenable than either of the two other possible interpretations, namely post-main-sequence evolution before or after mass transfer. Serious problems remain, however, even if my basic interpretation is correct.

A pre-main-sequence interpretation can best explain why both stars are above the main-sequence but are both detached; why stars around 1 $M_\odot$ (in some, like SS Boo, the two stars are even less massive) can appear 'evolved', when the galaxy is not old enough for them to have evolved off the main sequence; and why one star is losing mass at a very rapid rate even though it is definitely not filling its Roche lobe. This interpretation can best explain the very strong H and K emission, which we associate with youthful chromospheric activity; the contact components in *bona fide* Algol-like systems are also subgiants, quite similar in spectral type and radius, but do not display this anomalously strong H and K emissions. And it has the least difficulty in explaining the fact that the more 'evolved' component can be either the more or the less massive one.

There are two additional pieces of evidence, not yet published, which support the pre-main-sequence interpretation. First, *JHKL* photometry of six of these systems shows that there is an infra-red excess of about 0^m5, probably in the cooler component.

Furthermore, *UBV* photometry suggests the possibility of a slight ultra-violet excess in some of the systems. Then we have an energy distribution which cannot be fitted by any one spectral type. A similar situation exists with the T Tau stars, although the infra-red excess in the T Tau stars is larger, about 2^m or more at *L*. Thus the energy distribution is consistent with my claim that the systems are young, but not so extremely young as the T Tau stars themselves. The effective temperature derived from consideration of the entire spectrum from *U* to *L* is around 1000 K cooler than that derived from the $B - V$ colour index alone, or the spectral type, which invariably is quite close to K0. A K0 star would be too hot for the cooler component to be on the Hayashi track for its mass, but an effective temperature about 1000 K cooler is acceptable. I have also mentioned before (p. 143) Montle's result that the average distance of these binaries from the galactic plane is about 80 parsecs, suggestive of an age of about 6 to 10×10^7 yrs. The average distance from the plane of Algol-like systems with a total mass less than 2.75 $M_\odot$ is about 155 parsecs, suggestive of an age of about 3 to 7×10^8 yrs. The dispersion of velocities perpendicular to the galactic plane $\langle Z^2 \rangle^{1/2}$, is about 12.5 km s$^{-1} \pm 2$ km s^{-1} for the RS CVn systems, suggestive of an age of about 1 to 3×10^8 yrs, and 15.5 km s$^{-1} \pm 4$ km s^{-1} for the Algol-like systems.

There seems to me to be many reasons why we should blame the wave-like distortions of the light-curve on an uneven distribution of surface brightness on the cooler star. One side of the cooler star seems to be fainter than the other. Why this should be so is unclear. The fainter side of the cooler star slowly changes its orientation with respect to the other star, executing one complete migration approximately every 10 yr. There is a good correlation between the orientation of the fainter side and the large period increases and decreases observed in the (O − C) diagram. This correlation is best explained as a result of mass loss, actually angular-momentum loss from the fainter side, since the binary speeds up when the fainter side is on the trailing hemisphere, and slows down when the fainter side is on the leading hemisphere. Eugene Milone and I have found the same correlation in RT Lac, another RS CVn-type system. In my paper I supposed that the fainter side was darkened by sunspots and that flare activity accompanying these spots might account for the outflow of matter. However, another interpretation should be explored. The energy equivalent of the required rate of angular-momentum loss is quite large compared to the luminosity of the cooler star. Perhaps we should say that the fainter side is fainter *because* it is losing mass, and only ask why it is losing mass. Using very simple assumptions, I found that 10^{-6} $M_\odot$ of mass lost each year at an ejection velocity of 3000 km s^{-1} could account for the observed period changes. If there is a strong magnetic field, the mass could be lost from the ends of long magnetic 'spokes'. A lower rate of mass loss and/or velocity of ejection would then give the same rate of loss of angular momentum.

Oliver: My list was not meant to be complete – just a sample of seven probable members of this class. Essentially all double-line late-type systems showing H and K emission in their spectra could be included. Virtually all systems of that type listed by Popper seem to belong to the class – we know of at least two dozen members right now. Perhaps there could be some argument about RT Lac.

Smak: I believe that the pre-main-sequence explanation is a very promising one, even if you do have to drop out those systems for which the mass ratios are inconsistent...

Hall: No, you don't have to. The mass loss itself gets over that problem – if it's at anything like the rate I claim. If the mass loss goes on for any length of time, the K0 IV star will soon become the less massive – certainly by the time it reaches the main sequence. That's how I get around the problem of the (presently) more massive secondary.

Smak: Then you might find that the primary went through the same phase that the secondary is now in, some time before.

Hall: That's the best way to think of the situation at present, but perhaps the primary and secondary evolve differently right from the beginning.

Smak: If we consider the contraction phase, it is quite possible that the secret lies more deeply, namely in the early history of the system.

Hall: I think these stars are suffering the effects of an extremely traumatic birth. We should not think of a simple 'interaction effect': the systems are suffering from their past history.

Plavec: Dr. Hall seems to believe that if H and K emission is observed in Algol-type systems it is inevitably associated with the disk around the primary component. This is only partially true. It's true that when the emission lines of the ring are extremely strong you observe also double or single emission in H and K. But, for example, the spectrum of U Sge discloses a reversal inside the deep absorption of K lines of the secondary spectrum. But the subgiants themselves do show emission of the same type as these R CVn systems. The difference only might be in the intensity of the line, and I would say that in those systems H and K emissions are particularly strong. But basically, they are no different from subgiants in Algol systems.

Hall: I have to reply to this. The Wilson-Bappu effect, and the apparent universality of its application, would tell you that any subgiant between late G and early K in spectral type must show some H and K emission. It would seem to me that the H and K emission arising from the subgiant component itself in Algol-like systems is probably just what would be expected from the Wilson-Bappu effect. The emission from the K0 IV component in the RS CVn systems, in contrast, is very much stronger. I can thank Dr. Popper for making this clear to me (although I don't want to hold him responsible for the things I say). About a year ago I was trying to interpret the RS CVn systems as post-main-sequence objects, i.e., similar to the Algol-like systems. Within that framework I could not easily understand the H and K emission and therefore wanted to dismiss it as simply Wilson-Bappu emission. But Dr. Popper emphasized to me how very much stronger the emission was in the RS CVn systems.

Batten: In the spectrum of the secondary component of U Cep, even though the emission is not strong enough to come above the continuum, it is definitely much stronger than you would expect from the Wilson-Bappu effect.

Popper: When you talked of an average value of z, did you make any allowance for the fact that those two groups of stars have systematically different luminosities?

Hall: Incompleteness can certainly have an important effect on the mean value derived for z. We have done two things. First we plotted the value of z versus $d \cos b$, thereby producing a side view or cross-section of the galaxy, so to speak. Examination of the mean z within various vertical sections assured us that it remained more or less constant as a function of $z \cos b$. Since the sample becomes more incomplete with increasing $z \cos b$ and since the mean value of z was not obviously correlated with $z \cos b$, we concluded tentatively that our mean value for z was not grossly in error. Second, we noticed that a concentration of interstellar absorption towards the galactic plane, along with the fact that obscuration contributes to incompleteness, would have the effect of exaggerating the derived mean value of z.

This doesn't completely answer your questions but, because the effect of incompleteness on the statistics is a crucial question, Montle is now working to remove the effect explicitly, in a more formal way. Until this is done, we do not want to publish any results.

D. B. Wood: As I listened to the first few days of this Symposium, the following idea struck me. We are at a stage of understanding the photometry of eclipsing stars that is comparable to that reached in the study of planetary motions in the 16th Century. We find ourselves with the necessity of inventing 'epicycles' to explain our observations – that is, we need gas streams, bumps, hot spots, and what have you. Instead of calling upon these various devices to explain deviations from the model of eclipsing stars, it seems particularly obvious that we should change the model. In this context, let me show you a specific example of what rectification does.

Rectification treats stars as ellipsoids with concentric isophotes. In reality, however, the isophotes on such a star will be distorted toward one end, except at phases 0°, 90°, 180° and 270°. Thus rectification will leave behind an apparent 'hot spot'. The effect of this spot will be to increase system brightness away from quadrature, until this brighter area is eclipsed. This phenomenon will look very much like what Dr. Walter has shown us. The amplitude of this effect would depend strongly upon the amount of distortion of the stars.

Walter: This is not quite clear. The hot spot is attributed to the primary component of Algol-type systems. The primary star is very nearly spherical, and there is no problem with the distribution of gravity darkening and limb darkening.

D. B. Wood: Well, another problem with rectification, especially applied to the Algol-type system, is that there's only one distortion parameter. If you have a very distorted cool star and a nearly spherical hot star, the distortion is averaged over the whole system. The effect of rectification is thus to treat the hot star as though it is more distorted than it really is.

Popper: I think the only way one could resolve this controversy would be for someone like Dr. Wood to use his techniques to analyze Dr. Walter's observations. Then Dr. Wood and Dr. Walter could compare their results and see whether, as a matter of fact, there were any outstanding differences. It appears that Dr. Walter has allowed for first-order effects of the difference in shapes of the components. I think that without some sort of comparison of that kind, further discussion as to whether Dr. Walter's technique is adequate will not be useful at this time.

Herczeg: Many of us participated in the non-recorded session dealing with β Lyr the other day. Although we certainly did not solve any of the major problems concerning this famous system, we were able to point out 4 or 5 current problems, the solution of which may be of some importance for future investigations. I will try to give a summary of these problems:

(1) We had the impression that the H/He ratio in the system is still a rather uncertain quantity although underabundance of hydrogen has been proposed by several authors. Perhaps we should try to re-examine this question using the best spectra and the best theory even if strong disturbances in the system make an exact solution almost impossible.

(2) It would be desirable to obtain a new photometric solution based on the methods of light-curve synthesis. The question of the secondary eclipse – how much light and of what colour is lost at phase $0^P.5$ – is of particular interest.

(3) The primary eclipse should be observed and reobserved faithfully (in several colours), not only at a few points but if possible rather continuously covered throughout its whole width, since permanent or recurrent features may shed light on the nature of the eclipsing body.

(4) We have quite a few sets of polarimetric observations that should be assembled and rediscussed under the following aspect: do they confirm any of the published models? We find many quotations and remarks in the literature to this effect but a thorough and critical comparison is still wanting.

(5) There was consensus among participants that publication (or making accessible) of the OAO observations of β Lyr in the near future would help our efforts considerably and is highly desirable.

Perhaps I am allowed to move that Drs. Batten and Sahade, organizers of the last year's co-ordinated program, should contact people responsible for the OAO observation and convey this question of ours to them.

Underhill: You forget, to obtain an abundance ratio you have to solve a serious problem in radiative transfer in a way that will be valid for a stream of helium.

Herczeg: We only want approximate data in order to decide, eventually, whether the B8 star is normal or peculiar.

Devinney: Perhaps some spectral scanning of β Lyr would be useful.

Underhill: A lot of OAO material is now being placed in a data bank. When I go back to Goddard, I will make inquiries for you.

(After the close of the Symposium, Dr. Underhill sent a copy of the following memorandum from Dr. C. D. Wende of NASA, with the request that it be included in the Proceedings. The memorandum was dated 20 September, 1972.)

The following is the current status of OAO-2 data in the National Space Science Data Center:
SAO Data: The final, revised version of the Smithsonian Celescope star catalog is expected to arrive today. This catalog will be available both on magnetic tape and on microfilm. The documen-

tation appears to be complete, and these data should be available from the NSSDC for general requests by late October.

WEP Data: The NSSDC has received approximately 135 reduced data tapes out of an expected total of 400 (*note*: these are arriving at the expected rate). Other WEP data, expected later, include microfilm copies of the contents of the reduced data tapes, microfilmed plots of the spectrometer data, and magnetic tapes of the reduced spectrometer data. Note that the reduced data tapes contain reduced (i.e., instrumentally corrected) photometer data, but unreduced spectrometer data. Data on the reduced data tapes (and the corresponding microfilm) are in chronological order by the date of observation; observations of specific objects must first be located in a catalog of observations, and then these observations may be found in the data tapes/microfilms. User documentation is being written by Wisconsin, but the completion date is as yet unknown. This documentation is the pacing item as far as release of these data are concerned.

Scientists desiring copies of these data should write to:

> *National Space Science Data Center*
> *code 601.4*
> *Goddard Space Flight Center*
> *Greenbelt, Maryland, 20771.*

Scientists who are not U.S. citizens also have access to these data by writing:

> *World Data Center A, for Rockets and Satellites*
> *Code 601*
> *Goddard Space Flight Center*
> *Greenbelt, Maryland, 20771, U.S.A.*

Persons desiring these data need not wait until its availability is formally announced, but may write to one of the above addresses, their names will be filed, and they will be notified as soon as the data are released.

References

Catalano, S. and Rodonò, M.: 1967, *Mem. Soc. Astr. Ital.* **38**, 395.
Eggen, O. J.: 1955, *Publ. Astron. Soc. Pacific* **67**, 315.
Hiltner, W. A.: 1947, *Astrophys. J.* **106**, 481.
Hummer, D. G. and Mihalas, D.: 1973, *Astrophys. J.* (in press).
Mihalas, D., Hummer, D. G., and Conti, P.: 1972, *Astrophys. J. Letters* **175**, L99.
Morton, D. C., Jenkins, E. B., and Macy, W. W.: 1972, *Astrophys. J.* **177**, 235.
Plavec, M.: 1967, *Bull. Astron. Inst. Czech.* **18**, 334.
Popper, D. M.: 1970, in K. Gyldenkerne and R. M. West (eds.), *Mass Loss and Evolution in Close Binaries,* Copenhagen Univ. Press, p. 13.
Struve, O.: 1946, *Ann. Astrophys.* **9**, 1.
Swings, P.: 1948, *Ann. Astrophys.* **11**, 228.
Swings, P. and Struve, O.: 1940, *Astrophys. J.* **91**, 546.
Walborn, N. R.: 1971, *Astrophys. J. Suppl.* **23**, 257.

SUMMARY

JORGE SAHADE*

Instituto de Astronomía y Física del Espacio, Buenos Aires, Argentina

I have been assigned the task of presenting a summary of the Conference that has just finished. Actually during the Symposium the review papers have touched upon many items, some of them apparently not strictly connected with the subject of the meeting but helpful to clarify the picture related to close binaries and aimed at establishing differences with single stars, whenever possible. Therefore, my task would be somewhat difficult it if were not that all the review papers were extremely clear, very well given and fully discussed and that there were so many contributions from the floor presented without any pressure from time or otherwise. Furthermore most of what I would say has been already said. All this makes my task a lot easier as it appears clear that I should confine myself to offer only a few comments. Let me start by saying that Dr. Struve would have felt most pleased to have been present at this meeting and realize how many astronomers are now interested, strongly interested, in the field of close binaries and how much the field has and is expanding taking also up the efforts of researchers that work in the radio and X-ray regions and attracting the attention of those who are interested in problems of general relativity.

The interest in the field, at Struve's time, was so extremely limited that in 1957 I made in the *Publications of the Astronomical Society of the Pacific* an appeal to photoelectric observers to engage is some cooperative programs that required an urgent and simultaneous attack from the photometric as well as from the spectrographic side. No reply was received. How different things would have been ten to fifteen years later!

From the observational point of view the Conference has summarized our present knowledge in regard to the structure commonly found in interacting binaries with the presence of gaseous streams usually coming from the less massive component of the system, with the presence of gaseous envelopes around one of the components – usually around the star towards which the gaseous stream is directed – and with the presence of outer, more tenous, envelopes that surround the whole system.

Observational evidence has been presented that adds to the one already available and confirms the existence of an interaction between the stream and the gaseous envelope around one of the components or perhaps in some cases with the star's outer layers, depending partly on how thick or thin the envelope is. The result of this interaction has been described with the terminology of 'hot spot' – the fourth element in the structure of circumstellar matter – and whenever the conditions, velocity of the stream, density of the envelope, are appropriate, it gives rise to the production of X-radiation in the soft range of energies.

We have been further given the observational facts – some recently secured –

---

* Member of the Carrera del Investigador Científico, Consejo Nacional de Investigaciones Científicas y Técnicas, Argentina.

Batten (ed.), Extended Atmospheres and Circumstellar Matter in Spectroscopic Binary Systems, 286–291.

related to stars – either single or binary – with expanding and with extended envelopes.

From the theoretical point of view we have been exposed to up to date information as to the state of the art in regard to the points touched in the observational reviews. They were clear in stating the difficulties of the problems and the present possibilities.

The last review paper presented the evolutionary pictures that are now available for close binaries and made use of them to suggest an explanation for some specific objects.

Our Conference has been very fruitful in bringing together a good number of research workers in the field of close binaries and provoking some interesting discussions such as the ones that referred to the period changes and to the problem of dealing with contact configurations.

However, no new discussion has been made related to the physical conditions, to the parameters that characterize each one of the elements of the structure of the circumstellar matter in different objects and/or types of objects, to the physics of mass outflow and of the whole process of mass exchange.

From the observational point of view we have to give the theoreticians the values of the parameters that characterize the present configuration of a close binary including information on everything which is there in addition to the stars in order that the comparison between theory and observations be more meaningful.

Following this line of thought, quantitative spectrophotometric measurements have been programmed for β Lyrae in order to try to understand something about the phenomena involved and to find a correlation, if possible, between the amount of mass being ejected from the primary component, the intensity of the emission lines, the variable features in the velocity and light curves and the amount of material that occupies the expanding envelope that surrounds the whole system.

Quantitative measures of the kind are badly needed and should be carried out in as many selected systems as possible. Particularly through Struve's pioneering work, which has provided us with a qualitative understanding of close binary phenomena, we know which systems deserve preferent and systematic attention. The problem is not simple but the outcome will certainly be rewarding.

In particular, spectrographic and photometric observations should be made simultaneously, should cover as long an interval of time as possible with a good coverage of one cycle and of different cycles, photometry should be done with interferometric filters in appropriately chosen wavelengths to avoid contamination with emission features, the coverage in energy range should be extended so as to obtain information in several energy ranges.

In this way we may perhaps come to understand systems like W Serpentis the light curve of which has defied so far an interpretation and whose spectrum in the photographic region does not give information about either star.

For the extension of the range in which observations of particular objects ought to be made, let us hope that Anne Underhill will succeed in including among the NASA projects a satellite with adequate instrumentation that will help to solve or gather more information on different problems. Perhaps a resolution from this Conference might be helpful.

Another important point is the following. Research on several types of peculiar objects has disclosed in many cases that the peculiarity can be traced to and explained in terms of the binary nature of the objects. There certainly must exist quite a number of additional peculiar objects that if investigated adequately will prove to be binaries and will extend our knowledge and help our understanding of evolution of close binaries. The stars that were discussed by Mrs. Peters and by Anne Underhill are two promising cases that may open new avenues of research in the field of peculiar close binary objects.

In this context Plavec's suggestion of some shell stars being perhaps binaries is to me a very good one. We have been studying peculiar binaries that show readily to be so for their velocity and light characteristics. But there certainly must be a number of binaries whose nature will be perhaps only disclosed after careful analysis of the spectral behavior as the characteristics – mass outflow and interaction – connected with the particular stage of evolution in which they are may mask or distort the stellar features in a very drastic way.

Perhaps I should remind you that to me – I know that somebody in this room may disagree with me – there is a fairly good evidence that suggests that the V/R variables are binaries with periods of a few hours. The components of such systems must be very different from main-sequence stars.

We used to say that the evolution of close binaries must be different from the evolution of single stars. And we are finding more and more objects that confirm that statement.

The secondary components of β Lyrae, V 453 Scorpii and ε Aurigae must be very peculiar objects indeed that find no counterpart among single objects. They are probably massive, more massive than the primaries, relatively small and are surrounded by thick envelopes. Suggestions have been made that they are black holes. It would certainly be terribly important to find evidence for black holes and in the case of close binaries we should be looking for at least two of the indicators that go with a black hole, namely, X-radiation and the particular type of light variation at primary minimum if there is no thick envelope around the black hole or the inclination is adequate. Neither of the three objects are known to be X-ray sources but they should be tried at different phases and particularly within the phase interval where we may expect to observe the stream colliding with the circumstellar envelope.

At the Elsinore Conference on Close Binaries I mentioned the three epochs we can distinguish in the study of close binaries. We could perhaps add the present epoch which seems to be characterized by a haste in trying to be the first one in identifying a black hole (black holes are stylish today as magnetic fields were some years ago) and, as a result, too many stars have been assigned to black hole identifications. It is very good to point out the possibility but one should try to see whether other alternatives are possible.

Going back to the objects I just mentioned, let me remind you that the spectrum of the secondary component of V453 Scorpii displays only emission lines that shift in antiphase with respect to the absorption lines from the primary star. The spectrum of the secondary component of β Lyrae may behave the same way. It did so at the time of Belopolsky and of Curtiss but at present the emission certainly comes from different sources and the interpretation of the feature is still not totally clear; the investigation under way by Batten and myself may clarify the situation.

In the same context, let me say that Hutchings' interpretation of the spectrum of the star from the surroundings of which the X-radiation of Cygnus X-1 comes from is very important. Such a spectrum appears to be quite similar to the one of the optical counterpart of Sco X-1 which, in turn, looks similar to that of a WR object and has been described as similar as that of an old nova.

It is just possible that we may be dealing with similar – not identical – kind of objects in the case of Cyg X-1 and of V453 Sco, although the mass assigned to the peculiar component of V453 Sco is much larger than in the case of Cyg X-1, if my memory is not failing me. Further investigation on the two objects is certainly needed.

In his talk Plavec gave arguments against the identification of some components of close binaries as black holes, and other arguments could still be added in certain cases, such as the need for a hot star, in the case of β Lyrae, to provide the excitation required to produce the emission lines of H and He I.

No reference has been made in this Conference to the fact that in double-lined early type binaries it is sometimes found that at least part of the envelope emits continuous radiation which veils the stellar lines of the star surrounded by the envelope giving rise to the phenomena of variable line intensity known in systems like UW Canis Majoris.

The complications in the light curve of W Serpentis, which appears to have some similarity with that of HD 187399 presented during the Conference, may find their explanation in the presence of a thick envelope around one of the components. After all the spectrum of W Serpentis is essentially a shell spectrum as it is the case in V367 Cygni.

The idea of the 'disk' as suggested by Huang for β Lyrae and ε Aurigae has provided a word which seems to be adequate for describing the rather flat circumstellar envelopes surrounding one component in Algol-type and other systems. But in objects like HR 2142, if Mrs Peters interpretation is correct, the circumstellar envelope must not be flat but occupy quite a volume. Similarly, in Wolf-Rayet stars, a large percentage of which – perhaps even 100% – are binaries, the envelope where the WR spectrum is formed must surround the whole star. For this reason, I prefer not to use the word disk to describe one of the elements of the circumstellar matter around binaries but to call it simply 'circumstellar envelope', disk being a particular case of such envelopes.

In regard to the 'outer envelope' (Batten's cloud), mention has been made in this Conference, of the densities involved. However, there are cases like those of υ Sagittarii and W Serpentis where forbidden lines appear in the spectrum, and that of γ_2 Velorum in the UV spectrum of which there appears an intercombination line displaced the same amount as the shell lines of He I; in these cases the densities of the outer observable layers of the envelope must be much lower, of the order of 10^6, as in the symbiotic stars.

Perhaps because a short reference was made in a Conference four years ago no mention has been made here to the existence of expanding envelopes around close binaries and the velocities involved. Here again figures on the rate of mass loss should be highly desirable as we keep repeating one or two values which were estimated several years ago.

Typical velocities of the expanding envelopes are near 200 km s^{-1} in the case of β Lyrae, 300 in the case of υ Sagittarii, 700 in the case of HD 47129 and 1300 km s^{-1} in γ_2 Velorum.

By the way, mention has been made in one of the sessions of the importance of radiation pressure in defining the motion of the stream. HD 47129 provides a good example of such a fact, as in such case there is a strong deflection of the stream away from the O star.

In the course of the Conference a suggestion has been made for the existence of stellar 'lumps'. Hall's suggestion seems to be difficult to accept and it appears preferable to think in terms of colliding streams. But in order to be able to decide or to come to a different interpretation one has first to know more about the observational facts, particularly, the behaviour with time of the feature that has been interpreted as indicating a possible 'lump' in the star.

I mentioned that I found the discussion on period changes extremely interesting. Let me add that I thought that Hall's model sounds very promising. But the idea should perhaps be checked by trying to see whether there is any correlation with the density of the circumstellar envelope as suggested by the spectra in each particular case.

Since Struve's time we have made quite some progress in our knowledge of close binaries through more refined photometric techniques, through the computations related to evolution, through the discovery of X-ray sources and through the work of many spectroscopists who were able to ascertain the binary nature of a number of objects which at Struve's time were very enigmatic. In regard to evolution, let me remind you that the idea that explains the Algol paradox was suggested by Crawford and Kraft and was readily accepted by Struve.

We have now advanced a great deal in our understanding of possible evolution in close binaries and Plavec is going in the right direction in trying to match theory with observations. So far the computations have taken for granted the mass exchange proposition and have felt somewhat satisfied in establishing the mass configuration and the position in the H − R diagram during evolution. The comparison with observations have made it clear that non-conservative cases should be taken up, as Plavec has shown, and this seems to be reasonable. Since the stage of rapid mass outflow from the more massive component of the system involves the loss of a substantial amount of mass in a relatively short time one would expect that there will be large velocities of outflow and, therefore, that mass will be lost to the system.

Much work has still to be done, both from the theoretical as well as from the observational side, until we understand the phase of rapid mass loss and, until we are able to describe each stage, the physics involved and can fully compare the result of the computations with the actual observed configuration. Then we will be able to answer questions like when is the 'disk' or 'circumstellar envelope' formed, whether during the fast or the slow phase of evolution, and other similar questions.

Plavec has interpreted υ Sagittarii and the binaries X-ray sources as objects undergoing second mass exchange. Actually β Lyrae, V453 Scorpii and perhaps the whole group of stars that several years ago I classed together with β Lyrae, with perhaps the exception of HD 47129, are most probably undergoing second mass exchange.

Since I mention υ Sagittarii let me state that although everybody talks quite seriously of υ Sgr as being a helium-rich star, I feel that such a conclusion from a spectrum which has strong H emission and an outer envelope that displays lines of H and He I, is at least subject to doubt.

Since I am an observational astronomer I have stressed the type of observations that should be carried out at the present time, namely, simultaneous photometric and spectrographic investigations, quantitative spectrophotometric analysis and study of peculiar objects. What I have said also points towards our needs from the theoretical point of view. Commision 42 should perhaps discuss and agree at the next General Assembly on a number of objects that should preferentially be observed.

To put an end to these comments I would say that for me personally the Conference has been very valuable and inspiring indeed, and look forward to our next meeting where the results from the Parksville interaction between astronomers will be apparent.

Let me finish my words by saying that it was particularly interesting for me to find that my old friend Su-Shu Huang has developed such a skill to find good figurative comparisons to illustrate his descriptions and explanations. After his talk I have come to the conclusion that if I want to understand the dynamics of gaseous streams I should take up smoking!